Advances in Experimental Medicine and Biology

Cell Biology and Translational Medicine

Volume 1212

Subseries Editor
Kursad Turksen

More information about this subseries at http://www.springer.com/series/15838

Kursad Turksen

Editor

Cell Biology and Translational Medicine, Volume 6

Stem Cells: Their Heterogeneity, Niche and Regenerative Potential

 Springer

Editor
Kursad Turksen (Retired)
Ottawa Hospital Research Institute
Ottawa, ON, Canada

ISSN 0065-2598   ISSN 2214-8019   (electronic)
Advances in Experimental Medicine and Biology
ISSN 2522-090X   ISSN 2522-0918   (electronic)
Cell Biology and Translational Medicine
ISBN 978-3-030-32825-2   ISBN 978-3-030-32823-8   (eBook)
https://doi.org/10.1007/978-3-030-32823-8

This Springer imprint is published by the registered company Springer Nature Switzerland AG.
The registered company address is: Gewerbestrasse 11, 6330 Cham, Switzerland

Preface

In this next volume in the Cell Biology and Translational Medicine series, we continue to explore the potential utility of stem cells in regenerative medicine. Chapters in this volume cover a range of topics in the area, including developmental aspects and role of the stem cell niche, function of decellularized scaffolds, and the potential of iPS cells in tissue and organ regeneration. Collectively, the chapters continue to cover crucial aspects in translational studies of tissue and organ regeneration and restoration of function in clinical settings.

I remain very grateful to Peter Butler, Editorial Director, and Meran Lloyd-Owen, Senior Editor, for their support of this series from its inception and for helping to foster its success until now. I would like, in addition, to take the opportunity to welcome Gonzalo Cordova as the Associate Editor of the series and acknowledge his support.

I would also like to acknowledge and thank Sara Germans-Huisman, Assistant Editor, for her outstanding efforts in helping to get this volume to the production stages.

A special thank you also goes to Rathika Ramkumar and Abinay Subramaniam for their outstanding efforts in the production of this volume.

Finally, sincere thanks are due to the contributors not only for their support of the series but also for their insight and effort to capture both the advances and remaining obstacles in their areas of research. I trust readers will find their contributions as interesting and helpful as I have.

Ottawa, ON, Canada   Kursad Turksen

Contents

Adv Exp Med Biol – Cell Biology and Translational Medicine (2019) 6: 1–29
https://doi.org/10.1007/5584_2019_350
© Springer Nature Switzerland AG 2019
Published online: 10 March 2019

Addressing Variability and Heterogeneity of Induced Pluripotent Stem Cell-Derived Cardiomyocytes

Sherri M. Biendarra-Tiegs, Frank J. Secreto, and Timothy J. Nelson

Abstract

Induced pluripotent stem cells (iPSCs) offer great promise in the areas of disease modeling, basic research, drug development, and regenerative medicine. Much of their value comes from the fact that they can be used to create otherwise inaccessible cell types, such as cardiomyocytes, which are genetically matched to a patient or any other individual of interest. A consistent issue plaguing the iPSC platform, however, involves excessive variability exhibited in the differentiated products. This includes discrepancies in genetic, epigenetic, and transcriptional features, cell signalling, the cell types produced from cardiac differentiation, and cardiomyocyte functionality. These properties can result from both the somatic source cells and environmental conditions related to the derivation and handling of these cells. Understanding the potential sources of variability, along with determining which factors are most relevant to a given application, are essential in advancing iPSC-based technologies.

S. M. Biendarra-Tiegs
Department of Molecular Pharmacology and Experimental Therapeutics, Mayo Clinic, Rochester, MN, USA

Program for Hypoplastic Left Heart Syndrome-Center for Regenerative Medicine, Mayo Clinic, Rochester, MN, USA

F. J. Secreto
Program for Hypoplastic Left Heart Syndrome-Center for Regenerative Medicine, Mayo Clinic, Rochester, MN, USA

Department of Internal Medicine, Division of General Internal Medicine, Mayo Clinic, Rochester, MN, USA

T. J. Nelson (✉)
Department of Molecular Pharmacology and Experimental Therapeutics, Mayo Clinic, Rochester, MN, USA

Program for Hypoplastic Left Heart Syndrome-Center for Regenerative Medicine, Mayo Clinic, Rochester, MN, USA

Department of Cardiovascular Medicine, Mayo Clinic, Rochester, MN, USA

Department of Pediatric and Adolescent Medicine, Division of Pediatric Cardiology, Mayo Clinic, Rochester, MN, USA

Department of Internal Medicine, Division of General Internal Medicine, Mayo Clinic, Rochester, MN, USA
e-mail: nelson.timothy@mayo.edu

Keywords

Cardiomyocytes · Cellular microenvironment · Differentiation · Induced pluripotent stem cells · Phenotypic variability · Stem cell heterogeneity

Abbreviations

AP	action potential
APD	action potential duration
cGMP	current Good Manufacturing Practice
CNV	copy number variation
DEG	differentially expressed gene
EB	embryoid body
ECM	extracellular matrix
ESA	etoposide sensitivity assay
ESC	embryonic stem cell
ESC-CM	embryonic stem cell-derived cardiomyocyte
FACS	fluorescence-activated cell sorting
iPSC	induced pluripotent stem cell
iPSC-CM	induced pluripotent stem cell-derived cardiomyocyte
mtDNA	mitochondrial DNA
SNV	single nucleotide variation
XCI	X-chromosome inactivation

1   Introduction

Since the original discovery of human induced pluripotent stem cells (iPSCs) in 2007 (Takahashi et al. 2007; Yu et al. 2007), this revolutionary technology has been adopted in a wide variety of settings including disease modeling, drug discovery and toxicity testing, and cell-based therapies (Musunuru et al. 2018; Yoshida and Yamanaka 2017). Much of the power of this platform lies in the ability to produce iPSCs from individuals with unique genetic backgrounds and subsequently differentiate them into a myriad of cell types, including those difficult to obtain by other means, such as cardiomyocytes. This is highly relevant to the pursuit of individualized medicine, including autologous cell-based therapies. Furthermore, the ability to capture variation in the human population and study genotype-phenotype relationships has led to the use of iPSCs in modeling a wide variety of diseases. For example, iPSC-derived cardiomyocytes (iPSC-CMs) have been applied to model diseases including long QT syndrome (Moretti et al. 2010), other

channelopathies such as Timothy syndrome (Yazawa et al. 2011), cardiomyopathies (Wyles et al. 2016), ventricular tachycardia (Jung et al. 2012), mitochondrial diseases including Barth syndrome (Wang et al. 2014), and even structural heart disease (Hrstka et al. 2017).

However, while the natural variation captured by iPSCs and their differentiated cardiomyocyte progeny is of great value, heterogeneity and variability of these cellular populations can also be problematic. Ideally, it would be advantageous to produce iPSC-CMs in a reliable and consistent manner. In terms of disease modeling, it is important to be able to understand the relationship between genotype and phenotype without confounding variables distorting the results. Furthermore, for disease modeling it is often desirable to have cells that closely resemble a particular cardiomyocyte subtype, including cells resembling those found in the ventricular or atrial chambers of the heart (Marczenke et al. 2017). This is also vital for reproducibility of the results, an ongoing concern in scientific research (Osterloh and Mulelane 2018). In terms of drug development applications, it is important to be working with iPSC-CMs of a sufficient developmental state to behave in a manner that is predictive of human heart tissue. Cardiomyocyte subtype is a concern in this setting, as well (Denning et al. 2016). In terms of cell-based therapies, it is critical to minimize any safety concerns such as genetically abnormal cells (Merkle et al. 2017). There has also been concern that mixed subtypes of cardiomyocytes could lead to arrhythmias (Liu et al. 2018).

In this review, we discuss sources and types of variability and heterogeneity in pluripotent stem cells and differentiated cardiomyocytes, as well as approaches which have been proposed to aid in retaining pertinent genetic variability while maximizing consistency. While the focus will be on human iPSC-CMs, much of what has been learned from embryonic stem cells (ESCs) or even murine pluripotent stem cells can be applied to the iPSC-CM platform. Ultimately, the goal will be to determine what facets of heterogeneity or variability are or are not permissible to a particular application, and tailor cell production or the study design accordingly.

2 Intrinsic and Acquired Variation in Pluripotent Stem Cell Lines

2.1 Genetic and Epigenetic Abnormalities

One potentially troubling source of variation in iPSC lines is abnormalities at the genetic or epigenetic level. Such variations have been discovered between different iPSC lines, different passages of the same iPSC line, or even different subpopulations within an iPSC culture. These include aneuploidy, chromosomal rearrangement, sub-chromosomal copy number variation (CNV), single nucleotide variation (SNV), variable X-chromosome inactivation, and aberrant DNA methylation. Any of these could potentially result in unexpected cellular properties such as acquisition or disappearance of disease-related phenotypes (Liang and Zhang 2013; Nguyen et al. 2013).

Large-scale studies have provided much insight into the frequency and specific natures of these abnormalities. For example, karyotypic analysis of over 1,700 iPSC and ESC cultures from 97 investigators in 29 labs revealed that for both cell types, approximately 12% of cultures were abnormal. In terms of the types of abnormalities, there were both similarities and differences between iPSCs and ESCs. Trisomy 12 was predominant for both, partial gain of chromosome 12 and trisomy 20q were also seen in both, trisomy 8 was more common for iPSCs than for ESCs, an additional chromosome X was more common in female ESCs, and trisomy 17 was only seen in ESCs (Taapken et al. 2011). In another study, 12 of 38 (~32%) of ESC lines and 13 of 66 (~20%) of iPSC lines had chromosomal aberrations, with 6 iPSC lines having a full trisomy of chromosome 1, 3, 9, or 12 (Mayshar et al. 2010). Rate of aneuploidy has also been shown to increase with higher passage (Mayshar et al. 2010). It has been suggested that karyotypic abnormalities could be derived from culture adaptation, were present in the parental cells, or arose from selective pressure during the reprogramming

process (Mayshar et al. 2010). Studies have indeed demonstrated that chromosomal abnormalities are in some cases present in the original somatic cells (Vitale et al. 2012). Reprogramming method and culturing substrate were not found to have a notable role in some studies (Mayshar et al. 2010; Taapken et al. 2011), while some other studies have found that certain passaging methodologies can lead to chromosomal abnormalities (Mitalipova et al. 2005). Karyotypic abnormalities can have functional consequences, as well. For example, spontaneously differentiated normal and abnormal ESC lines demonstrated differences in expression of differentiation-related genes and different propensities for particular lineages (Fazeli et al. 2011).

CNVs have been observed in pluripotent stem cells as well. For example, one study applied single-cell array-based comparative genomic hybridization to reveal notable fractions of both somatic cells and ESCs with diverse megabase-scale chromosomal abnormalities. The authors identified replication break fork collapse and breakage-induced replication as a potential cause, possibly a result of sub-optimal culture conditions (Jacobs et al. 2014). Genomic analysis of 58 iPSC lines from 10 laboratories revealed CNVs that were donor-specific and others that varied between lines from the same donor. There were some genomic loci that were frequently affected, suggesting a basis in the reprogramming process. In some cases, the deletion of tumor suppressors or duplication of cell growth-related genes suggested a survival or proliferative advantage (Salomonis et al. 2016). Another study similarly concluded that CNVs were produced in the reprogramming process and provided a selective advantage, but also found more CNVs in early-passage iPSCs (Hussein et al. 2011). An evaluation of 711 cell lines from 301 healthy individuals reported lower levels of genetic aberrations than had been detected in some previous studies, likely because the authors also had access to donor-matched reference samples and were thus able to identify germline copy number variations. Most of the aberrations they did find were unique to individual iPSC lines, but some alterations were found in multiple cell lines from the same donor. The number of these alterations was not associated with passage number, donor age, gender, or the results of the quality control assay PluriTest (Kilpinen et al. 2017). The prevalence of SNVs has also been investigated, and one study identified between 1058 and 1808 heterozygous SNVs in each iPSC line examined, with 50% of these being synonymous changes. Since the SNVs were not shared between iPSC lines from the same donor, the abnormalities were deemed likely to have resulted from the reprogramming process (Cheng et al. 2012).

Epigenetic differences between iPSC lines, including variable levels of aberrant DNA methylation, have also been described. For example, a genomic analysis of 58 iPSC lines from 10 laboratories showed that while ESCs and iPSCs were generally indistinguishable at the level of global gene expression, there were notable differences in methylation profiles (Salomonis et al. 2016). Differences in DNA methylation have also been found for iPSCs derived from distinct source cell types (neonatal dermal fibroblasts, adult dermal fibroblasts, and CD34$^+$ cells from peripheral blood mononuclear cells) via different reprogramming technologies, with this heterogeneity reduced after prolonged culture to a more ESC-like DNA methylation state (Tesarova et al. 2016). Conversely, another group found that epigenetic patterns of different iPSC lines were similar to each other and to ESCs, regardless of source cell, although in that case the same reprogramming approach was used for all the iPSC lines. There was some random aberrant hypermethylation observed at early passages, but this was decreased with additional passaging (Nishino and Umezawa 2016). Transcriptional profiling of 317 human iPSC lines from 101 individuals revealed transcriptional variability in Polycomb repressive complex 2 (PRC2) and H3K27me3 targets, which appeared to be independent of genetic background, suggesting that the reprogramming process could be the source. In this same study, some genes showed allelic imbalance while others demonstrated biallelic expression. These patterns were in some cases consistent within individuals,

but different across individuals (Carcamo-Orive et al. 2017). This mixture of genes with monoallelic or biallelic expression had previously been seen for both iPSCs (Pick et al. 2009) and ESCs (Kim et al. 2007). One study reported low frequency loss of imprinting in some iPSC lines, which was stable in culture (Hiura et al. 2013). These epigenetic differences can have functional relevance, since it has been shown that epigenetic features can be used to identify iPSC lines with particular differentiation capacities and perhaps even maturation capacity (Nishizawa et al. 2016). For example, histone modifications H3K27me3 and H3K4me3 at lineage-associated and pluripotency genes in ESCs influence developmental potential towards particular lineages (Hong et al. 2011).

The X-chromosome status of female iPSCs and ESCs has also been an area of extensive characterization (Wutz 2012). This was originally described by Silva et al., who showed that ESCs tend to lose XIST RNA expression during culture, leading to three different classes of cells. Class I is pre X-chromosome inactivation (XCI) with a capacity to recapitulate XCI upon differentiation. Class II cells show elevated XIST-positive cells and XCI status. Class III cells have lost XIST expression but still have an inactivated X-chromosome which is not reactivated upon differentiation. Some of these class III lines demonstrate poor spontaneous differentiation in embryoid bodies (EBs) (Silva et al. 2008). Analysis of dozens of iPSC lines has shown notable variation in XIST expression, H3K27me3, and XCI status (Geens et al. 2016; Mayshar et al. 2010; Salomonis et al. 2016). Single cell-derived iPSC clones from the same donor show various states of XCI right after clonal isolation, with both pre- and post-XCI cells within individual colonies (Andoh-Noda et al. 2017). Other studies have similarly noted a mixture of cells with different XCI status in the same passage or even the same colony (Geens et al. 2016; Tanasijevic et al. 2009). In some cases XCI is acquired over time, with no reactivation with repeated passaging (Andoh-Noda et al. 2017). In other cases erosion of XCI in culture has been reported, with this being a stable condition that cannot be restored by differentiation or reprogramming (Geens et al. 2016; Mekhoubad et al. 2012). One of these studies which reported some reactivation of X-chromosomes additionally noted that clusters of genes in certain chromosomal areas were being reactivated sooner than those in others (DeBoever et al. 2017). Another study, while reporting activation of X-chromosomes in some iPSCs upon reprogramming, did not note any correlation of XCI status to passage number, culture, conditions, or reprogramming method (Bruck and Benvenisty 2011). X-chromosome status can have functional implications for these iPSCs. Loss of XIST in female iPSCs is correlated with upregulation of X-linked oncogenes, downregulated tumor suppressors, accelerated growth rate in vitro, and poorer differentiation in teratomas (Anguera et al. 2012). Developmental genes are also differentially methylated in female iPSC lines with different XIST expression and XCI status (Salomonis et al. 2016).

2.2 Contribution of Source Cells to iPSC Properties

The main sources of genomic, epigenetic, and transcriptional variation between different iPSC lines remains a major question within the field, although a number of studies have helped to provide insight. Transcriptional profiling of 317 human iPSC lines from 101 individuals revealed that ~50% of genome-wide expression variability could be explained by the variation across individuals (Carcamo-Orive et al. 2017). It was even possible to identify expression quantitative trait loci contributing to this variation, which could be conducive to studying variants identified in genome-wide association studies (Carcamo-Orive et al. 2017; DeBoever et al. 2017). Other variables such as donor age, body mass index, sex, ancestry, reprogramming batch and technician, RNA preparation technician, Sendai virus lot, and reprogramming cell source influenced expression variation for only a small number of genes. There were, however,

differences in the degree of similarity between iPSC lines derived from the same individual, with Polycomb targets playing a major role in non-genetic variability both within and between individuals (Carcamo-Orive et al. 2017). When genomic analysis was performed on 58 cell lines from 10 laboratories, donor, sex, reprogramming technology, and originating laboratory, but not passage number, were major driving covariates in mRNA, miRNA, and methylation profiling. In regards to methylation profiling, cell of origin played a contributing role but there was no clear connection to differences in somatic methylation profiles (Salomonis et al. 2016). According to an analysis of 711 lines from 301 healthy individuals, between 5 and 46% of variation in iPSC phenotypes including genome-wide assays, protein immunostaining, differentiation capacity, and cellular morphology was due to differences between individuals, and this donor variance was primarily due to genetic differences (Kilpinen et al. 2017).

Other studies have likewise found that genetic differences between individuals are a major contributing factor to variation between cell lines, such as in mRNA levels, splicing, and imprinting (Rouhani et al. 2014). For example, three iPSC clones from the same individual could not be distinguished by transcriptional profiling and functional pathway analysis, and were distinct from ESCs and iPSCs from different donors. These differences between unique donors were retained after differentiation to all three germ layers in embryoid bodies (Schuster et al. 2015). Generated isogenic ESC and iPSC lines have been shown not to have significantly different gene expression in either an undifferentiation or differentiated state, and have little difference in methylations profiles while undifferentiated (Mallon et al. 2014).

On the other hand, the role of epigenetic memory derived from the somatic cell of origin in the variability between cell lines has been more controversial. It is known that different somatic cells have distinct epigenetic profiles, even for the same cell type from different locations. For example, genome-wide DNA methylation and transcriptome data on matched pairs of dural and scalp fibroblasts showed strong epigenetic memory based on sampling location. More epigenetic variability was observed with age, especially for the scalp-derived cells (Ivanov et al. 2016). However, it has also been found that epigenetic memory is not necessarily a major contributor to transcriptional variation (DeBoever et al. 2017; Rouhani et al. 2014). One study examined matched iPSCs from fibroblasts and blood from multiple donors and observed that lines from the same donor were highly transcriptionally and epigenetically similar, but that different donors had specific transcriptome and methylation patterns that contribute to distinct differentiation capacities (Kyttala et al. 2016). Similarly, in another study very few differences in DNA methylation states or gene expression patterns were detected between iPSCs derived from lymphoblastoid cell lines and from fibroblasts. Again, genetic variation was found to be the largest contributor to differences between different cell lines (Burrows et al. 2016). In fact, if variation between individuals is not corrected for appropriately, transcriptional differences between iPSCs and ESCs and between iPSCs from different somatic tissues of origin seem much larger than in actuality (Rouhani et al. 2014). When considering genetically matched ESC and iPSC lines, genetic background had a larger impact on transcriptional variation than either somatic origin or Sendai virus reprogramming method (Choi et al. 2015). Putting genetic contributions aside, however, iPSCs derived from murine ventricular cardiomyocytes demonstrate a higher propensity to spontaneously differentiate into ventricular-like cardiomyocytes than genetically matched ESCs or iPSCs from tail-tip fibroblasts. This was thought to potentially be due to distinct transcriptomes and DNA methylation, including at promoters of cardiac genes (Xu et al. 2012). Cardiac differentiation efficiency has also been shown to be higher for cardiac progenitor cell-derived iPSCs than for fibroblast-derived iPSCs, possibly due to differential methylation at the NKX2-5 promoter. However, these epigenetic differences decreased with passaging and there were no significant differences in morphology, calcium handling, or electrophysiology of the resulting cardiomyocytes. Moreover, these cells had a similar therapeutic effect in a murine myocardial infarction model (Sanchez-Freire et al. 2014). In order to address this source of variability, our laboratory devised an approach to negate the influence of somatic origin on methylation and transcriptional profiles of the resultant iPSCs, via comparison of murine iPSC clones against a standardized gene expression profile. Expression levels of two pluripotency genes, Oct4 and Zfp42, were identified to indicate increased cardiogenicity regardless of cell source or reprogramming strategy, thus allowing a way to address clonal variability (Hartjes et al. 2014).

The role of somatic cell source in cellular aberrations is another potential concern. Blood-derived iPSCs have been found to be less likely to acquire aberrant DNA methylations than iPSCs from other somatic sources (Nishizawa et al. 2016). In terms of genomic aberrations, it has been determined that an average iPSC line has two CNVs that are not apparent in the originating fibroblasts, although by using more sensitive techniques it can be seen that at least 50% of those CNVs are actually low frequency somatic genomic variants in the parental fibroblasts which are revealed due to the clonal origin of iPSCs. It has been estimated that about 30% of fibroblasts have somatic CNVs (Abyzov et al. 2012). Examinations of SNVs in murine iPSCs have also suggested that most mutations occur prior to reprogramming (albeit at very low allele frequency) and are captured by the clonal nature of iPSCs, although some mutations can occur later on (Li et al. 2015; Young et al. 2012). One study identified 4 somatic mutation classes: clonal, subclonal (which would have arisen during reprogramming or culturing), UV-damage mutations, and copy number alterations. Most point mutations were found to be in areas of repressed chromatin and thus not influence gene expression in iPSCs, although subclonal mutations were associated with altered gene expression to a greater degree. Furthermore, over a third of the genes overlapped by copy number alterations had altered expression. Still, mutations that did not influence gene expression in iPSCs could still potentially have effects in differentiated tissues, so they should not necessarily be discounted. As for the UV-damage mutations, these were found in ~50% of iPSCs from skin fibroblast. However, the number of mutations in cancer genes was not significantly different than what would be expected by random chance (D'Antonio et al. 2018).

It has become apparent in the past few years that mutations in mitochondrial DNA (mtDNA) can also vary across iPSC lines. Studies have shown that that individual fibroblasts can carry unique mutations, and that mutations in iPSCs can be found in very low levels in parental fibroblasts (and thus may not even be detectable when analyzing whole tissue). These mutations may even be homoplasmic or present in high heteroplasmy. iPSCs from older adults have been reported to exhibit more mtDNA mutations than those derived from younger individuals, and even blood-derived iPSC lines may harbor mitochondrial mutations. These mutations can subsequently lead to defects in metabolic function and respiration (Kang et al. 2016). It was previously shown that while somatic murine cells with high mtDNA mutation load can be reprogrammed to iPSCs, the resultant cells have slower proliferation and differentiation defects (Wahlestedt et al. 2014). Our laboratory reported that low levels of mtDNA mutations in fibroblasts, even from healthy individuals, are detectable following reprogramming into iPSCs. While cardiac differentiation potential was not impacted by mtDNA mutations, this could lead to impaired mitochondrial respiration in iPSC-CMs. Additionally, we observed that a subset of iPSC clones derived from patients diagnosed with mitochondrial disease exhibit low levels of mtDNA heteroplasmy, and thus do provide a representative model system (Perales-Clemente et al. 2016).

3   The Dynamic Transcriptional State of Pluripotent Stem Cells

Of course, the ultimate uses of iPSCs typically involve the differentiation of these cells into a somatic cell type such as cardiomyocytes. The iPSCs must thus be receptive to developmental cues at the time of initiation of differentiation, and progress fully down a desired trajectory. Therefore, in addition to genetic and epigenetic properties, transcriptional heterogeneity of iPSCs at the time of initiation can have a notable impact upon how the cells respond to those cues (Fig. 1).

Recently, it was discovered that iPSC cultures contain two subtypes of cells which differ in morphology, cell-matrix and cell-cell adhesion, pluripotency, and gene expression. Both of these can differentiate into all 3 germ layers, but have different propensities towards these different germ layers when undergoing spontaneous differentiation (Yu et al. 2018a). Along these lines, it had previously been shown that murine ESCs can be described by one of two transcriptional states, and that DNA methylation plays an important role in maintaining these states (Singer et al. 2014). The Nucleosome Remodeling and Deacetylation (NuRD) complex was also found to modulate transcriptional heterogeneity and the expression of pluripotency of genes in murine ESCs, thus controlling response to differentiation signals (Reynolds et al. 2012). These findings suggest that even within a single culture, different cells can have distinct responses to the same differentiation cues. Moreover, they represent a sampling of a much broader body of work describing heterogeneity in pluripotent and development factors, as well as signalling molecules (Singh 2015).

In particular, studies have focused on the concept of heterogenous pluripotency factor gene expression. For example, it has been reported that some ESCs exhibit high Nanog expression levels, while others display levels considerably lower than expected, with the latter group being particularly prone to undergo spontaneous differentiation. Transitions from the high to the low state were modeled to be rare and stochastic, while transitions in the opposite direction were predicted to be frequent (Kalmar et al. 2009). Model simulations have further shown that low-Nanog cells act as an intermediate state to reduce the barrier of transition in the differentiation process (Yu et al. 2018b). However, it has also been suggested that Nanog heterogeneity

Fig. 1 Sources of variation in iPSCs at the time of initiation of differentiation. Differences in genomes, epigenomes, and transcriptomes between iPSC lines, or even different cells within the same culture, can influence response to differentiation cues. Endogenous signalling and cell cycle position have also been found to exert a noticeable effect upon differentiation trajectories

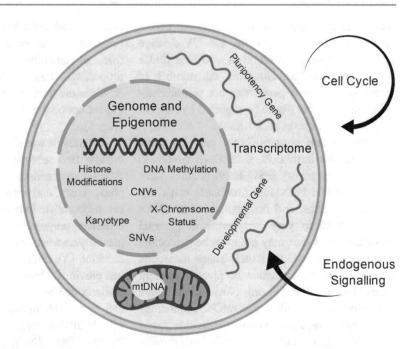

could be due to specific *in vitro* culture conditions and may not be functionally significant (Smith 2013). The reporters used in such studies can also disturb the cell states they are intended to model. For example, genetic reporters for Nanog can influence behavior of pluripotency-related positive feedback loops and lead to a bifurcation that results in heterogeneous Nanog expression (Smith et al. 2017). Still, some researchers continue to assert that these Nanog fluctuations are real and functionally relevant, based on the fact that similar fluctuations are not seen for some other pluripotency genes, mathematical models have supported bimodal distribution of Nanog, and single-cell RNA seq in murine ESCs suggests bimodal expression (Yu et al. 2018b).

Developmental genes have also been reported to exhibit heterogeneous expression in pluripotent stem cells. For example, Hes1 is a developmental factor that regulates cell proliferation and differentiation in embryogenesis, and along with its downstream gene targets, demonstrates an oscillating expression pattern in murine ESCs. High levels promote mesodermal differentiation whereas low levels promote neural differentiation through modulation of Notch signalling and the

cell cycle (Kobayashi and Kageyama 2010; Kobayashi et al. 2009). Modeling approaches have also shown that intrinsic noise in the Hes1 gene regulatory network could explain heterogeneity in murine ESC differentiation (Sturrock et al. 2013). In human ESCs, a Wnt reporter has been used as a read-out of heterogeneity in endogenous Wnt signalling activity, even in cells with similar expression of pluripotency markers. Moreover, the level of Wnt signalling activity in pluripotent stem cells correlates with lineage propensity in differentiation, with high Wnt expression promoting endoderm and cardiac differentiation, and low Wnt enhancing neuroectodermal differentiation (Blauwkamp et al. 2012; Paige et al. 2010).

Interestingly, studies have found that when pluripotent stem cells are exposed to signals which induce differentiation, they activate developmental pathways in an asynchronous manner. Recent evidence has suggested that this is linked to the cell cycle (Dalton 2015). It has been observed that G1 cells are more responsive to differentiation cues, which could help explain heterogeneity in expression of developmental factors. It is possible that developmental genes are transcriptionally primed in G1 and/or that some pluripotency markers could be diminished in G1, and that a favorable epigenetic and nuclear architecture state in G1 promotes activation of developmental programs (Dalton 2015; Singh et al. 2014). One group used the FUCCI reporter system to show that differentiation capacity of ESCs and iPSCs varies throughout the cell cycle, with early G1 cells having a propensity towards endoderm/mesoderm and late G1 cells tending to differentiate towards neuroectoderm. They found that cells in G2/S/M, on the other hand, responded poorly to differentiation signals. The authors further focused on the differences between early and late G1, and ultimately found that in early G1, level of cyclin D is low, so Smad2/3 can bind and activate endoderm genes, whereas in late G1 cyclin D is high and CDK4/6 is activated and phosphorylates Smad2/3, thus preventing nuclear entry (Pauklin and Vallier 2014). Using the FUCCI system combined with fluorescence-activated cell sorting (FACS) and RNA-seq, other researchers found that heterogeneous expression of developmental regulators in cells which also express pluripotency genes is linked to cell cycle position, with increased expression of these regulators in G1. Major changes in global 5-hydroxymethylcytosine, namely upregulation of 5-hydroxymethylation in G1, were linked to both cell-cycle progression and expression of developmental factors. G1 was seen to be a window of time when the cells could respond to external differentiation signals via gene activation, possibly due to chromatin being in a more permissive state (Singh et al. 2014). Interestingly, it has been reported that when ESCs and iPSCs are cultured with DMSO, this activates the retinoblastoma (Rb) protein, increases proportion of cells in early G1 phase, and improves differentiation efficiency across all germ layers. Such culture manipulation has been used to differentiate cardiomyocytes from an ESC line predicted to be impaired in mesodermal differentiation ability (Chetty et al. 2013). A recent study describing a comparison of various cell cycle inhibitors in human pluripotent stems ultimately identified nocodazole as an efficient and non-toxic means to synchronize these cells in the G2/M phase. This may provide a valuable framework for further investigation into the relationship between cell cycle and differentiation (Yiangou et al. 2019).

Over the past several years, single-cell transcriptional profiling has provided additional insight into heterogeneity of pluripotent stem cells. One such study reported that heterogeneity in expression levels for a number of pluripotency genes was greater in iPSCs than ESCs, and that significant cell-to-cell variability exists even in cells positive for Tra-1-60 and SSEA-4 (Narsinh et al. 2011). From single cell RNA-seq analysis, it has been observed that genes with a higher coefficient of variation in human and murine ESCs form co-expression clusters and partly explain bivalency of gene expression. This data aligns with the idea that pluripotent stem cells alternate between different transient and reversible cell states, although this does not appear to involve lineage priming since genes with a high coefficient of variation were not shown to be enriched for any particular biological process (Mantsoki et al. 2016). Moving forward, it has been suggested that integrative network models—namely gene network models involving epigenetic, transcriptional, and signaling information—from single cell data will be very important for better understanding how self-renewal and differentiation are regulated (Espinosa Angarica and Del Sol 2016).

4 Extrinsic Influences on Pluripotency and Differentiation

Given the extent of reported heterogeneity between pluripotent stem cells, numerous studies have scrutinized the role of extrinsic factors on their properties. One study examining gene expression profiles of 66 iPSC lines found that the lines clustered together according to laboratory and study of origin (Mayshar et al. 2010). Likewise, a reanalysis of microarray gene expression data from seven labs showed strong correlation between gene expression signatures and lab of origin for both ESCs and iPSCs (Newman and

Cooper 2010). Part of this could be due to the culture conditions used. For example, prior to the availability of commercially produced media designed specifically for maintaining pluripotent stem cells, labs employed a variety of in-house developed cell culture media types. Notably, though, a comparison involving eight reported in-house culture methods and two widely available commercial medias (mTeSR1 and STEMPRO) demonstrated that the commercial medias were superior in supporting the maintenance of pluripotent stem cells (International Stem Cell Initiative C 2010). More recently, single-cell RNA-seq of murine ESCs showed enhanced heterogeneity of pluripotency and differentiation marker gene expression for cells cultured in serum as compared to serum free conditions. The most variable of these genes had distinct chromatin state signatures (Guo et al. 2016). A comparison of defined media and media with serum, enzymatic and mechanical passaging, and feeder-free and mouse embryonic fibroblast (MEF) substrates for iPSC and ESC culture demonstrated differences in genomic stability for these different conditions, with more genetic instability in particular for cells subjected to single-cell enzymatic passaging with Accutase (Garitaonandia et al. 2015). Oxygen levels have also been shown to influence properties of pluripotent stem cells, with hypoxic conditions resulting in increased expression of pluripotency markers, reduced chromosomal abnormalities, and reduced transcriptional heterogeneity for ESCs (Forsyth et al. 2008; Lim et al. 2011).

Furthermore, both high and low pH can influence pluripotency of murine and human ESCs, although reports regarding the effects of lactate levels have been more conflicting (Chen et al. 2010; Gupta et al. 2017). One of these reports showed that media acidification due to accumulation of lactic acid from high culture density leads to DNA damage and genomic alterations in ESCs grown on feeders, even over the course of a single passage. This was not seen for a feeder-free system, however (Jacobs et al. 2016). The presence of other metabolites in media can also influence pluripotency. For example, secreted factors from cell culture of ESCs lead to decreased

pluripotency marker expression in a system of multiplexed culture chambers (Titmarsh et al. 2013). Build-up of metabolites in media can also influence pluripotency in murine ESCs by priming them for differentiation (Yeo et al. 2013). Some of these effects can be addressed via perfusion culture (Gupta et al. 2017; Yeo et al. 2013). In addition to increasing metabolite levels in media, high density culture results in a higher proportion of cells in G1 (Jacobs et al. 2016; Laco et al. 2018; Wu et al. 2015) and thus could potentially influence differentiation capacity.

Notably, one group found that levels of Wnt fluctuate according to the cell cycle and that higher-density pluripotent stem cells exhibited more cell death and required lower doses of the GSK3β inhibitor CHIR99021 to induce cardiac differentiation. Conversely, cultures consisting of a greater percentage of cells in S/G2/M, along with exhibiting high expression of NANOG and OCT4a, demonstrated an increased propensity for undergoing cardiac differentiation. The authors were therefore able to increase efficiency of more confluent cultures by decreasing concentration of CHIR99021. Ultimately, they discovered that CHIR99021 treatment increased expression of Cyclin D1, promoted cell-cycle progression, and increased genetic instability from acidified media in high-density culture, ultimately leading to cell death. Lower confluence along with increased S/G2/M phase enhanced expression of Wnt inhibitors TCF7L1/2, so those less dense cultures required more CHIR in order to induce higher β-catenin levels via GSK3β inhibition, and ultimately achieve suppression of TCF7L1/2. This then allowed sufficient activation of Wnt target gene expression. Variations in TCF7L1/2 levels and/or cell cycle could thus lead to different differentiation results for the same CHIR concentration. Interestingly, mesoderm development was not found to be as affected by confluency and cell cycle as was full progression to cardiac differentiation (Laco et al. 2018).

In terms of embryoid body (EB) differentiations, outputs can also be influenced by culture conditions, namely colony and EB sizes. Gata6 and Pax6 expression are both impacted by colony size, with higher input Gata6/Pax6 being connected to more endoderm gene expression. Conversely, there is enhanced mesoderm and cardiac induction at larger EB sizes (Bauwens et al. 2008). Interestingly, the same group later found that efficient cardiac differentiation in EBs is promoted by endogenous extra-embryonic endoderm-like cells which are influenced by aggregate size (Bauwens et al. 2011). A comparison of EB and monolayer cardiac differentiation demonstrated more efficient cardiac differentiation and maturation, as well as homogeneity in cell structure, for the monolayer differentiations (Jeziorowska et al. 2017). Extracellular matrix (ECM) can also potentially influence differentiation ability. For murine ESCs, collagen I and III were individually correlated with decreased cardiac differentiation efficiency, but increased differentiation efficiency when combined. Similar findings were found for the combination of high fibronectin, Wnt2a, and Activin A, suggesting that interactions between growth factors and ECM signalling pathways could modulate stem cell fate (Flaim et al. 2008).

5   The Diverse Nature of Cardiac Differentiations from Pluripotent Stem Cells

5.1   Heterogeneous Cell Populations Resulting from Cardiac Differentiation

Even once iPSCs are successfully differentiated to a cardiac fate, there is still a wide range of variability and heterogeneity in the resultant cell populations. One aspect of this is that cardiac differentiations typically produce a combination of cardiomyocytes and non-cardiomyocytes at varied proportions. There is evidence that these non-cardiomyocytes can impact properties of the cardiomyocytes themselves. For instance, one study reported that when non-cardiomyocytes were removed from a EB-based differentiation of ESCs, development/maturation of electrophysiology and calcium handling were stunted, but these phenotypes were rescued when non-cardiomyocytes were added back (Kim et al.

2010). A second group likewise found that non-cardiomyocytes had an effect upon iPSC-CM electrophysiology and contractility, although they observed optimal properties in several parameters around ~70% cardiomyocytes (Iseoka et al. 2018). However, it is possible that these effects could be cell line-dependent. A study with murine embryonic stem cell-derived cardiomyocytes (ESC-CMs) showed that one line had shortened action potential duration (APD) associated with purification of cardiomyocytes (αMHC+ cells), but another line had a slightly prolonged APD and increased action potential (AP) maximum upstroke velocity when cultured using the same conditions (Hannes et al. 2015). Beyond functional properties, one of these studies also reported an increased proportion of cardiomyocytes expressing ventricular versus atrial myosin light chain for co-cultures with higher cardiomyocyte purity (Iseoka et al. 2018). In an earlier report, it was also found that contaminating non-cardiomyocytes release NRG-1β, which can promote development of working-type (ventricular and atrial) cardiomyocytes (Zhu et al. 2010). Interestingly, one study discovered that BRAF-mutant fibroblast-like cells from cardiac-directed iPSC differentiation promote cardiomyocyte hypertrophy phenotypes via TGFβ paracrine signaling, and that examining purified cardiomyocytes could mask the contributions of non-cardiomyocytes to cardiomyocyte disease processes (Josowitz et al. 2016). This suggests that non-cardiomyocytes may be particularly relevant to some disease modeling applications. There is still much to be learned regarding the interactions between cardiomyocytes and non-cardiomyocytes derived from iPSCs, and this is likely to be an ongoing focus of investigation.

Regarding the cardiomyocytes themselves, there can also be heterogeneity between cultures and within the same culture in when it comes to of properties typically associated with atrial, ventricular, or nodal/pacemaker cardiomyocyte subtypes. Traditionally, it has been asserted that cardiac differentiations produce a heterogeneous population of these subtypes with distinct functional and molecular properties. In order to facilitate phenotyping of hiPSC-CMs, Kane, et al. have recently proposed a semi-quantitative system for wholistically classifying cardiomyocytes into specific subtypes based on a variety of parameters including AP morphology, gene/protein marker expression, cell morphology, calcium transients, and conduction (Kane and Terracciano 2017).

In terms of electrophysiology, there can be significant variability in APs both between distinct clusters of ESC-CMs and even within the same cluster, with individual clusters frequently having multiple types of APs (Vestergaard et al. 2017; Zhu et al. 2016). Some groups have used signal processing and machine learning to develop platforms to evaluate and classify the electrophysiology of ESC-CMs, and subsequently demonstrated that most cultures exhibit multiple AP phenotypes and even display a continuum of properties between different AP morphologies (Gorospe et al. 2014). However, various studies use different parameters to categorize AP profiles as atrial-like, ventricular-like, or nodal-like, with some researchers questioning whether chamber specificity can be determined via AP morphologies alone (Du et al. 2015; Kane et al. 2016).

One study examined the concordance between electrophysiology and expression of the proposed pacemaker markers HCN4 and Isl1 at Days 40 and 60 of differentiation by acquiring APs of single cells optically, then assessing protein expression via immunofluorescence in the same cell. The researchers saw that HCN4 expression was higher in the cells with pacemaker-like APs initially but that differences decreased with downregulation of HCN4 over time. Conversely, Isl1 expression was initially not different for cells with different AP profiles, but became statistically higher in electrophysiologically pacemaker-like versus ventricular-like cells over time. Therefore, they deemed that neither protein marker was sufficient to identify pacemaker-like cells. Interestingly, they saw that differences in AP properties of the collective groups between Day 40 and 60 seemingly reflected an increase in ventricular- and atrial-like cardiomyocytes, suggesting that subtype may not be determined by Day 40 (Yechikov et al. 2016). Other studies

have also found that subtype classification by AP morphology is influenced by time in culture. In one case it was reported that time in culture lead to a transition from nodal-like to ventricular-like APs, with a transient atrial-like phenotype appearing between Days 57–70. That group also performed flow cytometry analysis of cTnT (cardiomyocyte marker), HCN3 (nodal marker), MYL2 (ventricular marker), and MYL7 (atrial marker), which further supported a transition from nodal to atrial/ventricular-like phenotypes from Day 30 to Day 60. Both approaches also revealed some cells with intermediate phenotypes, and ultimately led to the conclusion that AP profiles could not be categorized into three distinct groups (Ben-Ari et al. 2016).

It is possible that culture conditions or microenvironment could further have an impact upon AP properties. For example, AP morphologies of iPSC-CMs seeded at different densities demonstrate distinct distributions, with these differences seemingly not due to gap junction conductance (Du et al. 2015). It has also been observed that similar APs can be found in local regions within clusters of ESC-CMs, with a continuous gradient of AP shapes between regions with distinct AP profiles (Zhu et al. 2016).

5.2　The Quest for Pure Cardiomyocyte Populations

This heterogeneity has led to the development of a variety of approaches to purify cardiomyocytes from the cardiac differentiation process and to enrich for or specifically differentiate cardiomyocytes with the properties associated with a particular cardiomyocyte subtype. These efforts would also aid in addressing variability in cellular distributions between independent differentiations. An early approach to enrich for cardiomyocytes was to use a Percoll density gradient with centrifugation, but this could only enrich to 40–70%(Ban et al. 2017). Mitochondrial staining via the TMRM dye was also proposed fairly early on, but later studies showed that this approach could not robustly discriminate cardiomyocytes early in differentiation from non-cardiomyocytes and undifferentiated ESCs (Elliott et al. 2011). Other proposed solutions have included expression of a drug resistance gene or fluorescent reporter gene driven by a cardiomyocyte reporter (followed by drug treatment or FACS), but these have the caveat of needing to genetically modify the cells (Ban et al. 2017).

One of the more common, non-invasive approaches is antibody-based enrichment via fluorescent activated cell sorting (FACS) or magnetic-activated cell sorting. Multiple papers from 2011 reported the identification of SIRPA (CD172a) and VCAM1 (CD106) as iPSC and ESC-derived cardiomyocyte cell-surface markers, respectively (Dubois et al. 2011; Elliott et al. 2011; Uosaki et al. 2011). While these markers can be useful, it should be kept in mind that they are not completely specific or selective. One of these papers reported that ~71% of NKX2-5 eGFP$^+$ ESC-CMs express VCAM1 and ~85% express SIRPA at Day 14 of differentiation, with only ~37% being dual-positive. Furthermore, only ~67% of VCAM+SIRPA+ cells were also eGFP+, and eGFP+SIRPA+ cells had higher expression of endothelial and smooth muscle markers (Elliott et al. 2011). Later reports have shown that VCAM1 is more highly expressed at earlier stages of differentiation (before Day 25) and that SIRPA expression exists as a continuum, which makes gating based on that alone to be difficult (Veevers et al. 2018).

Another promising non-invasive approach is to take advantage of metabolic differences between cardiomyocytes and non-cardiomyocytes. Differences in glucose and lactate metabolism between non-cardiomyocytes and cardiomyocytes from murine and human pluripotent stem cells allow for cardiomyocyte enrichment in glucose-depleted media with supplementation of lactate (Tohyama et al. 2013). A subsequently-developed protocol involving glucose- and glutamine-depleted media plus lactose was shown to also kill pluripotent stem cells remaining after differentiation (Tohyama et al. 2016). Other methods such as molecular beacons to label cardiomyocyte-specific mRNAs, miRNA-based enrichment, and microfluidic

systems are still in relatively early stages of development, but may prove to be useful in the future (Ban et al. 2017).

Likewise, numerous different approaches have been pursued in order to isolate cardiomyocytes with properties of a specific cardiomyocyte subtype. These have included an SLN reporter for atrial-like cardiomyocytes, a cGATA6 reporter for nodal-like cardiomyocytes, and an MLC-2v reporter for ventricular-like cardiomyocytes (Bizy et al. 2013; Josowitz et al. 2014; Zhu et al. 2010). A molecular beacon approach has been investigated in this context as well, namely the use of molecular beacons targeting *Irx4* mRNA in murine ESCs to select for ventricular-like cardiomyocytes. However, a high load of molecular beacons per cell were needed in order to achieve significant signal (Ban et al. 2015). Another group recently used an ESC line for which GFP expression was driven by the MYL2 promoter in order to screen for cell-surface markers of ventricular cardiomyocytes. They found that a CD77+/CD200- population was >97+ cTNI+ with 65% expression MYL2-GFP, allowing for selection of a nearly pure cardiomyocyte population which was enriched for ventricular-like cells. While this approach worked well for the ESC lines they tested, the two iPSC lines they attempted to use interestingly had little-to-no CD77 expression. This enrichment approach was amenable to both EB and monolayer-based differentiations, but with somewhat less efficiency in the monolayer differentiation (Veevers et al. 2018). Other researchers took a unique approach where instead of trying to sort out specific subpopulations, they aimed to identify them *in situ*. To that end, they used subtype-specific promoters (MLC-2v, SLN, and SHOX2) to express a voltage-sensitive fluorescent protein in iPSC-CMs for subtype-specific optical AP recordings (Chen et al. 2017).

Other groups have taken a more developmental biology-informed approach and thereby developed differentiation protocols tailored to the production of particular cardiomyocyte subtypes. Initial work with neonatal rat ventricular myocytes and murine ESCs showed that overexpression of *Tbx18* or *Isl1* transcription factors was associated with development of the nodal subtype (Dorn et al. 2015; Kapoor et al. 2013). Inhibition of NRG-1β/ErbB signalling can also enhance the proportion of nodal-like cells, as can co-modulation of BMP, RA, and FGF signalling pathways (Protze et al. 2017; Zhu et al. 2010). Interestingly, iPSCs co-cultured with the visceral endoderm-like cell line END-2 produced primarily nodal-like cells, as well (Schweizer et al. 2017).

Protocols have also been proposed for the targeted production of working-type cardiomyocytes. There have been a couple protocols that involved modulation of canonical Wnt signalling by the small molecule IWR-1 in order to produce ventricular-like cardiomyocytes from ESCs and iPSCs (Karakikes et al. 2014; Weng et al. 2014). Gremlin 2 has been reported to upregulate pro-atrial transcription factors and downregulate atrial fate-repressive transcription factors during the differentiation of murine ESCs via stimulation of JNK signaling (Tanwar et al. 2014). More studies, though, have focused on the role of retinoid signaling in atrial verus ventricular development from pluripotent stem cells. Protocols involving retinoic acid treatment can promote atrial-like phenotypes, whereas protocols which include treatment with a retinoic acid receptor antagonist can promote ventricular-like development (Devalla et al. 2015; Lemme et al. 2018; Zhang et al. 2011). A subsequent study showed that atrial and ventricular-like cardiomyocytes develop optimally from specific mesoderm populations (CD235a+/CYP26a1+ for ventricular-like and RALDH2+ for atrial-like), and that these different mesoderms can be specified with different concentrations of BMP4 and Activin A. The RALDH2+ mesoderm responds to retinol to thus make atrial-like cardiomyocytes, since only cells with ALDH expression can synthesize retinoic acid from retinol. Retinoic acid can specify both mesoderms to an atrial fate, but the RALDH2+ mesoderm is more efficient for the production of atrial-like cells. Conversely, without retinoid signalling the RALDH2+ mesoderm can produce ventricular-like cardiomyocytes, but at low efficiency. Importantly, this study also showed that differential cell

lines may have variable expression of endogenous Nodal/Activin A and that different cytokine lots can have different activity, and thus optimization of differentiation reagents is necessary (Lee et al. 2017). One group has even used a reporter for the atrial transcription factor COUP-TFII to further enrich atrial-like hESC-CMs from a retinoic acid-directed cardiac differentiation. Interestingly, though, they also found that COUP-TFII was not required for atrial specification of the hESCs, highlighting that the processes associated with development of different cardiomyocyte subtypes have not yet been fully elucidated (Schwach et al. 2017).

5.3 Phenotypic Variability of Cardiomyocytes

Beyond the consideration of different cardiomyocyte subtype-like populations arising from cardiac differentiation, there is also quite a bit of variability and heterogeneity in other aspects of pluripotent stem cell-derived cardiomyocyte properties. A number of studies have focused on evaluating electrophysiological properties in particular, which have been demonstrated to differ between differentiations (with different cell lines or differentiation protocols) for both mouse and human (Hannes et al. 2015; Pekkanen-Mattila et al. 2010). For example, one study reported that there was heterogeneity in electrophysiological phenotypes of ESC-CMs differentiated with two different methods, with approximately one third of cells demonstrating fairly mature electrophysiological properties (maximum diastolic potential < −70 mV and upstroke velocity >140 V/S) but others appearing more embryonic-like (Pekkanen-Mattila et al. 2010). Even cell lines with similar gene expression profiles at the pluripotent cell state can have distinct electrophysiological properties, which was the case for an ESC line and an iPSC line differentiated to cardiomyocytes in one particular study. This comparison revealed great differences in APs and sodium currents at Day 60 of differentiation, with higher sodium currents in the iPSC-CMs and differential responsiveness to lidocaine and tetrodotoxin. There was also variation in AP frequency and APD, as well as differences in subtype classification between the lines and as a function of time (Sheng et al. 2012). AP profiles can change in numerous ways as the cardiomyocytes undergo maturation with time in culture, due to ongoing development of multiple electrophysiological currents in terms of current density and properties (Sartiani et al. 2007). Such variability can have implications for the application of these cells, where it is often important to elucidate which differences are biologically meaningful. For example, there can be variability in APD and drug responses for iPSC-CMs from LQT3 patients. In response to this, one research group created an *in silico* model to identify plausible mechanisms, and henceforth identified currents with possible differences at baseline or in response to drug treatment (Paci et al. 2017). Culture conditions can have a profound effect upon electrophysiological properties of hiPSC-CMs, which can be a particularly important consideration for drug-screening applications. In one study, it was discovered that more drugs prolonged field potential duration of iPSC-CMs in serum-containing media than in serum-free media, with some drugs also inducing arrhythmias at lower concentrations in the serum-containing media. This was a result of the media formulation impacting both compound availability (dissolved drug concentrations were surprisingly lower in the serum-free media) and baseline electrophysiology (the cells in serum-containing media had longer field potential durations) (Schocken et al. 2018).

Much of the heterogeneity in pluripotent-stem cell-derived cardiomyocytes can be attributed to the maturation status of these cells. With changes in maturation come changes in numerous cardiomyocyte properties including cell morphology (size, shape, nucleation), gene expression, contractility (sarcomere organization, myosin light chain isoforms, troponin T isoforms), electrophysiology (ion channels, APs, cell-cell coupling, conduction velocity), calcium handling, metabolism (including mitochondrial maturity), and proliferation. Numerous different approaches

have been taken to modulate and enhance the maturity of these cells and there have been several informative reviews on this topic, including recent reviews by Scuderi et al. and Tu et al. (Scuderi and Butcher 2017; Tu et al. 2018).

One simple approach is to culture the cells for extended periods of time, even months. This can lead to changes in morphology, contractile properties, calcium handling, electrophysiology, and gene expression (Lundy et al. 2013). Culture substrate can also have a notable impact upon cardiomyocyte development, since the use of substrates with physiological stiffness, micro- or nano-patterned surfaces, incorporation of native cardiac extracellular matrix components, and culture in 3D scaffolds can be used to promote advanced maturation (Carson et al. 2016; Fong et al. 2016; Nunes et al. 2013; Ribeiro et al. 2015; Ruan et al. 2015; Tiburcy et al. 2017; Zhang et al. 2013). In an effort to even further mimic physiology, both electrical and mechanical stimulation have been used to promote cardiomyocyte maturation (Mihic et al. 2014; Nunes et al. 2013; Ruan et al. 2015, 2016; Shen et al. 2017). Even increasing the conductivity of the culture system can enhance maturation, for example through the incorporation of trace amounts of electrically conductive silicon nanowires into scaffold-free cardiac spheroids (Tan et al. 2015).

In addition to physical influences upon the cardiomyocytes and their development, chemical influences designed to mimic *in vivo* maturation factors can also be quite impactful. For example, both tri-iodo-L-thyronine and dexamethasone (thyroid and glucocorticoid hormones, respectively) can enhance multiple measures of cardiomyocyte maturation. The combination of the two with a Matrigel mattress protocol is able to promote development of a T-tubule network (Parikh et al. 2017), which has historically been a bottleneck in the maturation of these cells (Scuderi and Butcher 2017). Some miRNAs are also able to impact cardiomyocyte maturation, as was found to be the case for overexpression of Let-7 miRNA family members (Kuppusamy et al. 2015).

A recent examination of cardiac differentiation from human pluripotent stem cells via single-cell RNA-seq was able to provide great insight into the transcriptional heterogeneity of the cells arising from this differentiation process, in particular revealing the role of HOPX in late stages of cardiac maturation (Friedman et al. 2018). A second recent study applied both single-cell RNA-seq and bulk RNA-seq over the course of a cardiac differentiation of iPSCs, and thereby identified distinct subpopulations of cardiomyocytes which were enriched for specific cardiac transcription factors and represented distinct maturation states. Through a variety of follow-up experiments, the authors furthermore found evidence that two of these transcription factors, *NR2F2* and *HEY2*, can promote atrial and ventricular transcriptional and electrophysiological phenotypes, respectively (Churko et al. 2018). Both of these studies provide a wealth of new information, and the continued use of single-cell RNA-seq will likely provide additional insight into the heterogeneity of iPSC-CMs and allow for generation of new hypotheses regarding how to better control the output of the differentiation process.

6 Looking Ahead: Approaches to Improve Consistency and Reproducibility

6.1 Improving and Validating the Starting Material

Overall, the creation of iPSCs from somatic cells and subsequently differentiation of these cells into cardiomyocytes involves taking mosaic cells from genetically diverse individuals and subjecting them to a wide variety of procedures and environmental conditions over the course of several months. Furthermore, there are no universally defined standards for these processes and culture conditions, which can differ markedly between groups or even individuals (Fig. 2). It should therefore not be surprising that different batches of iPSCs and iPSC-CMs demonstrate considerable variability, and that heterogeneity can even be present within a single population.

Moving forward, there will continue to be a need to reduce undesired variability within the iPSC-derived cardiomyocyte platform, in order to highlight true biological differences that are relevant for the given application. A component of this will be ensuring that the starting iPSCs are of high quality and meet a certain set of desired standards such as pluripotency and differentiation capacity. It has been suggested that one approach to this would be to choose cellular starting material that is less likely to have accumulated mutations or abnormalities (such as multipotent stem cells) (Silva et al. 2015). Quality control assays are also very useful in this regard. The teratoma assay is an established gold-standard for the capacity of pluripotent stem cells to differentiate into all three germ layers. However, it has been shown that it is not necessarily sufficient as a stand-alone means of evaluating pluripotent stem cell quality. For example, one study found that 45/46 evaluated cell lines could form teratomas with all three germ layers, yet 23 of those cell lines had contamination, karyotypic abnormalities, or features suggestive of spontaneous differentiation in culture (Salomonis et al. 2016). It has also been reported that murine iPSCs can demonstrate differences in cardiogenic potential despite a lack of variability in teratoma formation (Hartjes et al. 2014). In response to this, one group created a quantitative scorecard (TeratoScore) based on gene expression data from *in vivo* cell types in order to differentiate pluripotent stem cell-derived teratomas from malignant tumors. This approach could even differentiate between normal and abnormal karyotype (Avior et al. 2015).

A variety of other types of assays are now available to provide additional quality information on pluripotent stem cells. For instance, one group established an unbiased approach to evaluate colony morphology of human pluripotent stem cells using automated live-cell, label-free imaging and analysis algorithms (Kato et al. 2016). The PluriTest was created in order to evaluate pluripotency based on gene expression profiles, using both a "pluripotency score" and a "novelty score", which quantifies how different the gene expression profile of the sample is from the historic data used by the algorithm (Muller et al. 2011). The ScoreCard assay also uses gene

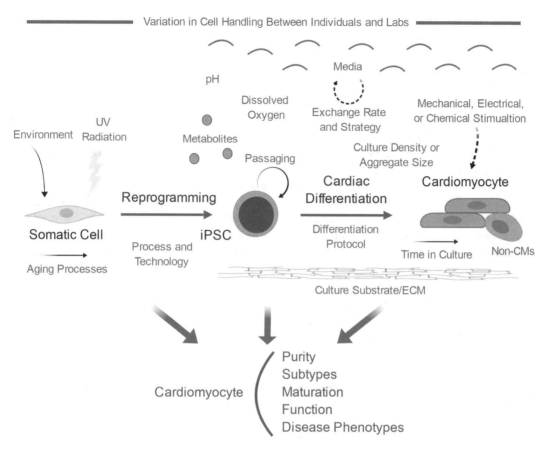

Fig. 2 Role of extrinsic factors in cellular properties. Numerous environmental and technical factors have been shown to influence the molecular and functional properties of somatic cells, iPSCs, and iPSC-derived cardiomyocytes. Altogether, these can ultimately modulate the final cardiomyocyte product and impact its utility for the desired applications

expression signatures, but to determine differentiation capacity (Tsankov et al. 2015). The Etoposide Sensitivity Assay (ESA) developed by our group takes advantage of the fact that pluripotent stem cells are hypersensitive to the topoisomerase inhibitor etoposide, and thus can be used to distinguish good quality iPSC clones from malignant teratocarcinoma clones. This is in contrast to PluriTest, which was shown to be unable to distinguish pluripotent teratocarcinoma cell lines or those with a considerable amount of spontaneous differentiation (Secreto et al. 2017).

Another study, though, highlighted the fact that some of these assays may not be sufficient in isolation. The authors profiled 18 cell lines which had variation in endogenous pluripotency gene expression and other properties, including those which were only partially reprogrammed and had low SSEA4 expression. However, this variability did not have any bearing on other criteria for evaluating pluripotency such as teratoma formation and the PluriTest assay. However, for the lines which did fulfill the most stringent pluripotency criteria, there was low interclonal and inter-individual variability. The authors concluded that thorough analyses of pluripotency are necessary and that proper characterization is vital to be able to distinguish differences between individuals from disease-associated differences (Vitale et al. 2012).

Recently, the International Stem Cell Initiative performed a detailed comparison of several different quality control assays via blinded analyses by independent experts in iPSCs and ESCs in four laboratories. Of the four methods the evaluated (PluriTest, 'Spin EB' system plus adapted lineage ScoreCard method, histological assessment of teratomas, and TeratoScore assessment of teratomas), all could be used to show pluripotency and each provided some information about differentiation potential. The authors suggested that the particular approach should be chosen based on the final application of the cells. For example, they asserted that a teratoma assay would be vital for cells that are intended for clinical purposes, since only that approach could evaluate both pluripotency and malignant potential (International Stem Cell I 2018).

There has been much focus on figuring out how to maintain pluripotent stem cell quality and consistency through the culture system used, for example by developing fully defined and integration-free conditions for iPSC reprogramming, and implementing pluripotent stem cell culture systems using chemically-defined media, attachment surfaces, and splitting reagents. These approaches have the potential to minimize the batch-to-batch variation that can be seen in culture systems which use serum-containing and Matrigel for cell attachment (Chen et al. 2011). One study reported the creation of current Good Manufacturing Practice (cGMP)-compliant iPSC lines for clinical purposes and tested differentiation capacity into cell types from all 3 germ layers. The authors proposed that the creation of repositories of well-characterized iPSC lines that could be expected to respond predictably to standard differentiation protocols. However, this must be taken with the caveat that they based their assertion that cGMP iPSC lines behave in a predictable manner on only two iPSC lines (Rao et al. 2018). Several chemically-defined differentiation conditions for pluripotent stem cells involving chemically defined medium and small molecules have also been described and subsequently shown to produce reproducible differentiation efficiency across 10+ iPSC lines differentiated repeatedly at multiple passages (Burridge et al. 2014; Lian et al. 2015, 2017). Some researchers even used

an albumin-free and chemically-defined medium for ventricular- and atrial-directed differentiations (using either retinoic acid or a retinoic acid inhibitor). They were able to achieve higher efficiency, higher cardiomyocyte yield, and lower inter-experimental variation as compared to differentiations performed using a B27-supplemented medium (Pei et al. 2017).

Significant effort has been devoted in general to the creation of new and improved cardiac differentiation protocols, with the aim of increasing efficiency, yield, and reproducibility. For instance, one early study involved the optimization of >45 different variables for cardiac differentiation of iPSCs and ESCs (Burridge et al. 2011). However, variability between cell lines can hinder the pursuit of universal differentiation protocols, and in some cases modified protocols have been created to enhance cardiac differentiation of specific cell lines which respond poorly to standard protocols (Hrstka et al. 2017; Yassa et al. 2018). Some suggestions have been made as to how to address this challenge in a more systematic way. One group applied a cytokine screening strategy to optimize cardiac output for murine ESC lines with differences in endogenous signaling of Activin/Nodal and BMP (Kattman et al. 2011).

Another group created a high-throughput platform to screen pluripotent stem cells in different microenvironments in order to optimize colony size, cell density, media composition, and substrate, and ultimately quantify endogenous signaling pathways and differentiation bias. They found that endogenous signaling is a major source of variability in how cells respond to exogenous induction conditions, and could therefore use their system to improve differentiation of difficult cell lines, including along a cardiac lineage (Nazareth et al. 2013). A different group focused on addressing the challenges associated with cardiac differentiation which can be posed by high density monolayer culture. They found that using rapamycin (mTOR inhibition) and CHIR99021 together improved efficiency and yield by reducing p53- and DNA damage-depended apoptosis in high density culture through reduction of p53 accumulation and mitochondrial ROS production. A similar effect could potentially be achieved by hypoxia and control over the nutrients present, instead of rapamycin treatment (Qiu et al. 2017).

6.2 Prioritizing Robust Study Designs and Cellular Manufacturing

Although the variable features of pluripotent stem cell-derived cardiomyocytes can pose a challenge, it should also be kept in mind that these may reflect true biological differences between individuals, an advantage of the iPSC system. For example, not only is beat rate variation present in iPSC-CMs, but heart rate variation can be observed *in vivo* (Binah et al. 2013). Additionally, an examination of ECGs from over 12,000 subjects undergoing routine medical exams for occupation purposes furthermore revealed natural variation in QT interval, down to 335 ms (Gallagher et al. 2006). In the case of diseased populations, electrophysiological recordings from sinus rhythm and chronic atrial fibrillation patients have been shown to exhibit inter-subject variability in AP morphology. Mathematical modeling has been used to determine possible causes of this, and revealing that variability in several different ion currents could modulate variability in APD and triangulation (Sanchez et al. 2014).

Ideally, studies involving iPSC-CMs should be carefully designed in order to highlight relevant differences between healthy and disease states, without being overshadowed by other inter-individual differences. It has been posited that variability among small cohorts of iPSCs could lead to inaccurate conclusions due to inherent differences arising from genetic variability (Kyttala et al. 2016). A number of studies have suggested that it is preferable to utilize cell lines from more individuals, rather than generating and studying multiple iPSC lines from the same individual, in order to differentiate disease mechanisms from the effect of genetic background in disease modeling (Burrows et al. 2016; Rouhani et al. 2014; Schuster et al. 2015).

A recent study concluded that use of more than one clone per individual can actually negatively impact the robustness of findings for transcriptionally-focused studies. Since differences between individuals play large roles in transcriptional variance, comparison of unrelated individuals, as generally done in disease modeling studies, will result in some differentially expressed genes (DEGs) that are not relevant to the disease of interest. They found that using more than one clone per individual actually increased spurious DEGs. While the use of multiple clones per individual can increase sensitivity (although not more than using more individuals), there is a larger loss in specificity. When multiple clones must be used per individual, the authors suggested using analysis methods that take into account the interdependence of the samples, such as an R package that they developed. The choice of controls was found to be another significant issue. An analysis of 77 studies published in 2016 showed that 79% of them used only unrelated controls. This is notable since very few spurious DEGs were found for the comparison of isogenic clones as opposed to a comparison between unrelated individuals. The authors suggested using two clones per individual with a mixed-models approach in order to obtain similar results to the use of isogenic controls, with at least 3 individuals per group in order to reduce false positives. When single clones from unrelated individuals are used, they suggested having at least 4 individuals per group, although having more than 6–7 per group did not improve performance (Germain and Testa 2017).

Other studies have likewise found the choice (or lack-thereof) of controls in iPSC disease modeling studies to be an issue. An analysis of 117 studies revealed that the median and average number of controls in such studies were only 1 and 1.6, respectively, and did not generally account for age, gender, or ethnicity. These authors suggested use of at least 3 controls from 3 separate subjects which are matched for such demographic factors. They proposed that these should be from unaffected family members whenever possible, and when not possible, as many as 12 or more individual donor lines should be used for controls since it has previously been reported that differences between iPSC and ESCs are negligible when that many lines are evaluated. Furthermore, for differentiated cells it may be essential to compare cells which have been in culture for similar amount of time, in order to reduce effect of maturation-related differences (Johnson et al. 2017).

Some researchers have suggested that when participants in iPSC studies are selected based on the presence or absence of polygenic disease the patients may be genetically heterogeneous and phenotypically variable, thus decreasing statistical power to detect the differences between cases and controls. Instead, they suggested the selection of patients with a known genetic variant with high penetrance and large effect size, or patients with high polygenic risk based on common genetic variants. They proposed an ideal study design with 4 different groups: patients with and without the disease penetrant variant or high polygenic risk, and controls with and without the same. The use of family members as controls could also help to control for genetic heterogeneity (Hoekstra et al. 2017). Finally, regardless of the number of clones per individual used in a study, it may be necessary to produce and screen multiple iPSC lines for chromosomal, nuclear gene, and mtDNA defects, any of which could potentially lead to misleading phenotypes (Kang et al. 2016).

While choice of the number and identity of samples and controls is highly relevant for disease modeling studies involving iPSCs, different considerations exist for the production of iPSCs and iPSC-CMs for therapeutic uses, high-throughput drug screening purposes, or other such applications. In these cases, improving consistency and quality in the cell manufacturing processes is of particular concern. To that end, various approaches to automating aspects of pluripotent stem cell culture have been investigated, including automated approaches to iPSC cell reprogramming, cell seeding, medium changes, passaging, differentiation, imaging, and harvesting (Konagaya et al. 2015; Kowalski et al. 2012; Paull et al. 2015; Serra et al. 2010). Some of these efforts have indeed been shown to

reduce well-to-well, plate-to-plate, and line-to-line variability (Crombie et al. 2017; Kowalski et al. 2012; Paull et al. 2015). Automated versus manual cell handling approaches have even been shown to differentially influence expression of pluripotency and differentiation marker expression in iPSCs (Archibald et al. 2016). In addition to adaptation of existing methods to automated approaches, ongoing improvements are being made to the methods themselves. For example, there was a recent report demonstrating that using dextran sulfate during cell seeding was able to control aggregate size and reduce heterogeneity and variability in suspension cultures of pluripotent stem cells. This is due to the fact that greater homogeneity in aggregates allows for more control over nutrient gradients and the prevention of large aggregate formation, in which the cells tend to lose pluripotency and undergo increased apoptosis (Lipsitz et al. 2018). Suspension culture methods will become increasingly valuable as a means to more efficiently produce large batches of pluripotent stem cell-derived cardiomyocytes, and thus improved methods in this area will also be of notable impact.

From a broader standpoint, generating cellular products requires overcoming variability in starting materials, reagents, microenvironment, and stochastic variability. Silva et al. have made some suggestions to help overcome these challenges, such as developing approaches for standardized comparative evaluation of cell product quality during the production process, and the need for robust and scalable standardized platforms for selection, purification, and validation of iPSCs (Silva et al. 2015). French et al. have provided an overview of the types of physical standards (reference materials) which will aid in improving reproducibility and consistency in the creation of differentiated cells from pluripotent stem cells. "Product" reference materials are representative of the product and can aid in evaluating its identity or potency. For example, these could be samples of specific batches, pooled populations from multiple batches, or other cell populations that are biologically equivalent in relevant properties. On the other hand, "method" reference materials, such as fixed cells or RNA samples, can be used to validate and define criteria for particular assays and perform necessary calibrations (French et al. 2015).

Building upon these ideas, Lipsitz et al. have been proponents of using quality-by-design principles to design cell manufacturing processes, an approach already commonly used by small-molecule pharmaceutical manufacturers. Quality-by-design integrates both scientific knowledge and risk analysis and involves product and process description, characterization, design, and monitoring. It also highlights the need to understand desired characteristics of the end product, attributes that influence safety and efficacy, and what parameters influence those attributes. In the case of pluripotent stem cell-derived cardiomyocytes, potency is now often evaluated via electrophysiological read-outs, but force-of-contraction assays may be of value for applications where they cells are intended to be used as a therapy and ultimately act as new heart tissue. One of the major issues for the use of pluripotent stem cell-derived cardiomyocytes in clinical applications is purity, since nodal cells and non-cardiomyocytes could potentially promote arrhythmias, and undifferentiated pluripotent stem cells can lead to teratomas. Ultimately, these authors highlighted a need to understand the influence of various factors such as dissolved oxygen, pH, metabolic by-products, and media exchange rate and strategy, and then monitor and control them if necessary, with the ability to reduce or at least understand the effect of variability in reagents being of equal importance (Lipsitz et al. 2018).

7   Final Remarks: Strategically Matching Approach to Application

Despite the challenges associated with the derivation and use of iPSC-derived cardiomyocytes, they have proven to be an extremely powerful platform in basic and translational science. Not only have these cells been used to model a wide variety of cardiac diseases, but they show great promise for drug safety testing and have demonstrated efficacy in large animal pre-clinical models (Gao et al. 2018; Ishida et al. 2018; Liu et al. 2018; Musunuru et al. 2018; Yoshida and Yamanaka 2017). Moreover, while there is still much to learn, there has been a progressively detailed understanding of how variability and heterogeneity in the iPSC-CM platform arises, and a number of proposed approaches to further enhance desired characteristics in the final cellular products. However, it should be recognized that it may not be feasible to validate and optimize all possible parameters and cellular properties for every single cell line. Therefore, it will become increasingly necessary to define which criteria are most important for a given study or application, and produce iPSC-CMs with those considerations in mind. This approach of developing purpose-built iPSC-CM products will require identifying which cellular features or functionalities are needed to achieve the ultimate purpose, and choosing the materials, protocols, and quality control measures based on those desired end properties. For instance, disease modeling relies heavily on careful selection of both the patient and control cell lines in order to uncover disease phenotypes and differentiate those from other aspects of inter-individual or inter-clonal variability. Conversely, for drug screening applications, the particular cell lines used may not matter as much as being able to achieve high batch-to-batch consistency.

There must also be an element of being able to balance risk versus benefit, which becomes particularly relevant for more translational or clinical applications. As one reflection of this, it may be appropriate to pursue the first clinical trials of iPSC-CMs in patient populations with severe disease and limited alternative treatment options, since those patients have the potential for achieving the greatest potential benefit despite the risks of a novel therapeutic modality. Then as quality control criteria for achieving maximal safety and efficacy are established, the number of suitable disease indications may grow.

Currently, the systems are in place to successfully create purpose-built iPSC-CMs for many applications. In fact, some of the challenges associated with these cells can actually be considered assets in expanding their versatility. For example, non-cardiomyocytes can be leveraged to modulate cardiomyocyte properties or reveal disease phenotypes, genetic variability can be used to recapitulate diverse populations *in vitro*, and the ability of these iPSC-CMs to display properties of varied cardiac subtypes and maturation states means that the cells can act as models of distinct regions of the heart across developmental stages. The future will only aid in refining these cells and expanding their utility. Therefore, as the field continues to discover what factors impact cardiomyocyte differentiation, purity, and ultimate phenotypes, as well as develop additional means by which to evaluate and control these factors, the potential of this platform will only grow.

Acknowledgements This work was supported by the Todd and Karen Wanek Family Program for Hypoplastic Left Heart Syndrome at Mayo Clinic. Figures were created with BioRender.

Conflict of interest T.J.N and Mayo Clinic have licensed reprogramming technology to ReGen Theranostics, Inc. in Rochester, MN.

Ethical approval The authors declare that this article does not contain any studies with human participants or animals.

References

Abyzov A et al (2012) Somatic copy number mosaicism in human skin revealed by induced pluripotent stem cells. Nature 492:438–442. https://doi.org/10.1038/nature11629

Andoh-Noda T et al (2017) Differential X chromosome inactivation patterns during the propagation of human induced pluripotent stem cells. Keio J Med 66:1–8. https://doi.org/10.2302/kjm.2016-0015-OA

Anguera MC et al (2012) Molecular signatures of human induced pluripotent stem cells highlight sex differences and cancer genes. Cell Stem Cell 11:75–90. https://doi.org/10.1016/j.stem.2012.03.008

Archibald PRT, Chandra A, Thomas D, Chose O, Massourides E, Laabi Y, Williams DJ (2016) Comparability of automated human induced pluripotent stem cell culture: a pilot study. Bioprocess Biosyst Eng 39:1847–1858. https://doi.org/10.1007/s00449-016-1659-9

Avior Y, Biancotti JC, Benvenisty N (2015) TeratoScore: assessing the differentiation potential of human pluripotent stem cells by quantitative expression analysis of teratomas. Stem Cell Rep 4:967–974. https://doi.org/10.1016/j.stemcr.2015.05.006

Ban K et al (2015) Non-genetic purification of ventricular cardiomyocytes from differentiating embryonic stem cells through molecular beacons targeting IRX-4. Stem Cell Rep 5:1239–1249. https://doi.org/10.1016/j.stemcr.2015.10.021

Ban K, Bae S, Yoon YS (2017) Current strategies and challenges for purification of cardiomyocytes derived from human pluripotent stem cells. Theranostics 7:2067–2077. https://doi.org/10.7150/thno.19427

Bauwens CL, Peerani R, Niebruegge S, Woodhouse KA, Kumacheva E, Husain M, Zandstra PW (2008) Control of human embryonic stem cell colony and aggregate size heterogeneity influences differentiation trajectories. Stem Cells 26:2300–2310. https://doi.org/10.1634/stemcells.2008-0183

Bauwens CL et al (2011) Geometric control of cardiomyogenic induction in human pluripotent stem cells. Tissue Eng Pt A 17:1901–1909. https://doi.org/10.1089/ten.tea.2010.0563

Ben-Ari M et al (2016) Developmental changes in electrophysiological characteristics of human-induced pluripotent stem cell derived cardiomyocytes. Heart Rhythm 13:2379–2387. https://doi.org/10.1016/j.hrthm.2016.08.045

Binah O, Weissman A, Itskovitz-Eldor J, Rosen MR (2013) Integrating beat rate variability: from single cells to hearts. Heart Rhythm 10:928–932. https://doi.org/10.1016/j.hrthm.2013.02.013

Bizy A et al (2013) Myosin light chain 2-based selection of human iPSC-derived early ventricular cardiac myocytes. Stem Cell Res 11:1335–1347. https://doi.org/10.1016/j.scr.2013.09.003

Blauwkamp TA, Nigam S, Ardehali R, Weissman IL, Nusse R (2012) Endogenous Wnt signalling in human embryonic stem cells generates an equilibrium of distinct lineage-specified progenitors. Nat Commun 3:1070. https://doi.org/10.1038/ncomms2064

Bruck T, Benvenisty N (2011) Meta-analysis of the heterogeneity of X chromosome inactivation in human pluripotent stem cells. Stem Cell Res 6:187–193. https://doi.org/10.1016/j.scr.2010.12.001

Burridge PW et al (2011) A Universal system for highly efficient cardiac differentiation of human induced pluripotent stem cells that eliminates interline variability. PLoS One 6:ARTN e18293. https://doi.org/10.1371/journal.pone.0018293

Burridge PW et al (2014) Chemically defined generation of human cardiomyocytes. Nat Methods 11:855–860. https://doi.org/10.1038/Nmeth.2999

Burrows CK, Banovich NE, Pavlovic BJ, Patterson K, Gallego Romero I, Pritchard JK, Gilad Y (2016) Genetic variation, not cell type of origin, underlies the majority of identifiable regulatory differences in iPSCs. PLoS Genet 12:e1005793. https://doi.org/10.1371/journal.pgen.1005793

Carcamo-Orive I et al (2017) Analysis of transcriptional variability in a large human iPSC library reveals genetic and non-genetic determinants of heterogeneity. Cell Stem Cell 20:518–532 e519. https://doi.org/10.1016/j.stem.2016.11.005

Carson D et al (2016) Nanotopography-induced structural anisotropy and sarcomere development in human cardiomyocytes derived from induced pluripotent stem cells. ACS Appl Mater Interfaces 8:21923–21932. https://doi.org/10.1021/acsami.5b11671

Chen XL, Chen A, Woo TL, Choo ABH, Reuveny S, SKW O (2010) Investigations into the metabolism of two-dimensional colony and suspended microcarrier cultures of human embryonic stem cells in serum-free media. Stem Cells Dev 19:1781–1792. https://doi.org/10.1089/scd.2010.0077

Chen G et al (2011) Chemically defined conditions for human iPSC derivation and culture. Nat Methods 8:424–429. https://doi.org/10.1038/nmeth.1593

Chen ZF et al (2017) Subtype-specific promoter-driven action potential imaging for precise disease modelling and drug testing in hiPSC-derived cardiomyocytes. Eur Heart J 38:292–301. https://doi.org/10.1093/eurheartj/ehw189

Cheng L et al (2012) Low incidence of DNA sequence variation in human induced pluripotent stem cells generated by nonintegrating plasmid expression. Cell Stem Cell 10:337–344. https://doi.org/10.1016/j.stem.2012.01.005

Chetty S, Pagliuca FW, Honore C, Kweudjeu A, Rezania A, Melton DA (2013) A simple tool to improve pluripotent stem cell differentiation. Nat Methods 10:553–556. https://doi.org/10.1038/nmeth.2442

Choi J et al (2015) A comparison of genetically matched cell lines reveals the equivalence of human iPSCs and ESCs. Nat Biotechnol 33:1173–1181. https://doi.org/10.1038/nbt.3388

Churko JM et al (2018) Defining human cardiac transcription factor hierarchies using integrated single-cell heterogeneity analysis. Nat Commun 9:ARTN 4906. https://doi.org/10.1038/s41467-018-07333-4

Crombie DE et al (2017) Development of a modular automated system for maintenance and differentiation of adherent human pluripotent stem cells. SLAS Discov 22:1016–1025. https://doi.org/10.1177/2472555217696797

Dalton S (2015) Linking the cell cycle to cell fate decisions. Trends Cell Biol 25:592–600. https://doi.org/10.1016/j.tcb.2015.07.007

D'Antonio M et al (2018) Insights into the mutational burden of human induced pluripotent stem cells from an integrative multi-omics approach. Cell Rep 24:883–894. https://doi.org/10.1016/j.celrep.2018.06.091

DeBoever C et al (2017) Large-scale profiling reveals the influence of genetic variation on gene expression in human induced pluripotent stem cells. Cell Stem Cell 20:533–546 e537. https://doi.org/10.1016/j.stem.2017.03.009

Denning C et al (2016) Cardiomyocytes from human pluripotent stem cells: from laboratory curiosity to industrial biomedical platform. Biochim Biophys Acta 1863:1728–1748. https://doi.org/10.1016/j.bbamcr.2015.10.014

Devalla HD et al (2015) Atrial-like cardiomyocytes from human pluripotent stem cells are a robust preclinical model for assessing atrial-selective pharmacology. Embo Mol Med 7:394–410. https://doi.org/10.15252/emmm.201404757

Dorn T et al (2015) Direct Nkx2-5 transcriptional repression of Isl1 controls cardiomyocyte subtype identity. Stem Cells 33:1113–1129. https://doi.org/10.1002/stem.1923

Du DTM, Hellen N, Kane C, Terracciano CMN (2015) Action potential morphology of human induced pluripotent stem cell-derived cardiomyocytes does not predict cardiac chamber specificity and is dependent on cell density. Biophys J 108:1–4. https://doi.org/10.1016/j.bpj.2014.11.008

Dubois NC et al (2011) SIRPA is a specific cell-surface marker for isolating cardiomyocytes derived from human pluripotent stem cells. Nat Biotechnol 29:1011–U1082. https://doi.org/10.1038/nbt.2005

Elliott DA et al (2011) NKX2-5(eGFP/w) hESCs for isolation of human cardiac progenitors and cardiomyocytes. Nat Methods 8:1037–1040. https://doi.org/10.1038/nmeth.1740

Espinosa Angarica V, Del Sol A (2016) Modeling heterogeneity in the pluripotent state: a promising strategy for improving the efficiency and fidelity of stem cell differentiation. Bioessays 38:758–768. https://doi.org/10.1002/bies.201600103

Fazeli A et al (2011) Altered patterns of differentiation in karyotypically abnormal human embryonic stem cells. Int J Dev Biol 55:175–180. https://doi.org/10.1387/ijdb.103177af

Flaim CJ, Teng D, Chien S, Bhatia SN (2008) Combinatorial signaling microenvironments for studying stem cell fate. Stem Cells Dev 17:29–39. https://doi.org/10.1089/scd.2007.0085

Fong AH et al (2016) Three-dimensional adult cardiac extracellular matrix promotes maturation of human induced pluripotent stem cell-derived cardiomyocytes. Tissue Eng Pt A 22:1016–1025. https://doi.org/10.1089/ten.tea.2016.0027

Forsyth NR, Kay A, Hampson K, Downing A, Talbot R, McWhir J (2008) Transcriptome alterations due to physiological normoxic (2% O-2) culture of human embryonic stem cells. Regen Med 3:817–833. https://doi.org/10.2217/17460751.3.6.817

French A et al (2015) Enabling consistency in pluripotent stem cell-derived products for research and development and clinical applications through material standards. Stem Cells Transl Med 4:217–223. https://doi.org/10.5966/sctm.2014-0233

Friedman CE et al (2018) Single-cell transcriptomic analysis of cardiac differentiation from human PSCs reveals HOPX-dependent cardiomyocyte maturation.

Cell Stem Cell 23:586–598 e588. https://doi.org/10.1016/j.stem.2018.09.009

Gallagher MM et al (2006) Distribution and prognostic significance of QT intervals in the lowest half centile in 12,012 apparently healthy persons. Am J Cardiol 98:933–935. https://doi.org/10.1016/j.amjcard.2006.04.035

Gao L et al (2018) Large cardiac muscle patches engineered from human induced-pluripotent stem cell-derived cardiac cells improve recovery from myocardial infarction in swine. Circulation 137:1712–1730. https://doi.org/10.1161/CIRCULATIONAHA.117.030785

Garitaonandia I et al (2015) Increased risk of genetic and epigenetic instability in human embryonic stem cells associated with specific culture conditions. PLoS One 10:ARTN e0118307. https://doi.org/10.1371/journal.pone.0118307

Geens M et al (2016) Female human pluripotent stem cells rapidly lose X chromosome inactivation marks and progress to a skewed methylation pattern during culture. Mol Hum Reprod 22:285–298. https://doi.org/10.1093/molehr/gaw004

Germain PL, Testa G (2017) Taming human genetic variability: transcriptomic meta-analysis guides the experimental design and interpretation of iPSC-based disease modeling. Stem Cell Rep 8:1784–1796. https://doi.org/10.1016/j.stemcr.2017.05.012

Gorospe G, Zhu RJ, Millrod MA, Zambidis ET, Tung L, Vidal R (2014) Automated grouping of action potentials of human embryonic stem cell-derived cardiomyocytes. IEEE Trans Biomed Eng 61:2389–2395. https://doi.org/10.1109/Tbme.2014.2311387

Guo GJ et al (2016) Serum-based culture conditions provoke gene expression variability in mouse embryonic stem cells as revealed by single-cell analysis. Cell Rep 14:956–965. https://doi.org/10.1016/j.celrep.2015.12.089

Gupta P, Hourigan K, Jadhav S, Bellare J, Verma P (2017) Effect of lactate and pH on mouse pluripotent stem cells: importance of media analysis. Biochem Eng J 118:25–33. https://doi.org/10.1016/j.bej.2016.11.005

Hannes T et al (2015) Electrophysiological characteristics of embryonic stem cell-derived cardiomyocytes are cell line-dependent. Cell Physiol Biochem 35:305–314. https://doi.org/10.1159/000369697

Hartjes KA, Li X, Martinez-Fernandez A, Roemmich AJ, Larsen BT, Terzic A, Nelson TJ (2014) Selection via pluripotency-related transcriptional screen minimizes the influence of somatic origin on iPSC differentiation propensity. Stem Cells 32:2350–2359. https://doi.org/10.1002/stem.1734

Hiura H et al (2013) Stability of genomic imprinting in human induced pluripotent stem cells. BMC Genet 14:32. https://doi.org/10.1186/1471-2156-14-32

Hoekstra SD, Stringer S, Heine VM, Posthuma D (2017) Genetically-informed patient selection for iPSC studies of complex diseases may aid in reducing cellular heterogeneity. Front Cell Neurosci 11:164. https://doi.org/10.3389/fncel.2017.00164

Hong SH, Rampalli S, Lee JB, McNicol J, Collins T, Draper JS, Bhatia M (2011) Cell fate potential of human pluripotent stem cells is encoded by histone modifications. Cell Stem Cell 9:24–36. https://doi.org/10.1016/j.stem.2011.06.002

Hrstka SC, Li X, Nelson TJ, Wanek Program Genetics Pipeline G (2017) NOTCH1-dependent nitric oxide signaling deficiency in hypoplastic left heart syndrome revealed through patient-specific phenotypes detected in bioengineered cardiogenesis. Stem Cells 35:1106–1119. https://doi.org/10.1002/stem.2582

Hussein SM et al (2011) Copy number variation and selection during reprogramming to pluripotency. Nature 471:58–62. https://doi.org/10.1038/nature09871

International Stem Cell I (2018) Assessment of established techniques to determine developmental and malignant potential of human pluripotent stem cells. Nat Commun 9:1925. https://doi.org/10.1038/s41467-018-04011-3

International Stem Cell Initiative C et al (2010) Comparison of defined culture systems for feeder cell free propagation of human embryonic stem cells. In Vitro Cell Dev Biol Anim 46:247–258. https://doi.org/10.1007/s11626-010-9297-z

Iseoka H et al (2018) Pivotal role of non-cardiomyocytes in electromechanical and therapeutic potential of induced pluripotent stem cell-derived engineered cardiac tissue. Tissue Eng Pt A 24:287–300. https://doi.org/10.1089/ten.tea.2016.0535

Ishida M et al (2018) Transplantation of human induced pluripotent stem cell-derived cardiomyocytes is superior to somatic stem cell therapy for restoring cardiac function and oxygen consumption in a porcine model of myocardial infarction. Transplantation 103(2):291–298. https://doi.org/10.1097/TP.0000000000002384

Ivanov NA et al (2016) Strong components of epigenetic memory in cultured human fibroblasts related to site of origin and donor age. PLoS Genet 12:e1005819. https://doi.org/10.1371/journal.pgen.1005819

Jacobs K, Mertzanidou A, Geens M, Nguyen HT, Staessen C, Spits C (2014) Low-grade chromosomal mosaicism in human somatic and embryonic stem cell populations. Nat Commun 5:4227. https://doi.org/10.1038/ncomms5227

Jacobs K et al (2016) Higher-density culture in human embryonic stem cells results in DNA damage and genome instability. Stem Cell Rep 6:330–341. https://doi.org/10.1016/j.stemcr.2016.01.015

Jeziorowska D et al (2017) Differential sarcomere and electrophysiological maturation of human iPSC-derived cardiac myocytes in monolayer vs. aggregation-based differentiation protocols. Int J Mol Sci 18:ARTN 1173. https://doi.org/10.3390/ijms18061173

Johnson AA, Andrews-Pfannkoch C, Nelson TJ, Pulido JS, Marmorstein AD (2017) Disease modeling studies using induced pluripotent stem cells: are we using enough controls? Regen Med 12:899–903. https://doi.org/10.2217/rme-2017-0101

Josowitz R et al (2014) Identification and purification of human induced pluripotent stem cell-derived atrial-like cardiomyocytes based on sarcolipin expression. PLoS One 9:ARTN e101316. https://doi.org/10.1371/journal.pone.0101316

Josowitz R et al (2016) Autonomous and non-autonomous defects underlie hypertrophic cardiomyopathy in BRAF-mutant hiPSC-derived cardiomyocytes. Stem Cell Rep 7:355–369. https://doi.org/10.1016/j.stemcr.2016.07.018

Jung CB et al (2012) Dantrolene rescues arrhythmogenic RYR2 defect in a patient-specific stem cell model of catecholaminergic polymorphic ventricular tachycardia. Embo Mol Med 4:180–191. https://doi.org/10.1002/emmm.201100194

Kalmar T, Lim C, Hayward P, Munoz-Descalzo S, Nichols J, Garcia-Ojalvo J, Martinez Arias A (2009) Regulated fluctuations in nanog expression mediate cell fate decisions in embryonic stem cells. PLoS Biol 7:e1000149. https://doi.org/10.1371/journal.pbio.1000149

Kane C, Terracciano CMN (2017) Concise review: criteria for chamber-specific categorization of human cardiac myocytes derived from pluripotent stem cells. Stem Cells 35:1881–1897. https://doi.org/10.1002/stem.2649

Kane C, Du DTM, Hellen N, Terracciano CM (2016) The fallacy of assigning chamber specificity to iPSC cardiac myocytes from action potential morphology. Biophys J 110:281–283. https://doi.org/10.1016/j.bpj.2015.08.052

Kang E et al (2016) Age-related accumulation of somatic mitochondrial DNA mutations in adult-derived human iPSCs. Cell Stem Cell 18:625–636. https://doi.org/10.1016/j.stem.2016.02.005

Kapoor N, Liang WB, Marban E, Cho HC (2013) Direct conversion of quiescent cardiomyocytes to pacemaker cells by expression of Tbx18. Nat Biotechnol 31:54–62. https://doi.org/10.1038/nbt.2465

Karakikes I et al (2014) Small molecule-mediated directed differentiation of human embryonic stem cells toward ventricular cardiomyocytes. Stem Cells Transl Med 3:18–31. https://doi.org/10.5966/sctm.2013-0110

Kato R et al (2016) Parametric analysis of colony morphology of non-labelled live human pluripotent stem cells for cell quality control. Sci Rep 6:34009. https://doi.org/10.1038/srep34009

Kattman SJ et al (2011) Stage-specific optimization of activin/nodal and BMP signaling promotes cardiac differentiation of mouse and human pluripotent stem cell lines. Cell Stem Cell 8:228–240. https://doi.org/10.1016/j.stem.2010.12.008

Kilpinen H et al (2017) Common genetic variation drives molecular heterogeneity in human iPSCs. Nature 546:370–375. https://doi.org/10.1038/nature22403

Kim KP et al (2007) Gene-specific vulnerability to imprinting variability in human embryonic stem cell lines. Genome Res 17:1731–1742. https://doi.org/10.1101/gr.6609207

Kim C et al (2010) Non-cardiomyocytes influence the electrophysiological maturation of human embryonic stem cell-derived cardiomyocytes during differentiation. Stem Cells Dev 19:783–795. https://doi.org/10.1089/scd.2009.0349

Kobayashi T, Kageyama R (2010) Hes1 regulates embryonic stem cell differentiation by suppressing Notch signaling. Genes Cells 15:689–698. https://doi.org/10.1111/j.1365-2443.2010.01413.x

Kobayashi T, Mizuno H, Imayoshi I, Furusawa C, Shirahige K, Kageyama R (2009) The cyclic gene Hes1 contributes to diverse differentiation responses of embryonic stem cells. Genes Dev 23:1870–1875. https://doi.org/10.1101/gad.1823109

Konagaya S, Ando T, Yamauchi T, Suemori H, Iwata H (2015) Long-term maintenance of human induced pluripotent stem cells by automated cell culture system. Sci Rep 5:ARTN 16647. https://doi.org/10.1038/srep16647

Kowalski MP, Yoder A, Liu L, Pajak L (2012) Controlling embryonic stem cell growth and differentiation by automation: enhanced and more reliable differentiation for drug discovery. J Biomol Screen 17:1171–1179. https://doi.org/10.1177/1087057112452783

Kuppusamy KT et al (2015) Let-7 family of microRNA is required for maturation and adult-like metabolism in stem cell-derived cardiomyocytes. P Natl Acad Sci U S A 112:E2785–E2794. https://doi.org/10.1073/pnas.1424042112

Kyttala A et al (2016) Genetic variability overrides the impact of parental cell type and determines iPSC differentiation potential. Stem Cell Rep 6:200–212. https://doi.org/10.1016/j.stemcr.2015.12.009

Laco F et al (2018) Unraveling the inconsistencies of cardiac differentiation efficiency induced by the GSK3 beta inhibitor CHIR99021 in human pluripotent stem cells. Stem Cell Rep 10:1851–1866. https://doi.org/10.1016/j.stemcr.2018.03.023

Lee JH, Protze SI, Laksman Z, Backx PH, Keller GM (2017) Human pluripotent stem cell-derived atrial and ventricular cardiomyocytes develop from distinct mesoderm populations. Cell Stem Cell 21:179–194.e4. https://doi.org/10.1016/j.stem.2017.07.003

Lemme M et al (2018) Atrial-like engineered heart tissue: an in vitro model of the human atrium. Stem Cell Rep 11:1378–1390. https://doi.org/10.1016/j.stemcr.2018.10.008

Li C et al (2015) Genetic heterogeneity of induced pluripotent stem cells: results from 24 clones derived from a single C57BL/6 mouse. PLoS One 10:e0120585. https://doi.org/10.1371/journal.pone.0120585

Lian XJ et al (2015) Chemically defined, albumin-free human cardiomyocyte generation. Nat Methods 12:595–596. https://doi.org/10.1038/nmeth.3448

Liang G, Zhang Y (2013) Genetic and epigenetic variations in iPSCs: potential causes and implications for application. Cell Stem Cell 13:149–159. https://doi.org/10.1016/j.stem.2013.07.001

Lim HJ, Han JY, Woo DH, Kim SE, Kim SK, Kang HG, Kim JH (2011) Biochemical and morphological effects of hypoxic environment on human embryonic stem cells in long-term culture and differentiating embryoid bodies. Mol Cell 31:123–132. https://doi.org/10.1007/s10059-011-0016-8

Lin YS et al (2017) Heparin promotes cardiac differentiation of human pluripotent stem cells in chemically defined albumin-free medium, enabling consistent manufacture of cardiomyocytes. Stem Cells Transl Med 6:527–538. https://doi.org/10.5966/sctm.2015-0428

Lipsitz YY, Tonge PD, Zandstra PW (2018) Chemically controlled aggregation of pluripotent stem cells. Biotechnol Bioeng 115:2061–2066. https://doi.org/10.1002/bit.26719

Liu YW et al (2018) Human embryonic stem cell-derived cardiomyocytes restore function in infarcted hearts of non-human primates. Nat Biotechnol 36:597–605. https://doi.org/10.1038/nbt.4162

Lundy SD, Zhu WZ, Regnier M, Laflamme MA (2013) Structural and functional maturation of cardiomyocytes derived from human pluripotent stem cells. Stem Cells Dev 22:1991–2002. https://doi.org/10.1089/scd.2012.0490

Mallon BS, Hamilton RS, Kozhich OA, Johnson KR, Fann YC, Rao MS, Robey PG (2014) Comparison of the molecular profiles of human embryonic and induced pluripotent stem cells of isogenic origin. Stem Cell Res 12:376–386. https://doi.org/10.1016/j.scr.2013.11.010

Mantsoki A, Devailly G, Joshi A (2016) Gene expression variability in mammalian embryonic stem cells using single cell RNA-seq data. Comput Biol Chem 63:52–61. https://doi.org/10.1016/j.compbiolchem.2016.02.004

Marczenke M, Fell J, Piccini I, Ropke A, Seebohm G, Greber B (2017) Generation and cardiac subtype-specific differentiation of PITX2-deficient human iPS cell lines for exploring familial atrial fibrillation. Stem Cell Res 21:26–28. https://doi.org/10.1016/j.scr.2017.03.015

Mayshar Y et al (2010) Identification and classification of chromosomal aberrations in human induced pluripotent stem cells. Cell Stem Cell 7:521–531. https://doi.org/10.1016/j.stem.2010.07.017

Mekhoubad S, Bock C, de Boer AS, Kiskinis E, Meissner A, Eggan K (2012) Erosion of dosage compensation impacts human iPSC disease modeling. Cell Stem Cell 10:595–609. https://doi.org/10.1016/j.stem.2012.02.014

Merkle FT et al (2017) Human pluripotent stem cells recurrently acquire and expand dominant negative P53 mutations. Nature 545:229–233. https://doi.org/10.1038/nature22312

Mihic A et al (2014) The effect of cyclic stretch on maturation and 3D tissue formation of human embryonic stem cell-derived cardiomyocytes. Biomaterials 35:2798–2808. https://doi.org/10.1016/j.biomaterials.2013.12.052

Mitalipova MM et al (2005) Preserving the genetic integrity of human embryonic stem cells. Nat Biotechnol 23:19–20. https://doi.org/10.1038/nbt0105-19

Moretti A et al (2010) Patient-specific induced pluripotent stem-cell models for long-QT syndrome. N Engl J Med 363:1397–1409. https://doi.org/10.1056/NEJMoa0908679

Muller FJ et al (2011) A bioinformatic assay for pluripotency in human cells. Nat Methods 8:315–317. https://doi.org/10.1038/nmeth.1580

Musunuru K et al (2018) Induced pluripotent stem cells for cardiovascular disease modeling and precision medicine: a scientific statement from the american heart association. Circ Genom Precis Med 11:e000043. https://doi.org/10.1161/HCG.0000000000000043

Narsinh KH et al (2011) Single cell transcriptional profiling reveals heterogeneity of human induced pluripotent stem cells. J Clin Invest 121:1217–1221. https://doi.org/10.1172/JCI44635

Nazareth EJP et al (2013) High-throughput fingerprinting of human pluripotent stem cell fate responses and lineage bias. Nat Methods 10:1225–1231. https://doi.org/10.1038/Nmeth.2684

Newman AM, Cooper JB (2010) Lab-specific gene expression signatures in pluripotent stem cells. Cell Stem Cell 7:258–262. https://doi.org/10.1016/j.stem.2010.06.016

Nguyen HT, Geens M, Spits C (2013) Genetic and epigenetic instability in human pluripotent stem cells. Hum Reprod Update 19:187–205. https://doi.org/10.1093/humupd/dms048

Nishino K, Umezawa A (2016) DNA methylation dynamics in human induced pluripotent stem cells. Hum Cell 29:97–100. https://doi.org/10.1007/s13577-016-0139-5

Nishizawa M et al (2016) Epigenetic variation between human induced pluripotent stem cell lines is an indicator of differentiation capacity. Cell Stem Cell 19:341–354. https://doi.org/10.1016/j.stem.2016.06.019

Nunes SS et al (2013) Biowire: a platform for maturation of human pluripotent stem cell-derived cardiomyocytes. Nat Methods 10:781–787. https://doi.org/10.1038/nmeth.2524

Osterloh JM, Mulelane K (2018) Manipulating cell fate while confronting reproducibility concerns. Biochem Pharmacol 151:144–156. https://doi.org/10.1016/j.bcp.2018.01.016

Paci M, Passini E, Severi S, Hyttinen J, Rodriguez B (2017) Phenotypic variability in LQT3 human induced pluripotent stem cell-derived cardiomyocytes and their response to antiarrhythmic pharmacologic therapy: an in silico approach. Heart Rhythm 14:1704–1712. https://doi.org/10.1016/j.hrthm.2017.07.026

Paige SL, Osugi T, Afanasiev OK, Pabon L, Reinecke H, Murry CE (2010) Endogenous Wnt/Beta-catenin signaling is required for cardiac differentiation in human embryonic stem cells. PLoS One 5:e11134. https://doi.org/10.1371/journal.pone.0011134

Parikh SS et al (2017) Thyroid and glucocorticoid hormones promote functional T-tubule development in human-induced pluripotent stem cell-derived cardiomyocytes. Circ Res 121:1323–1330. https://doi.org/10.1161/Circresaha.117.311920

Pauklin S, Vallier L (2014) The cell-cycle state of stem cells determines cell fate propensity. Cell 156:1338. https://doi.org/10.1016/j.cell.2014.02.044

Paull D et al (2015) Automated, high-throughput derivation, characterization and differentiation of induced pluripotent stem cells. Nat Methods 12:885–892. https://doi.org/10.1038/Nmeth.3507

Pei F et al (2017) Chemical-defined and albumin-free generation of human atrial and ventricular myocytes from human pluripotent stem cells. Stem Cell Res 19:94–103. https://doi.org/10.1016/j.scr.2017.01.006

Pekkanen-Mattila M, Chapman H, Kerkela E, Suuronen R, Skottman H, Koivisto AP, Aalto-Setala K (2010) Human embryonic stem cell-derived cardiomyocytes: demonstration of a portion of cardiac cells with fairly mature electrical phenotype. Exp Biol Med 235:522–530. https://doi.org/10.1258/ebm.2010.009345

Perales-Clemente E et al (2016) Natural underlying mtDNA heteroplasmy as a potential source of intraperson hiPSC variability. EMBO J 35:1979–1990. https://doi.org/10.15252/embj.201694892

Pick M, Stelzer Y, Bar-Nur O, Mayshar Y, Eden A, Benvenisty N (2009) Clone- and gene-specific aberrations of parental imprinting in human induced pluripotent stem cells. Stem Cells 27:2686–2690. https://doi.org/10.1002/stem.205

Protze SI, Liu J, Nussinovitch U, Ohana L, Backx PH, Gepstein L, Keller GM (2017) Sinoatrial node cardiomyocytes derived from human pluripotent cells function as a biological pacemaker. Nat Biotechnol 35:56–68. https://doi.org/10.1038/nbt.3745

Qiu XX et al (2017) Rapamycin and CHIR99021 coordinate robust cardiomyocyte differentiation from human pluripotent stem cells via reducing p53-dependent apoptosis. J Am Heart Assoc 6:ARTN e005295. https://doi.org/10.1161/JAHA.116.005295

Rao MS et al (2018) Illustrating the potency of current Good Manufacturing Practice-compliant induced pluripotent stem cell lines as a source of multiple cell lineages using standardized protocols. Cytotherapy 20:861–872. https://doi.org/10.1016/j.jcyt.2018.03.037

Reynolds N et al (2012) NuRD suppresses pluripotency gene expression to promote transcriptional heterogeneity and lineage commitment. Cell Stem Cell 10:583–594. https://doi.org/10.1016/j.stem.2012.02.020

Ribeiro AJ et al (2015) Contractility of single cardiomyocytes differentiated from pluripotent stem cells depends on physiological shape and substrate stiffness. Proc Natl Acad Sci U S A 112:12705–12710. https://doi.org/10.1073/pnas.1508073112

Rouhani F, Kumasaka N, de Brito MC, Bradley A, Vallier L, Gaffney D (2014) Genetic background drives transcriptional variation in human induced pluripotent stem cells. PLoS Genet 10:e1004432. https://doi.org/10.1371/journal.pgen.1004432

Ruan JL et al (2015) Mechanical stress promotes maturation of human myocardium from pluripotent stem cell-derived progenitors. Stem Cells 33:2148–2157. https://doi.org/10.1002/stem.2036

Ruan JL et al (2016) Mechanical stress conditioning and electrical stimulation promote contractility and force Maturation of induced pluripotent stem cell-derived human cardiac tissue. Circulation 134:1557–1567. https://doi.org/10.1161/Circulationaha.114.014998

Salomonis N et al (2016) Integrated genomic analysis of diverse induced pluripotent stem cells from the progenitor cell biology consortium. Stem Cell Rep 7:110–125. https://doi.org/10.1016/j.stemcr.2016.05.006

Sanchez C et al (2014) Inter-subject variability in human atrial action potential in sinus rhythm versus chronic atrial fibrillation. PLoS One 9:ARTN e105897. https://doi.org/10.1371/journal.pone.0105897

Sanchez-Freire V et al (2014) Effect of human donor cell source on differentiation and function of cardiac induced pluripotent stem cells. J Am Coll Cardiol 64:436–448. https://doi.org/10.1016/j.jacc.2014.04.056

Sartiani L, Bettiol E, Stillitano F, Mugelli A, Cerbai E, Jaconi ME (2007) Developmental changes in cardiomyocytes differentiated from human embryonic stem cells: a molecular and electrophysiological approach. Stem Cells 25:1136–1144. https://doi.org/10.1634/stemcells.2006.0466

Schocken D et al (2018) Comparative analysis of media effects on human induced pluripotent stem cell-derived cardiomyocytes in proarrhythmia risk assessment. J Pharmacol Toxicol Methods 90:39–47. https://doi.org/10.1016/j.vascn.2017.11.002

Schuster J et al (2015) Transcriptome profiling reveals degree of variability in induced pluripotent stem cell lines: impact for human disease modeling. Cell Rep 17:327–337. https://doi.org/10.1089/cell.2015.0009

Schwach V et al (2017) A COUP-TFII human embryonic stem cell reporter line to identify and select atrial cardiomyocytes. Stem Cell Rep 9:1765–1779. https://doi.org/10.1016/j.stemcr.2017.10.024

Schweizer PA et al (2017) Subtype-specific differentiation of cardiac pacemaker cell clusters from human induced pluripotent stem cells. Stem Cell Res Ther 8:229. https://doi.org/10.1186/s13287-017-0681-4

Scuderi GJ, Butcher J (2017) Naturally engineered maturation of cardiomyocytes. Front Cell Dev Biol 5:50. https://doi.org/10.3389/fcell.2017.00050

Secreto FJ et al (2017) Quantification of etoposide hypersensitivity: a sensitive, functional method for assessing pluripotent stem cell quality. Stem Cells Transl Med 6:1829–1839. https://doi.org/10.1002/sctm.17-0116

Serra M et al (2010) Improving expansion of pluripotent human embryonic stem cells in perfused bioreactors through oxygen control. J Biotechnol 148:208–215. https://doi.org/10.1016/j.jbiotec.2010.06.015

Shen N et al (2017) Steps toward maturation of embryonic stem cell-derived cardiomyocytes by defined physical signals. Stem Cell Rep 9:122–135. https://doi.org/10.1016/j.stemcr.2017.04.021

Sheng XW, Reppel M, Nguemo F, Mohammad FI, Kuzmenkin A, Hescheler J, Pfannkuche K (2012) Human pluripotent stem cell-derived cardiomyocytes: response to TTX and lidocain reveals strong cell to cell variability. PLoS One 7:ARTN e45963. https://doi.org/10.1371/journal.pone.0045963

Silva SS, Rowntree RK, Mekhoubad S, Lee JT (2008) X-chromosome inactivation and epigenetic fluidity in human embryonic stem cells. Proc Natl Acad Sci U S A 105:4820–4825. https://doi.org/10.1073/pnas.0712136105

Silva M et al (2015) Generating iPSCs: translating cell reprogramming science into scalable and robust biomanufacturing strategies. Cell Stem Cell 16:13–17. https://doi.org/10.1016/j.stem.2014.12.013

Singer ZS et al (2014) Dynamic heterogeneity and DNA methylation in embryonic stem cells. Mol Cell 55:319–331. https://doi.org/10.1016/j.molcel.2014.06.029

Singh AM (2015) Cell cycle-driven heterogeneity: on the road to demystifying the transitions between "poised" and "restricted" pluripotent cell states. Stem Cells Int 2015:219514. https://doi.org/10.1155/2015/219514

Singh AM et al (2014) Cell-cycle control of developmentally regulated transcription factors accounts for heterogeneity in human pluripotent cells. Stem Cell Rep 2:398. https://doi.org/10.1016/j.stemcr.2014.02.009

Smith A (2013) Nanog heterogeneity: tilting at windmills? Cell Stem Cell 13:6–7. https://doi.org/10.1016/j.stem.2013.06.016

Smith RCG, Stumpf PS, Ridden SJ, Sim A, Filippi S, Harrington HA, MacArthur BD (2017) Nanog fluctuations in embryonic stem cells highlight the problem of measurement in cell biology. Biophys J 112:2641–2652. https://doi.org/10.1016/j.bpj.2017.05.005

Sturrock M, Hellander A, Matzavinos A, Chaplain MA (2013) Spatial stochastic modelling of the Hes1 gene regulatory network: intrinsic noise can explain heterogeneity in embryonic stem cell differentiation. J R Soc Interface 10:20120988. https://doi.org/10.1098/rsif.2012.0988

Taapken SM, Nisler BS, Newton MA, Sampsell-Barron TL, Leonhard KA, McIntire EM, Montgomery KD (2011) Karotypic abnormalities in human induced pluripotent stem cells and embryonic stem cells. Nat Biotechnol 29:313–314. https://doi.org/10.1038/nbt.1835

Takahashi K, Tanabe K, Ohnuki M, Narita M, Ichisaka T, Tomoda K, Yamanaka S (2007) Induction of pluripotent stem cells from adult human fibroblasts by defined factors. Cell 131:861–872. https://doi.org/10.1016/j.cell.2007.11.019

Tan Y et al (2015) Silicon nanowire-induced maturation of cardiomyocytes derived from human induced pluripotent stem cells. Nano Lett 15:2765–2772. https://doi.org/10.1021/nl502227a

Tanasijevic B, Dai B, Ezashi T, Livingston K, Roberts RM, Rasmussen TP (2009) Progressive accumulation of epigenetic heterogeneity during human ES cell culture. Epigenetics 4:330–338

Tanwar V et al (2014) Gremlin 2 promotes differentiation of embryonic stem cells to atrial fate by activation of the JNK signaling pathway. Stem Cells 32:1774–1788. https://doi.org/10.1002/stem.1703

Tesarova L, Simara P, Stejskal S, Koutna I (2016) The aberrant DNA methylation profile of human induced pluripotent stem cells is connected to the reprogramming process and is normalized during in vitro culture. PLoS One 11:e0157974. https://doi.org/10.1371/journal.pone.0157974

Tiburcy M et al (2017) Defined engineered human myocardium with advanced maturation for applications in heart failure modeling and repair. Circulation 135:1832–1847. https://doi.org/10.1161/CIRCULATIONAHA.116.024145

Titmarsh DM, Ovchinnikov DA, Wolvetang EJ, Cooper-White JJ (2013) Full factorial screening of human embryonic stem cell maintenance with multiplexed microbioreactor arrays. Biotechnol J 8:822–834. https://doi.org/10.1002/biot.201200375

Tohyama S et al (2013) Distinct metabolic flow enables large-scale purification of mouse and human pluripotent stem cell-derived cardiomyocytes. Cell Stem Cell 12:127–137. https://doi.org/10.1016/j.stem.2012.09.013

Tohyama S et al (2016) Glutamine oxidation is indispensable for survival of human pluripotent stem cells. Cell Metab 23:663–674. https://doi.org/10.1016/j.cmet.2016.03.001

Tsankov AM et al (2015) A qPCR ScoreCard quantifies the differentiation potential of human pluripotent stem cells. Nat Biotechnol 33:1182–1192. https://doi.org/10.1038/nbt.3387

Tu C, Chao BS, Wu JC (2018) Strategies for improving the maturity of human induced pluripotent stem cell-derived cardiomyocytes. Circ Res 123:512–514. https://doi.org/10.1161/CIRCRESAHA.118.313472

Uosaki H, Fukushima H, Takeuchi A, Matsuoka S, Nakatsuji N, Yamanaka S, Yamashita JK (2011) Efficient and scalable purification of cardiomyocytes from human embryonic and induced pluripotent stem cells by VCAM1 surface expression. PLoS One 6:ARTN e23657. https://doi.org/10.1371/journal.pone.0023657

Veevers J et al (2018) Cell-surface marker signature for enrichment of ventricular cardiomyocytes derived from human embryonic stem cells. Stem Cell Rep 11:828–841. https://doi.org/10.1016/j.stemcr.2018.07.007

Vestergaard ML et al (2017) Human embryonic stem cell-derived cardiomyocytes self-arrange with areas of different subtypes during differentiation. Stem Cells Dev 26:1566–1577. https://doi.org/10.1089/scd.2017.0054

Vitale AM, Matigian NA, Ravishankar S, Bellette B, Wood SA, Wolvetang EJ, Mackay-Sim A (2012) Variability in the generation of induced pluripotent stem cells: importance for disease modeling. Stem Cells Transl Med 1:641–650. https://doi.org/10.5966/sctm.2012-0043

Wahlestedt M, Ameur A, Moraghebi R, Norddahl GL, Sten G, Woods NB, Bryder D (2014) Somatic cells with a heavy mitochondrial DNA mutational load render induced pluripotent stem cells with distinct differentiation defects. Stem Cells 32:1173–1182. https://doi.org/10.1002/stem.1630

Wang G et al (2014) Modeling the mitochondrial cardiomyopathy of Barth syndrome with induced pluripotent stem cell and heart-on-chip technologies. Nat Med 20:616–623. https://doi.org/10.1038/nm.3545

Weng ZH et al (2014) A simple, cost-effective but highly efficient system for deriving ventricular cardiomyocytes from human pluripotent stem cells. Stem Cells Dev 23:1704–1716. https://doi.org/10.1089/scd.2013.0509

Wu JC, Fan YJ, Tzanakakis ES (2015) Increased culture density is linked to decelerated proliferation, prolonged G(1) phase, and enhanced propensity for differentiation of self-renewing human pluripotent stem cells. Stem Cells Dev 24:892–903. https://doi.org/10.1089/scd.2014.0384

Wutz A (2012) Epigenetic alterations in human pluripotent stem cells: a tale of two cultures. Cell Stem Cell 11:9–15. https://doi.org/10.1016/j.stem.2012.06.012

Wyles SP et al (2016) Modeling structural and functional deficiencies of RBM20 familial dilated cardiomyopathy using human induced pluripotent stem cells. Hum Mol Genet 25.254–265. https://doi.org/10.1093/hmg/ddv468

Xu H et al (2012) Highly efficient derivation of ventricular cardiomyocytes from induced pluripotent stem cells with a distinct epigenetic signature. Cell Res 22:142–154. https://doi.org/10.1038/cr.2011.171

Yassa ME, Mansour IA, Sewelam NI, Hamza H, Gaafar T (2018) The impact of growth factors on human induced pluripotent stem cells differentiation into cardiomyocytes. Life Sci 196:38–47. https://doi.org/10.1016/j.lfs.2018.01.009

Yazawa M, Hsueh B, Jia X, Pasca AM, Bernstein JA, Hallmayer J, Dolmetsch RE (2011) Using induced pluripotent stem cells to investigate cardiac phenotypes in Timothy syndrome. Nature 471:230–234. https://doi.org/10.1038/nature09855

Yechikov S, Copaciu R, Gluck JM, Deng W, Chiamvimonvat N, Chan JW, Lieu DK (2016) Same-single-cell analysis of pacemaker-specific markers in human induced pluripotent stem cell-derived cardiomyocyte subtypes classified by electrophysiology. Stem Cells 34:2670–2680. https://doi.org/10.1002/stem.2466

Yeo D et al (2013) Improving embryonic stem cell expansion through the combination of perfusion and Bioprocess model design. PLoS One 8:e81728. https://doi.org/10.1371/journal.pone.0081728

Yiangou L et al (2019) Method to synchronize cell cycle of human pluripotent stem cells without affecting their fundamental characteristics. Stem Cell Rep 12:165–179. https://doi.org/10.1016/j.stemcr.2018.11.020

Yoshida Y, Yamanaka S (2017) Induced pluripotent stem cells 10 years later: for cardiac applications. Circ Res 120:1958–1968. https://doi.org/10.1161/CIRCRESAHA.117.311080

Young MA et al (2012) Background mutations in parental cells account for most of the genetic heterogeneity of induced pluripotent stem cells. Cell Stem Cell 10:570–582. https://doi.org/10.1016/j.stem.2012.03.002

Yu J et al (2007) Induced pluripotent stem cell lines derived from human somatic cells. Science 318:1917–1920. https://doi.org/10.1126/science.1151526

Yu L et al (2018a) Low cell-matrix adhesion reveals two subtypes of human pluripotent stem cells. Stem Cell Rep 11:142–156. https://doi.org/10.1016/j.stemcr.2018.06.003

Yu P, Nie Q, Tang C, Zhang L (2018b) Nanog induced intermediate state in regulating stem cell differentiation and reprogramming. BMC Syst Biol 12:22. https://doi.org/10.1186/s12918-018-0552-3

Zhang QZ et al (2011) Direct differentiation of atrial and ventricular myocytes from human embryonic stem cells by alternating retinoid signals. Cell Res 21:579–587. https://doi.org/10.1038/cr.2010.163

Zhang DH, Shadrin IY, Lam J, Xian HQ, Snodgrass HR, Bursac N (2013) Tissue-engineered cardiac patch for advanced functional maturation of human ESC-derived cardiomyocytes. Biomaterials 34:5813–5820. https://doi.org/10.1016/j.biomaterials.2013.04.026

Zhu WZ, Xie YH, Moyes KW, Gold JD, Askari B, Laflamme MA (2010) Neuregulin/ErbB signaling regulates cardiac subtype specification in differentiating human embryonic stem cells. Circ Res 107:776–U258. https://doi.org/10.1161/Circresaha.110.223917

Zhu RJ, Millrod MA, Zambidis ET, Tung L (2016) Variability of action potentials within and among cardiac cell clusters derived from human embryonic stem cells. Sci Rep 6:ARTN 18544. https://doi.org/10.1038/srep18544

Adv Exp Med Biol – Cell Biology and Translational Medicine (2019) 6: 31–48
https://doi.org/10.1007/5584_2019_374
© Springer Nature Switzerland AG 2019
Published online: 16 April 2019

Direct Lineage Reprogramming in the CNS

Justine Bajohr and Maryam Faiz

Abstract

Direct lineage reprogramming is the conversion of one specialized cell type to another without the need for a pluripotent intermediate. To date, a wide variety of cell types have been successfully generated using direct reprogramming, both *in vitro* and *in vivo*. These newly converted cells have the potential to replace cells that are lost to disease and/or injury. In this chapter, we will focus on direct reprogramming in the central nervous system. We will review current progress in the field with regards to all the major neural cell types and explore how cellular heterogeneity, both in the starter cell and target cell population, may have implications for direct reprogramming. Finally, we will discuss new technologies that will improve our understanding of the reprogramming process and aid the development of more specific and efficient future CNS-based reprogramming strategies.

Keywords

Cellular reprogramming · Direct lineage conversion · Cellular heterogeneity · Neurological disease/Injury · Central nervous system

J. Bajohr and M. Faiz (✉)
Department of Surgery, University of Toronto, Toronto, Canada
e-mail: maryam.faiz@utoronto.ca

Abbreviations

6-OHDA	6-hydroxydopamine
Ascl1	achaete-scute family bHLH transcription factor 1
BAM factors	combination of the transcription factors *Ascl1, Brn2 and Mytl1*
Brn2	POU Class 3 Homeobox 2
CHAT	Choline O-Acetyltransferase
c-Myc	cellular Myc
CNP	2′,3′-Cyclic Nucleotide 3′ Phosphodiesterase
CNS	central nervous system
CRISPR	clustered regularly interspaced short palindromic repeats
CRISPRa	CRISPR activation
DAT	Dopamine transporter
DDC	DOPA Decarboxylase
Dlx2	Distal-Less Homeobox 2
DREADD	Designer Receptors Exclusively Activated by Designer Drugs
E47	transcription factor 3
Ezh2	Enhancer Of Zeste 2 Polycomb Repressive Complex 2 Subunit
Fezf2	FEZ Family Zinc Finger 2
Foxa2	Forkhead Box A2
FoxG1	forkhead box G1
GABA	Gamma-amino butyric acid
GLUT1	glucose transporter protein type 1
GRN	gene regulatory network
Hb9	Motor Neuron And Pancreas Homeobox 1
iPSC	induced pluripotent stem cell

Isl1	Insulin gene enhancer protein ISL-1
ITPR2	Inositol 1,4,5-Trisphosphate Receptor Type 2
Klf4	Kruppel Like Factor 4
Lhx3	LIM Homeobox 3
Lmx1a	LIM Homeobox Transcription Factor 1 Alpha
MBP	myelin basic protein
Mecom	MDS1 And EVI1 Complex Locus
miRNA	microRNA
MOL6	mature oligodendrocytes expressing Grm3 (Glutamate Metabotropic Receptor 3) and Jph4 (Junctophilin 4)
MS	multiple sclerosis
MyoD	myogenic differentiation 1
Myt1l	myelin transcription factor 1 like protein
NANOG	Nanog Homeobox
NeuroD1	Neurogenic Differentiation Factor 1
NFIA	Nuclear Factor I A
NFIB	Nuclear Factor I B
NG2 glia	Neural/glial antigen 2 expressing glial cells
Ngn2	Neurogenin 2
Nkx6.2	NK6 Homeobox 2
NSC	neural stem cell
NSPC	neural stem and progenitor cells
Nurr1	Nuclear receptor related 1 protein
OCT4	octamer-binding transcription factor 4
Olig1	Oligodendrocyte Transcription Factor 1
Olig2	Oligodendrocyte Transcription Factor 2
OPC	oligodendrocyte progenitor cell
Pax6	Paired Box 6
ROS	reactive oxygen species
S1 cortex	primary somatosensory cortex
sc RNA-seq	single cell RNA sequencing
Sox10	SRY-Box 10
Sox2	SRY-Box 2
Sox9	SRY-box 9
VMAT2	Vesicular monoamine transporter 2
VPA	valproic acid
Zfp536	Zinc Finger Protein 536

1   Introduction

Historically, it was believed that cell fate was fixed after the completion of development (Heins et al. 2002; Barker et al. 2018; Faiz and Nagy 2013; Vierbuchen and Wernig 2011). However, discoveries including cell fusion, somatic nuclear transfer, and most recently reprogramming to pluripotency (or the generation of induced pluripotent stem cells, iPSCs) have shown that cell fate is flexible (Faiz and Nagy 2013; Vierbuchen and Wernig 2011; Gurdon 1962; Chen et al. 2015; Graf and Enver 2009; Blau et al. 1983, 1985; Xie et al. 2004; Takahashi and Yamanaka 2006). In this review, we will focus on direct lineage reprogramming, which is the conversion of one specialized cell type to another (Graf and Enver 2009; Xu et al. 2015; Wang and Zhang 2018; Gascón et al. 2017a; Masserdotti et al. 2016) (Fig. 1). This was first demonstrated by Davis and colleagues, who showed that overexpression of *MyoD* resulted in the conversion of fibroblasts to myoblasts (Davis et al. 1987). More recently, a number of studies have demonstrated successful conversion of various other cell types, both *in vitro* and *in vivo* (for review, see (Barker et al. 2018; Chen et al. 2015; Xu et al. 2015; Wang and Zhang 2018; Gascón et al. 2017a; Masserdotti et al. 2016)). This ground-breaking technology has had a significant impact on the field of regenerative medicine, as directly reprogrammed cells could be used to replace those lost or damaged to disease or injury (Barker et al. 2018; Faiz and Nagy 2013; Chen et al. 2015; Graf and Enver 2009; Takahashi and Yamanaka 2006; Xu et al. 2015; Wang and Zhang 2018; Gascón et al. 2017a; Masserdotti et al. 2016).

Direct lineage conversion uses the delivery of specific factors to induce the conversion of cells without the need for a pluripotent intermediate (Graf and Enver 2009; Xu et al. 2015; Wang and Zhang 2018; Gascón et al. 2017a; Masserdotti et al. 2016). Typically, transcription factors have been used, but the feasibility of using small molecules (Hu et al. 2015; Li et al. 2015),

Fig. 1 Direct lineage reprogramming. Direct lineage reprogramming is the conversion of one specialized cell type (Cell A) to another (Cell B) without the need for a pluripotent intermediate. It can be initiated by a variety of methods (small molecules, microRNAs), but is typically achieved by the overexpression of transcription factors. Illustrated by Kayla Hoffman-Rogers

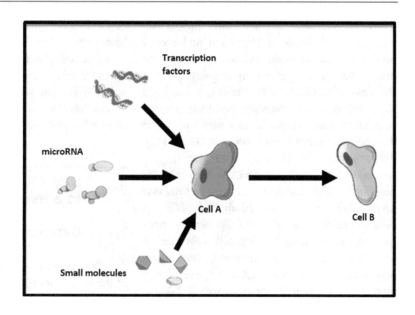

microRNAs (Yoo et al. 2011; Victor et al. 2014), and CRISPRa (Chakraborty et al. 2014; Black et al. 2016) (Fig. 1) has also been demonstrated. To date, most studies have identified reprogramming factors based on their role in specifying a target cell fate during development, and/or uniquely high gene expression in a target cell. For example, Najm and colleagues used microarray data from different central nervous system (CNS) cells to identify a pool of genes that were exclusively upregulated in oligodendrocytes (Najm et al. 2013). These genes were then tested for their ability to convert fibroblasts to oligodendrocytes (Najm et al. 2013).

Many studies have focused on identifying "core" factors that are needed for cellular conversion using a reductionist-additive approach (Ninkovic and Götz 2018). In this paradigm, one factor is removed at a time until the "necessary" factor(s) are found (Ninkovic and Götz 2018). Additional factors are then added back until a desired phenotype or efficiency is achieved (Ninkovic and Götz 2018). For example, following confirmation that a cocktail of eleven reprogramming factors was able to reprogram fibroblasts to motor neurons, Son and colleagues removed one transcription factor at a time and analyzed its effect on the conversion (Son et al.

2011). This allowed them to determine that *Ascl1* or *Lhx3* were crucial for fibroblast to neuron conversion (Son et al. 2011). Then, to determine the optimal combination of transcription factors for a motor neuron phenotype, they added back single transcription factors and identified a "core set" of seven (*Ascl1/Brn2/Myt1l/Lhx3/Hb9/Isl1/ Ngn2*) (*Son* et al. 2011). This approach suggests that an end goal is to achieve reprogramming with the smallest number of factors. Indeed, the seminal study by Davis and colleagues used only *MyoD* – highlighting the feasibility of a single factor for direct lineage reprogramming (Davis et al. 1987). One transcription factor for reprogramming may be favorable for future clinical applications, both in terms of feasibility of delivery and patient safety and tolerability. While it has been argued that single-factor reprogramming results in immature cell phenotypes (Morris 2016), neuronal reprogramming strategies using only one factor have resulted in the generation of mature and functional neurons, albeit at times with a slower maturation rate (Chanda et al. 2014; Zhu et al. 2018; Heinrich et al. 2010; Guo et al. 2014).

Interestingly, single-factor lineage reprogramming highlights the ability of certain reprogramming factors to behave as "pioneers"

(Ninkovic and Götz 2018). Pioneer factors can bind to closed areas of chromatin and recruit supporting transcription factors that may be needed to initiate the reprogramming process (Ninkovic and Götz 2018). Further, it has been suggested that the feasibility and efficiency of conversion using single factors may be due to their pioneer activity (Ninkovic and Götz 2018). For example, the pioneer factor *Ascl1*, may endogenously recruit other factors beneficial for fibroblast to neuron conversion, such as *Brn2* and *Mytl1* (Ninkovic and Götz 2018). This ability to bind to closed areas of chromatin demonstrates one way in which a starting cell state can be overridden; as in development, the genes regulating alternate cell fates are epigenetically repressed via chromatin modifications (Ninkovic and Götz 2018). Although a valuable insight into how reprogramming is initiated, many of the mechanisms that drive direct reprogramming have yet to be fully elucidated. It has been suggested that this is a complex process, dependent on many variables, including chromatin remodeling (Ninkovic and Götz 2018; Wapinski et al. 2017) and metabolic changes (Gascón et al. 2016, 2017b), amongst others (see (Xu et al. 2015; Wang and Zhang 2018; Gascón et al. 2017a, b; Masserdotti et al. 2015, 2016; Gascón et al. 2016) for a comprehensive review).

While the mechanisms of reprogramming remain unclear, the applicability of direct reprogramming technology is unmistakable. Direct lineage conversion has been used in many tissue systems and provides a novel therapeutic option for drug-resistant diseases or diseases with no current treatment options (Xu et al. 2015; Berninger 2010). In this review we will use the neural lineage as a model system to explore direct lineage reprogramming. Most studies have focused on direct reprogramming to neurons (reviewed in (Chen et al. 2015; Xu et al. 2015; Wang and Zhang 2018; Gascón et al. 2017a; Masserdotti et al. 2016)), because of the significant loss or injury to these cells in most neurological conditions. However, other neural lineage cells, for example, oligodendrocytes, may also be of interest. We will discuss the progress and current state of the field of direct lineage reprogramming with regards to all the major CNS cell types. We will explore how cellular heterogeneity, both in the starter cell population and the target cell type, may have implications for direct reprogramming. Finally, we will discuss new technologies that will improve our understanding of direct reprogramming and development of future conversion strategies.

2   Direct Reprogramming to a Neural Cell Fate

2.1   Overview

The first report of direct reprogramming to cells of the neural lineage used the transcription factor *Pax6* to convert astrocytes to neurons *in vitro* (Heins et al. 2002). Subsequent studies showed that the delivery of other transcription factors, such as *Ascl1* (Chanda et al. 2014), *Brn2* (Zhu et al. 2018) and *Ngn2* (Heinrich et al. 2010), could also convert astrocytes to neurons *in vitro*. Direct conversion has also been used to generate other neural cells such as oligodendrocytes (Najm et al. 2013; Yang et al. 2013; Mokhtarzadeh Khanghahi et al. 2018) and astrocytes (Caiazzo et al. 2015; Tian et al. 2016). It has also been shown that a wide variety of cell types, including those of a non-neural lineage, can be converted to the neural lineage. Fibroblasts and hepatocytes, two examples of non-neural cells, were successfully reprogrammed to neurons using a combination of *Brn2*/*Mytl1*/*Ascl1* (Vierbuchen et al. 2010; Marro et al. 2011). There are both advantages and disadvantages in using cells that belong to non-neural lineages as a source population for reprogramming. Veritably, it broadens the potential scope of direct reprogramming, as it does not limit choice of a starting cell type. Conversely, neural lineage cells, such as astrocytes, may already have relevant epigenetic marks and active transcription factors, that may result in easier reprogramming (Faiz and Nagy 2013; Ninkovic and Götz 2018). Thus, future studies must include a functional comparison of cells that are generated from neural versus non-neural starter populations.

Following initial *in vitro* studies, a number of reports demonstrated *in vivo* reprogramming in the brain and spinal cord. This is of particular interest for brain repair, as it enables the targeted generation of new cells at the site of injury and circumvents the need for transplantation of exogenous cells and the associated risks, namely immune-rejection and the potential for cell mutagenesis from long-term cell culture (Faiz and Nagy 2013; Xu et al. 2015; Gascón et al. 2017a). It also provides an alternative to strategies using endogenous neural stem cells that reside within the brain and spinal cord. Attempts to generate neurons from these neural stem cells have resulted in low differentiation into the proper mature neuronal phenotypes, and poor long-term survival (Barker et al. 2018; Gascón et al. 2017a; Arvidsson et al. 2002; Thored et al. 2007).

In 2005, Buffo and colleagues demonstrated for the first time in the CNS that the manipulation of transcription factors could alter cell fate *in vivo* (Buffo et al. 2005). They converted NG2 glia into cells of a neuronal phenotype by inhibiting the expression of *Olig2* (Buffo et al. 2005). This inhibition was achieved through the specific delivery of the dominant negative form of *Olig2* to NG2 glia (Buffo et al. 2005). *In vivo* direct conversion has now been shown in the healthy brains of both young and old mice (Niu et al. 2013; Rouaux and Arlotta 2013). Of clinical relevance, the success of *in vivo* direct reprogramming has also been demonstrated in a number of models of CNS injury and disease, including stroke (Faiz et al. 2015), stab wound injury (Chen et al. 2015; Guo et al. 2014; Heinrich et al. 2014), spinal cord injury (Su et al. 2014), Alzheimer's disease (Chen et al. 2015; Guo et al. 2014) and Parkinson's disease (Rivetti di Val Cervo et al. 2017). Interestingly, it has been suggested that aspects of the injured/diseased environment, such as the increase of beneficial growth factors, increased plasticity of glial cells and increased glycolysis may actually enhance the reprogramming process (Gascón et al. 2017a; Guo et al. 2014; Grande et al. 2013). These disease-induced changes could explain why some reprogramming paradigms

have encountered success in an injury context, but no conversion (or a significantly reduced conversion) was observed when the same transcription factor(s) were delivered to the uninjured brain (Heinrich et al. 2014; Grande et al. 2013). Conversely, it has also been noted that an increased production of reactive oxygen species (ROS) during injury could be deleterious to newly generated cells and explain the discrepancy in conversion success between *in vitro* and *in vivo* studies (Gascón et al. 2017a). A better understanding of the mechanisms that underlie each particular injury or disease model will allow for reprogramming strategies that are tailored and optimized for different applications.

2.2 Target Cell Type

Many neurological disorders or conditions have at their core, a significant loss or injury to the cells of the CNS. However, not all disorders implicate the same cells and as such, it is important to generate specific cell types that are needed for a particular disorder. The versatility of direct lineage reprogramming technology is clear – studies have shown the generation of all the main cell types of the CNS, including certain subtypes and progenitors.

2.2.1 Neurons

Generating Neurons In Vitro

Neurons are affected in a wide variety of neurological conditions, and thus direct lineage reprogramming strategies have mainly been focused on regenerating these cells. Since their seminal *Pax6* study, work from Magdalena Götz's lab has also demonstrated that a combination of *Ascl1/Dlx2* or *Ngn2* results in the conversion of astrocytes to GABAergic and glutamatergic neurons, respectively (Vierbuchen and Wernig 2011; Xu et al. 2015; Heinrich et al. 2010). Simultaneously, work done by Vierbuchen and colleagues established the ability of the combination of *Ascl1/Brn2/Mytl1* (referred to as BAM factors) to induce glutamatergic neurons from fibroblasts (Vierbuchen et al. 2010). The conversion of glial cells (both astrocytes and NG2 glia) to neurons using *NeuroD1* by Gong Chen's lab further demonstrated that transcription factors involved in later stages of neuronal development could also be used to regenerate neurons (Guo et al. 2014).

A number of other starter cell types have also been successfully converted to neurons. Non-neural cell types, such as pericytes, have been reprogrammed to glutamatergic and GABAergic cells (Karow et al. 2018) and the BAM factors have been used to reprogram hepatocytes to glutamatergic-like neuronal cells (Marro et al. 2011). Additionally, it has been shown that microglia can be converted to functional neurons with the delivery of *NeuroD1* alone (Matsuda et al. 2019).

Importantly, the type of neuron lost or affected in a particular disease is often of a specific subtype (i.e.: dopaminergic neurons in Parkinson's disease and motor neurons in Amyotrophic Lateral Sclerosis), and differs across various neurological conditions (Faiz and Nagy 2013; Chen et al. 2015; Xu et al. 2015; Wang and Zhang 2018; Masserdotti et al. 2016). As such, the generation of a random assortment of neuronal subtypes, or the ability to generate only one specific subtype would likely be of minimal therapeutic benefit. For example, generating cholinergic neurons in Alzheimer's disease is likely to confer more benefit than in Parkinson's disease, where dopaminergic neurons are needed. Direct reprogramming must therefore reliably generate subtype specific cell types appropriate for the neurological deficit in question (Faiz and Nagy 2013; Chen et al. 2015; Xu et al. 2015; Wang and Zhang 2018; Masserdotti et al. 2016). Accordingly, *in vitro* studies have shown the generation of dopaminergic (Rivetti di Val Cervo et al. 2017; Kim et al. 2011; Caiazzo et al. 2011; Sheng et al. 2012), motor (Son et al. 2011), serotonergic (Vadodaria et al. 2016), and cholinergic (Liang et al. 2018; Liu et al. 2013) neurons, amongst others (see (Masserdotti et al. 2016) for in depth review) using specific combinations of transcription factors.

In summary, direct lineage reprogramming *in vitro* is clearly feasible, customizable and reliable in generating new neurons. However, *in vitro* lineage conversion still requires transplantation into the brain.

Generating Neurons In Vivo

One of the most exciting features of direct reprogramming is the ability to target endogenous cells at their source. Thus, *in vivo* studies generating novel populations of neurons are of particular interest to the field. Work performed by a number of groups has shown the reliable generation of new neurons *in vivo* using direct reprogramming in healthy and injured environments, and has been extensively reviewed elsewhere.(Chen et al. 2015; Xu et al. 2015; Wang and Zhang 2018; Gascón et al. 2017a; Masserdotti et al. 2016) What is lacking and of significant interest however, is a systematic comparison of different transcription factors and delivery strategies in various models of disease and injury (Gascón et al. 2017a). Although the transcription factors used in these studies (*Sox2* (Niu et al. 2013; Heinrich et al. 2014), BAM factors (Torper et al. 2013), *NeuroD1* (Guo et al. 2014) and *Ascl1/Lmx1a/Nurr1* (Torper et al. 2015)) correspond to *in vitro* studies, there is variation with regards to the delivery system used. It has been proposed that the choice of delivery system may affect the reprogramming paradigm, as there is variance in their temporal kinetics (Gascón et al. 2017a). As such, clear conclusions on the "best" direct reprogramming paradigm for a particular starting cell type, target cell type or disease state cannot yet be made with certainty. Nonetheless, these newly generated neurons are capable of surviving, maturing and integrating into the pre-existing neural circuitry, as shown by electrophysiological and functional assays (Guo et al. 2014; Niu et al. 2013; Heinrich et al. 2014; Torper et al. 2013, 2015).

One hurdle that remains with regards to *in vivo* neuronal reprogramming is subtype specific neuronal regeneration. Success seen in *in vitro* studies of neuronal subtype generation has not been replicated to the same extent *in vivo,* even with the use of the same transcription factors (Chen et al. 2015; Xu et al. 2015; Wang and Zhang 2018; Gascón et al. 2017a; Masserdotti et al. 2016). The reasons for this are unclear, but as discussed above, could be attributed to the deleterious environment that results from injury (Gascón et al. 2017a). A more complex *in vivo* environment may require multiple transcription factors and/or a combination of both transcription factors and small molecules or microRNA to generate specific neuronal sub-types. In fact, Rivetti di Val Cervo and colleagues successfully obtained a novel population of dopaminergic neurons from astrocytes *in vivo* when they used a combination of both *NeuroD1/Ascl1/Lmx1a* and the microRNA miR218 (Rivetti di Val Cervo et al. 2017).

Neuron to Neuron Reprogramming

While most studies have focused on converting non-neuronal cells to neurons, reports of neuron to neuron reprogramming show that there is cell fate flexibility even within this population. In the cortex, Rouaux and Arlotta were able to successfully convert layer 2/3 callosal projection neurons into layer 5/6 corticofugal projections using only the transcription factor *Fezf2* (Rouaux and Arlotta 2013). More recently, Niu and colleagues used a combination of *Sox2/Nurr1/Foxa2/Lmx1a* paired with valproic acid (VPA) to reprogram striatal neurons to dopaminergic neurons (Niu et al. 2018). Interestingly, they showed that these newly induced dopaminergic cells arise directly from the striatal neurons, without passing through a progenitor stage (Niu et al. 2018). These studies beg the question of whether a shared identity (i.e. neuron) between the starting and target cell is an important consideration for easily generating specific neuronal subtypes *in vivo*.

2.2.2 Oligodendrocytes

Oligodendrocytes play crucial roles in maintaining proper cell signaling in the CNS and many diseases result from their widespread loss or malfunction. Oligodendrocyte death and subsequent de-myelination is characteristic to the pathology of multiple sclerosis (Lassmann et al. 2012; Reich et al. 2018; Sawcer et al. 2014), and a reduction in myelin is seen in multi-system atrophy (Burn and Jaros 2001). Oligodendrocytes have also been implicated in Alzheimer's disease. Although traditionally thought of as a grey matter disease, Alzheimer's disease presents with white matter disruption, impaired myelination patterns and decreased oligodendrocyte and oligodendrocyte progenitor gene expression (Desai et al. 2009, 2010; Roth et al. 2005). Interestingly, a mouse model of Alzheimer's disease showed that impaired myelination and decreased CNPase and MBP expression precedes the onset of tau and amyloid pathology (Desai et al. 2009). Finally, white matter damage and oligodendrocyte dysfunction have been proposed as a risk factor and predictor of stroke (Kuller et al. 2004) and of schizophrenia (Cassoli et al. 2015).

Given the significant implication of oligodendrocytes in disease, strategies to restore or replenish damaged or lost oligodendrocytes are needed. Yet, there is a clear disparity in the number of studies investigating direct reprogramming to neurons versus oligodendrocytes. Only three studies to date have looked at using direct reprogramming to generate new populations of oligodendrocytes and their precursors. Work done by Najm and colleagues, as well as Yang and colleagues produced oligodendrocyte progenitor cells (OPCs) and oligodendrocytes from fibroblasts *in vitro* using combinations of transcription factors involved in OPC development and oligodendrocyte function (*Sox10/Olig2/Nkx6.2* and *Sox10/Olig2/Zfp536*, respectively) (Najm et al. 2013; Yang et al. 2013). The oligodendrocytes generated from both these studies expressed characteristic OPC and oligodendrocyte markers and showed myelination capability in transplantation experiments (Najm et al. 2013; Yang et al. 2013). More recently, Khanghahi and colleagues reported that both *in vitro* and *in vivo* delivery of *Sox10* alone to astrocytes in cuprizone induced de-myelinated mice results in the generation of new oligodendrocyte-like cells (Mokhtarzadeh

Khanghahi et al. 2018). Cells transduced *in vitro* expressed markers of OPC and oligodendrocyte lineage and were transplanted into the corpus callosum of the cuprizone mice (Mokhtarzadeh Khanghahi et al. 2018).

Although promising, there are significantly fewer studies of oligodendrocyte reprogramming in comparison with neuronal reprogramming. Future work examining oligodendrocyte generation and the optimal factors involved are warranted.

2.2.3 Astrocytes

Most reprogramming studies have focused on astrocytes as the starter cell type, rather than the target cell type. Nonetheless, a few reports have shown the feasibility of generating astrocytes. From a pool of 14 genes involved in determining astrocyte fate, Caiazzo and colleagues found that the combination of *NFIA/NFIB/Sox9* could successfully reprogram fibroblasts to astrocytes (Caiazzo et al. 2015). Similarly, work by Tian and colleagues demonstrated that a cocktail of 6 small molecules generated functional astrocytes from fibroblasts (Tian et al. 2016). Given the recent knowledge that a subset of astrocytes, A2 cells, are neuroprotective and conducive to recovery following injury, future studies examining conversion of a starter cell to a beneficial astrocyte subtype, such as the A2 phenotype, may be of interest (Liddelow and Barres 2017; Liddelow et al. 2017; Zamanian et al. 2012; Toft-Hansen et al. 2011).

2.2.4 Stem/Progenitor Cells

To date, studies have generated both neural stem cells and glial progenitors that can give rise to mature neurons and glia. Researchers in the Wernig lab demonstrated that the use of two transcription factors, *FoxG1* and *Brn2,* that are important for neural stem cell (NSC) fate, could convert fibroblasts into tripotent NSCs (Lujan et al. 2012). Not only were these NSCs capable of further differentiation into functional neurons, astrocytes and oligodendrocytes, but they also demonstrated clear proliferative capacity, capable of being passaged up to 17 times, without loss of function (Lujan et al. 2012). Additionally, other studies by Han and colleagues, as well as Ring and colleagues have shown conversion of fibroblasts to NSCs using a combination of *Brn2/Sox2/Klf4/c-Myc/E47* and *Sox2* alone, respectively (Xu et al. 2015; Han et al. 2012; Ring et al. 2012). Similarly, astrocytes have also been successfully converted to NSPCs through delivery of both single factors (*OCT4, SOX2* or *NANOG*) (Corti et al. 2012) and combination of factors (*Foxg1/Sox2/Brn2*) (Ma et al. 2018). Finally, OPCs have also been generated in the Tesar lab using a combination of the transcription factors *Sox10/Olig2/Nkx6.2* (Najm et al. 2013). The pathophysiology of a particular disease may determine whether it is advantageous to induce progenitor populations or mature cell types. Given this, future studies comparing the functional outcomes of direct reprogramming to progenitors versus mature cells will be of interest.

2.2.5 Cell Heterogeneity

It is now clear that neurons are not the only cell type in the CNS with delineated subtypes comprising different roles, with recent work demonstrating intra-cellular differences in oligodendrocyte populations. Using single-cell RNA sequencing (sc RNA-seq), Marques and colleagues found unique transcriptome profiles according to the age and region of oligodendrocytes and progenitors in mice (Marques et al. 2016). In addition, they noted a novel population of cells (ITPR2+) hypothesized to be involved in periods of rapid myelination (Marques et al. 2016). Furthermore, they noted that varying regions of the CNS were associated with differing forms of mature oligodendrocytes, such as MOL6 oligodendrocytes specific to the S1 cortex and corpus callosum (Marques et al. 2016). Differences in mature oligodendrocytes in the CNS are also present in the context of disease. Work by Jäkel and colleagues showed that not only is there a similar heterogeneity of oligodendrocytes in humans, but that MS patients had a unique loss of certain mature oligodendrocyte populations (OLIG1+) when compared to controls (Jäkel et al. 2019). These findings will be of particular importance when considering transcription factor cocktails used to create functional, myelinating oligodendrocytes and how the transcription factor combinations may vary based on disease need.

2.3 Starting Cell Type

When performing direct lineage reprogramming, genetic systems with cell specific promoters can allow for targeting of a precise cellular population (Wang and Zhang 2018). Traditionally, starter cell populations have been chosen based on their developmental closeness to the target cell type (Gascón et al. 2017a; Masserdotti et al. 2016; Waddington 1957). However, developmental closeness may not be the only, or even "best" reason for choosing a particular starting cell type. In the context of injury or disease, it may be more relevant to choose a starting cell based on the role of that cell at the time of reprogramming. For example, cells that contribute to ongoing neuronal death and therefore disease pathology may be the most clinically relevant choice for reprogramming.

2.3.1 Developmental Closeness

The Waddington model, used to explain normal cell fate determination, denotes a linear differentiation and restriction pattern of cell type during development (Waddington 1957). It was hypothesized that more closely related starter and target cells would be easier to convert and reprogram (Gascón et al. 2017a; Masserdotti et al. 2016). Initial studies using starting cell populations that belonged to a non-neuronal lineage, such as fibroblasts, required a combination of transcription factors (Son et al. 2011; Vierbuchen et al. 2010; Kim et al. 2011; Caiazzo et al. 2011; Sheng et al. 2012) for conversion to neurons. In contrast, starting cell populations within the neural lineage, such as astrocytes, could be successfully converted with just one transcription factor (Gascón et al. 2017a; Heinrich et al. 2010; Guo et al. 2014; Niu et al. 2013). However, more recent work has demonstrated the use of single factors to convert non-neural cells neurons. Chanda and colleagues demonstrated the generation of neurons from fibroblasts using only *Ascl1*

(Chanda et al. 2014) and *MYT1L* alone has been shown to reprogram pericytes into mature cholinergic neurons (Liang et al. 2018). Given these findings, a new model of reprogramming has been proposed – the Cook's Island model (Masserdotti et al. 2016; Sieweke 2015). In this analogy, the starting cell is a boat and target cell types are the various islands to which the boat can travel (Masserdotti et al. 2016; Sieweke 2015). The boat may face various challenges or hurdles depending on the proximity of the island, but all are potentially accessible (Masserdotti et al. 2016; Sieweke 2015).

2.3.2 Cellular Heterogeneity

Cell heterogeneity within the starter cell population is of particular interest to direct lineage reprogramming. One specific subtype of astrocyte, for instance, may be especially conducive to generating a particular subtype of neuron (Fig. 2). Conversely, as we broaden our scope of potential cells that can be generated by reprogramming, it will be important that those generated are of the correct subtype for the particular disease or injury at hand.

Astrocytes

Recent work by Liddelow and colleagues has shown that astrocytes have at least two defined functional states in the context of disease/injury, termed A1 and A2 (Liddelow and Barres 2017; Liddelow et al. 2017). A1 cells are present in many disease states, including Alzheimer's Disease, Huntington's Disease, Parkinson's Disease and Multiple Sclerosis (Liddelow and Barres 2017; Liddelow et al. 2017). Furthermore, A1 astrocytes lose many normal astrocytic functions, such as phagocytic capacity and the promotion of synaptic formation and become toxic, killing neurons and oligodendrocytes, and impairing oligodendrocyte progenitor cell (OPC) differentiation (Liddelow and Barres 2017; Liddelow et al. 2017). Conversely, A2 cells are thought to be neuroprotective (Liddelow and Barres 2017). They upregulate a number of neurotrophic factors, cytokines and thrombospondins that may help repair and rebuild lost synapses (Liddelow and Barres 2017; Zamanian et al. 2012). In addition, it has also

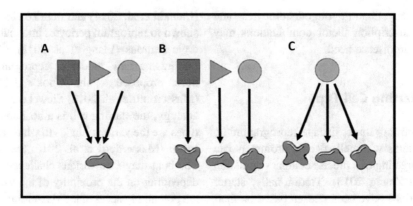

Fig. 2 Cellular heterogeneity. Heterogeneity of both the starting and target cell populations is an important consideration for direct lineage reprogramming. Diversity of the starting cell population may determine what types of target cell types are generated. All subsets of starting cells may give rise to one only type of target cell (**a**). Alternately, specific subtypes of starting cells may only give rise to specific subtypes of target cells (**b**). Or, only one type of starting cell may generate all target cell subtypes (**c**). Illustrated by Kayla Hoffman-Rogers

been postulated that there are many more astrocyte subtypes that have yet to be characterized (Liddelow and Barres 2017).

Given the heterogeneity of astrocytes, it may prove advantageous to reprogram astrocyte subtypes that are detrimental to disease outcome or progression (such as A1 cells), over reprogramming protective subtypes (such as A2 cells) that could lead to worse disease outcomes (Liddelow and Barres 2017; Liddelow et al. 2017; Zamanian et al. 2012; Toft-Hansen et al. 2011). Furthermore, it would be worthwhile investigating whether A1 neurotoxic astrocytes could be reprogrammed to their more beneficial A2 counterparts. This has been suggested in work done by Gong Chen's lab, which noted that astrocytes transduced in their *NeuroD1* mediated astrocyte to neuron paradigm showed a reduction in A1 gene expression prior to their conversion to neurons (Zhang et al. 2018).

Microglia
Microglia have also been shown to have at least two distinct subtypes, termed M1 and M2, with more recent work suggesting that many sub-classes, or even a continuum of microglial states may exist (Liddelow and Barres 2017;

Boche et al. 2013; Tang and Le 2016). These subtypes pertain to activation states that correspond with particular functions: M1 microglia are pro-inflammatory and potentially damaging to neighboring cells, whereas M2 microglia are involved in tissue repair and are immunosuppressive (Liddelow and Barres 2017; Boche et al. 2013). Interestingly, this activation pattern is thought to be dependent on the particular injury or disease state (Boche et al. 2013). In fact, work by Tang and Le have shown that changes in M1 and M2 microglia phenotype correspond to different stages of disease (Tang and Le 2016). This knowledge may be of particular relevance in future clinical applications of direct reprogramming, allowing for tailored paradigms based off disease progression.

2.4 Direct Reprogramming: Readouts

In order to ensure the clinical relevance and feasibility of direct reprogramming, there is a need to generate mature cells that can integrate into existing host circuitry, have long-term survival and perform proper cell functions (Barker et al.

2018; Xu et al. 2015; Wang and Zhang 2018; Gascón et al. 2017a; Berninger 2010; Yang et al. 2011). If the cells generated fail to meet these conditions, it is unlikely that they could be utilized as a novel therapy for neurological diseases.

2.4.1 Characterization of Target Cells

To characterize newly reprogrammed cells, many studies have examined the expression of cell type specific proteins and patterns of global gene expression that correspond to native cells (Barker et al. 2018; Faiz and Nagy 2013; Xu et al. 2015; Gascón et al. 2017a; Masserdotti et al. 2016; Yang et al. 2011). Some studies have also used a lack of gene/protein expression, of cells of unwanted lineages or of cells of the starting population to be indicative of proper cell conversion. For example, Niu and colleagues demonstrated that reprogrammed dopaminergic neurons expressed cell-type specific markers [DDC (DOPA Decarboxylase), VMAT2 (Vesicular monoamine transporter 2) and DAT (Dopamine transporter)], but also confirmed that the reprogrammed cells did not express markers associated with other neuronal subtypes (cholinergic or glutamatergic, using CHAT and GLUT1, respectively) (Niu et al. 2018).

Functional assays specific to the desired cell type are also important (Barker et al. 2018; Xu et al. 2015; Wang and Zhang 2018; Gascón et al. 2017a; Berninger 2010; Yang et al. 2011) (Fig. 3). For neurons, there are both general and subtype specific means of assessing neuronal function and integration (Yang et al. 2011). Patch-clamp recording can demonstrate whether reprogrammed neurons exhibit electrophysiological characteristics of neurons – their ability to fire action potentials and their synaptic patterns (Chen et al. 2015; Wang and Zhang 2018; Yang et al. 2011). Most studies to date have demonstrated mature, electrically active neurons, both *in vivo* and *in vitro*. The firing patterns of reprogrammed cells can also be compared to the expected firing patterns of native cells to assess similarity in function, as was done by Niu and colleagues (Niu et al. 2018). Furthermore, fluorescent reporters can be used to trace reprogrammed

cells and assess the extent of their integration into host circuitry (Chen et al. 2015; Torper et al. 2015). For example, Torper and colleagues noted that in a mouse model of Parkinson's disease, newly generated neurons did not migrate to alternate regions of the CNS, but rather integrated locally (Torper et al. 2015).

For oligodendrocytes, the ability to myelinate is key. To characterize the reprogrammed OPCs or oligodendrocytes, studies have used mouse models of demyelination or impaired myelination (Najm et al. 2013; Yang et al. 2013; Mokhtarzadeh Khanghahi et al. 2018; Chernoff 1981; Blakemore 1972; Matsushima and Morell 2001). The *Shiverer* mouse, for example, lacks myelin basic protein (MBP) and consequently, the ability to form compact myelin (Chernoff 1981). It has been used in studies such as those done by Najm and colleagues to demonstrate the generation of MBP+ myelin following transplantation of OPCs that were directly reprogrammed from fibroblasts (Najm et al. 2013).

2.4.2 Functional Outcomes

Ultimately, the goal of direct reprogramming is to be a clinical treatment option for neurological diseases. Therefore, it is crucial to employ animal models of disease to determine whether reprogramming can induce functional recovery, slow disease progression or reverse disease progression/impairments all together (Chen et al. 2015; Xu et al. 2015). To date, there has been limited study of the outcomes of reprogramming with regards to disease progression or prevention in disease models. The first report of functional recovery was by Rivetti di Val Cervo and colleagues in a unilateral 6-hydroxydopamine (6-OHDA) mouse model of Parkinson's disease (Rivetti di Val Cervo et al. 2017). Following astrocyte reprogramming to dopaminergic neurons, newly generated neurons were capable of rescuing gait deficits and dopamine-deficient circling behaviors (Rivetti di Val Cervo et al. 2017). A second report by Chen and colleagues, showed improvement in motor and fear memory deficits following ischemia (Chen et al. 2018).

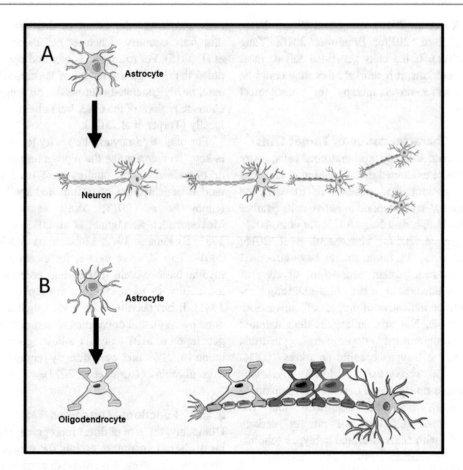

Fig. 3 Functional target cells. The goal of direct lineage reprogramming is to generate functional cells for repair or regeneration. Newly generated cells must recapitulate the function of their endogenous counterparts (blue neurons, red oligodendrocytes). After astrocyte to neuron conversion, new neurons (yellow) must fire action potentials, form synapses and integrate into the exiting host circuitry (blue neurons) (**a**). Similarly, new oligodendrocytes (yellow) must generate myelin and ensheath existing neurons like native oligodendrocytes (red) (**b**). Illustrated by Kayla Hoffman-Rogers

2.4.3 Application of New Technologies

Many exciting and novel technologies have recently emerged that will benefit our understanding of the reprogramming process and cellular outcomes. A new tool for analyzing reprogrammed cell identity is the CellNet database (Xu et al. 2015). It identifies gene regulatory networks (GRNs) in reprogrammed cells, and therefore enables confirmation of reprogramming factor expression in target cells and the comparison of gene expression profiles of experimental and naive cells (Xu et al. 2015; Cahan et al. 2014; Morris et al. 2014). Most striking, however, is the utility of CellNet in predicting how reprogramming paradigms could be improved, which is based on incorrect expression of GRNs (Xu et al. 2015; Morris et al. 2014).

In a pioneering study by Cadwell and colleagues, electrophysiological and single cell RNA-seq readouts were combined to create Patch-seq technology (Cadwell et al. 2016). This technique results in the simultaneous acquisition of cell transcriptomes (sc RNA-seq) and electrophysiological information (Patch-clamp readings), thereby correlating changes in cell function and the transcriptome within a single cell (Cadwell et al. 2016). This is of particular interest, as changes in cell function can be predicted based on particular transcriptome patterns or modifications (Cadwell et al. 2016). This technology has already been used to identify and predict populations of highly functional reprogrammed neurons generated from iPSCs (Bardy et al. 2016). The application of Patch-seq in direct reprogramming studies is thus greatly warranted, as it would enable a more tailored approach for the creation of specific cell types by identifying reprogramming factors that produce bonafide reprogrammed cells.

CRISPR is another new technology that can be used to determine genes that are involved in cell fate changes and therefore, elucidate optimal transcription factor combinations for direct reprogramming paradigms (Liu et al. 2018). This strategy has unveiled novel factor(s) that result in the conversion of fibroblasts to neurons, such as Ezh2/Mecom, which were not traditionally thought to be key proneural genes, (Liu et al. 2018). As the field strives for subtype specific cell generation, as well as tailorable and translatable therapies, utilizing the power of CRISPR may be of great interest.

Finally, optogenetics and pharmacogenetics provide novel means by which to target populations of cells and manipulate their activity (Deisseroth 2011; Amamoto and Arlotta 2014; Steinbeck and Studer 2015). This technology can be used to specifically analyze whether newly generated cells directly contribute to functional recovery (Amamoto and Arlotta 2014; Steinbeck and Studer 2015). It can also be employed to potentiate the activity of reprogrammed cells (Amamoto and Arlotta 2014; Steinbeck and Studer 2015). Indeed, in a study done by Dell'Anno and colleagues, a DREADD pharmacogenic system was used to selectively activate reprogrammed dopaminergic neurons to enhance their activity (Dell'Anno et al. 2014). Researchers noted that upon activation of these cells, dopamine levels were increased, even up to 5 weeks following reprogramming, supporting the use of chemogenetics as an adjunct strategy to direct reprogramming (Dell'Anno et al. 2014).

3 Conclusions and Future Directions

The field of reprogramming is still in its infancy. Although significant progress has been made in our understanding of direct reprogramming in the CNS since the pioneering work done by Heins and colleagues (2002), new research into the mechanisms underlying direct reprogramming will allow us to tailor better strategies for brain repair. Studies that will determine the optimal starting cell types that are needed for the generation of functional target cells, as well as experiments that systematically compare the efficacy of different reprogramming paradigms are needed. Moreover, research into the impact of cellular heterogeneity in reprogramming will result in better designed reprogramming strategies that cater to a specific disease or injury state. Our progress will only become faster with the implementation of the novel, cutting technologies, such as CellNet and Patch-seq. Given the integrative and multi-faceted nature of direct reprogramming, it seems only fitting to employ an interdisciplinary approach as we move forward.

References

Amamoto R, Arlotta P (2014) Development-inspired reprogramming of the mammalian central nervous system. Science (New York, NY) 343(6170):1239882. https://doi.org/10.1126/science.1239882

Arvidsson A, Collin T, Kirik D, Kokaia Z, Lindvall O (2002) Neuronal replacement from endogenous precursors in the adult brain after stroke. Nat Med 8 (9):963–970. https://doi.org/10.1038/nm747

Bardy C, van den Hurk M, Kakaradov B, Erwin JA, Jaeger BN, Hernandez RV, Eames T, Paucar AA, Gorris M, Marchand C et al (2016) Predicting the functional states of human iPSC-derived neurons with single-cell RNA-seq and electrophysiology. Mol Psychiatry 21(11):1573–1588. https://doi.org/10.1038/mp.2016.158

Barker RA, Götz M, Parmar M (2018) New approaches for brain repair—from rescue to reprogramming. Nature 557(7705):329. https://doi.org/10.1038/s41586-018-0087-1

Berninger B (2010) Making neurons from mature glia: a far-fetched dream? Neuropharmacology 58 (6):894–902. https://doi.org/10.1016/j.neuropharm.2009.11.004

Black JB, Adler AF, Wang H-G, D'Ippolito AM, Hutchinson HA, Reddy TE, Pitt GS, Leong KW, Gersbach CA (2016) Targeted epigenetic remodeling of endogenous loci by CRISPR/Cas9-based transcriptional activators directly converts fibroblasts to neuronal cells. Cell Stem Cell 19(3):406–414. https://doi.org/10.1016/j.stem.2016.07.001

Blakemore WF (1972) Observations on oligodendrocyte degeneration, the resolution of status spongiosus and remyelination in cuprizone intoxication in mice. J Neurocytol 1(4):413–426. https://doi.org/10.1007/BF01102943

Blau HM, Chiu CP, Webster C (1983) Cytoplasmic activation of human nuclear genes in stable heterocaryons. Cell 32(4):1171–1180

Blau HM, Pavlath GK, Hardeman EC, Chiu CP, Silberstein L, Webster SG, Miller SC, Webster C (1985) Plasticity of the differentiated state. Science (New York, NY) 230(4727):758–766

Boche D, Perry VH, Nicoll JAR (2013) Review: activation patterns of microglia and their identification in the human brain. Neuropathol Appl Neurobiol 39 (1):3–18. https://doi.org/10.1111/nan.12011

Buffo A, Vosko MR, Ertürk D, Hamann GF, Jucker M, Rowitch D, Götz M (2005) Expression pattern of the transcription factor Olig2 in response to brain injuries: implications for neuronal repair. Proc Natl Acad Sci U S A 102(50):18183–18188. https://doi.org/10.1073/pnas.0506535102

Burn DJ, Jaros E (2001) Multiple system atrophy: cellular and molecular pathology. Mol Pathol 54(6):419–426

Cadwell CR, Palasantza A, Jiang X, Berens P, Deng Q, Yilmaz M, Reimer J, Shen S, Bethge M, Tolias KF et al (2016) Electrophysiological, transcriptomic and morphologic profiling of single neurons using Patch-seq. Nat Biotechnol 34(2):199–203. https://doi.org/10.1038/nbt.3445

Cahan P, Li H, Morris SA, Lummertz da Rocha E, Daley GQ, Collins JJ (2014) CellNet: network biology applied to stem cell engineering. Cell 158 (4):903–915. https://doi.org/10.1016/j.cell.2014.07.020

Caiazzo M, Dell'Anno MT, Dvoretskova E, Lazarevic D, Taverna S, Leo D, Sotnikova TD, Menegon A, Roncaglia P, Colciago G et al (2011) Direct generation of functional dopaminergic neurons from mouse and human fibroblasts. Nature 476(7359):224–227. https://doi.org/10.1038/nature10284

Caiazzo M, Giannelli S, Valente P, Lignani G, Carissimo A, Sessa A, Colasante G, Bartolomeo R, Massimino L, Ferroni S et al (2015) Direct conversion of fibroblasts into functional astrocytes by defined transcription factors. Stem Cell Rep 4(1):25–36. https://doi.org/10.1016/j.stemcr.2014.12.002

Cassoli JS, Guest PC, Malchow B, Schmitt A, Falkai P, Martins-de-Souza D (2015) Disturbed macroconnectivity in schizophrenia linked to oligodendrocyte dysfunction: from structural findings to molecules. NPJ Schizophr 1:15034. https://doi.org/10.1038/npjschz.2015.34

Chakraborty S, Ji H, Kabadi AM, Gersbach CA, Christoforou N, Leong KW (2014) A CRISPR/Cas9-based system for reprogramming cell lineage specification. Stem Cell Rep 3(6):940–947. https://doi.org/10.1016/j.stemcr.2014.09.013

Chanda S, Ang CE, Davila J, Pak C, Mall M, Lee QY, Ahlenius H, Jung SW, Südhof TC, Wernig M (2014) Generation of induced neuronal cells by the single reprogramming factor ASCL1. Stem Cell Rep 3 (2):282–296. https://doi.org/10.1016/j.stemcr.2014.05.020

Chen G, Wernig M, Berninger B, Nakafuku M, Parmar M, Zhang C-L (2015) In vivo reprogramming for brain and spinal cord repair. eNeuro 2(5). https://doi.org/10.1523/ENEURO.0106-15.2015, https://www.ncbi.nlm.nih.gov/pmc/articles/PMC4699832/

Chen Y, Ma N, Pei Z, Wu Z, Do-Monte FH, Huang P, Yellin E, Chen M, Yin J, Lee G, Minier A, Hu Y, Bai Y, Lee K, Quirk G, Chen G (2018) Functional repair after ischemic injury through high effciency in situ astrocyte-to-neuron conversion. bioRxiv. https://doi.org/10.1101/294967

Chernoff GF (1981) Shiverer: an autosomal recessive mutant mouse with myelin deficiency. J Hered 72(2):128

Corti S, Nizzardo M, Simone C, Falcone M, Donadoni C, Salani S, Rizzo F, Nardini M, Riboldi G, Magri F et al (2012) Direct reprogramming of human astrocytes into neural stem cells and neurons. Exp Cell Res 318 (13–16):1528–1541. https://doi.org/10.1016/j.yexcr.2012.02.040

Davis RL, Weintraub H, Lassar AB (1987) Expression of a single transfected cDNA converts fibroblasts to myoblasts. Cell 51(6):987–1000

Deisseroth K (2011) Optogenetics. Nat Methods 8 (1):26–29. https://doi.org/10.1038/nmeth.f.324

Dell'Anno MT, Caiazzo M, Leo D, Dvoretskova E, Medrihan L, Colasante G, Giannelli S, Theka I, Russo G, Mus L et al (2014) Remote control of induced dopaminergic neurons in parkinsonian rats. J Clin Invest 124(7):3215–3229. https://doi.org/10.1172/JCI74664

Desai MK, Sudol KL, Janelsins MC, Mastrangelo MA, Frazer ME, Bowers WJ (2009) Triple-transgenic Alzheimer's disease mice exhibit region-specific abnormalities in brain myelination patterns prior to appearance of amyloid and tau pathology. Glia 57 (1):54–65. https://doi.org/10.1002/glia.20734

Desai MK, Mastrangelo MA, Ryan DA, Sudol KL, Narrow WC, Bowers WJ (2010) Early oligodendrocyte/myelin pathology in Alzheimer's disease mice constitutes a novel therapeutic target. Am J Pathol 177(3):1422–1435. https://doi.org/10.2353/ajpath.2010.100087

Faiz M, Nagy A (2013) Induced pluripotent stem cells and disorders of the nervous system: progress, problems, and prospects. Neuroscientist 19(6):567–577. https://doi.org/10.1177/1073858413493148

Faiz M, Sachewsky N, Gascón S, Bang KWA, Morshead CM, Nagy A (2015) Adult neural stem cells from the subventricular zone give rise to reactive astrocytes in the cortex after stroke. Cell Stem Cell 17(5):624–634. https://doi.org/10.1016/j.stem.2015.08.002

Gascón S, Murenu E, Masserdotti G, Ortega F, Russo GL, Petrik D, Deshpande A, Heinrich C, Karow M, Robertson SP et al (2016) Identification and successful negotiation of a metabolic checkpoint in direct neuronal reprogramming. Cell Stem Cell 18(3):396–409. https://doi.org/10.1016/j.stem.2015.12.003

Gascón S, Masserdotti G, Russo GL, Götz M (2017a) Direct neuronal reprogramming: achievements, hurdles, and new roads to success. Cell Stem Cell 21 (1):18–34. https://doi.org/10.1016/j.stem.2017.06.011

Gascón S, Ortega F, Götz M (2017b) Transient CREB-mediated transcription is key in direct neuronal reprogramming. Neurogenesis (Austin) 4(1): e1285383. https://doi.org/10.1080/23262133.2017.1285383

Graf T, Enver T (2009) Forcing cells to change lineages. Nature 462(7273):587–594. https://doi.org/10.1038/nature08533

Grande A, Sumiyoshi K, López-Juárez A, Howard J, Sakthivel B, Aronow B, Campbell K, Nakafuku M (2013) Environmental impact on direct neuronal reprogramming in vivo in the adult brain. Nat Commun 4:2373. https://doi.org/10.1038/ncomms3373

Guo Z, Zhang L, Wu Z, Chen Y, Wang F, Chen G (2014) In vivo direct reprogramming of reactive glial cells into functional neurons after brain injury and in an Alzheimer's disease model. Cell Stem Cell 14 (2):188–202. https://doi.org/10.1016/j.stem.2013.12.001

Gurdon JB (1962) Adult frogs derived from the nuclei of single somatic cells. Dev Biol 4(2):256–273. https://doi.org/10.1016/0012-1606(62)90043-X

Han DW, Tapia N, Hermann A, Hemmer K, Höing S, Araúzo-Bravo MJ, Zaehres H, Wu G, Frank S, Moritz S et al (2012) Direct reprogramming of fibroblasts into neural stem cells by defined factors. Cell Stem Cell 10 (4):465–472. https://doi.org/10.1016/j.stem.2012.02.021

Heinrich C, Blum R, Gascón S, Masserdotti G, Tripathi P, Sánchez R, Tiedt S, Schroeder T, Götz M, Berninger B (2010) Directing astroglia from the cerebral cortex into subtype specific functional neurons. PLoS Biol 8(5): e1000373. https://doi.org/10.1371/journal.pbio.1000373

Heinrich C, Bergami M, Gascón S, Lepier A, Viganò F, Dimou L, Sutor B, Berninger B, Götz M (2014) Sox2-mediated conversion of NG2 glia into induced neurons in the injured adult cerebral cortex. Stem Cell Rep 3 (6):1000–1014. https://doi.org/10.1016/j.stemcr.2014.10.007

Heins N, Malatesta P, Cecconi F, Nakafuku M, Tucker KL, Hack MA, Chapouton P, Barde Y-A, Götz M (2002) Glial cells generate neurons: the role of the transcription factor Pax6. Nat Neurosci 5(4):308–315. https://doi.org/10.1038/nn828

Hu W, Qiu B, Guan W, Wang Q, Wang M, Li W, Gao L, Shen L, Huang Y, Xie G et al (2015) Direct conversion of normal and Alzheimer's disease human fibroblasts into neuronal cells by small molecules. Cell Stem Cell 17(2):204–212. https://doi.org/10.1016/j.stem.2015.07.006

Jäkel S, Agirre E, Falcão AM, van Bruggen D, Lee KW, Knuesel I, Malhotra D, Ffrench-Constant C, Williams A, Castelo-Branco G (2019) Altered human oligodendrocyte heterogeneity in multiple sclerosis. Nature 23:1. https://doi.org/10.1038/s41586-019-0903-2

Karow M, Camp JG, Falk S, Gerber T, Pataskar A, Gac-Santel M, Kageyama J, Brazovskaja A, Garding A, Fan W et al (2018) Direct pericyte-to-neuron reprogramming via unfolding of a neural stem cell-like program. Nat Neurosci 21(7):932–940. https://doi.org/10.1038/s41593-018-0168-3

Kim J, Su SC, Wang H, Cheng AW, Cassady JP, Lodato MA, Lengner CJ, Chung C-Y, Dawlaty MM, Tsai L-H et al (2011) Functional integration of dopaminergic neurons directly converted from mouse fibroblasts. Cell Stem Cell 9(5):413–419. https://doi.org/10.1016/j.stem.2011.09.011

Kuller LH, Longstreth WT, Arnold AM, Bernick C, Bryan RN, Beauchamp NJ (2004) Cardiovascular Health Study Collaborative Research Group. White matter hyperintensity on cranial magnetic resonance imaging: a predictor of stroke. Stroke 35(8):1821–1825. https://doi.org/10.1161/01.STR.0000132193.35955.69

Lassmann H, van Horssen J, Mahad D (2012) Progressive multiple sclerosis: pathology and pathogenesis. Nat Rev Neurol 8(11):647–656. https://doi.org/10.1038/nrneurol.2012.168

Li X, Zuo X, Jing J, Ma Y, Wang J, Liu D, Zhu J, Du X, Xiong L, Du Y et al (2015) Small-molecule-driven direct reprogramming of mouse fibroblasts into functional neurons. Cell Stem Cell 17(2):195–203. https://doi.org/10.1016/j.stem.2015.06.003

Liang X-G, Tan C, Wang C-K, Tao R-R, Huang Y-J, Ma K-F, Fukunaga K, Huang M-Z, Han F (2018) Mytl1 induced direct reprogramming of pericytes into cholinergic neurons. CNS Neurosci Ther 24(9):801–809. https://doi.org/10.1111/cns.12821

Liddelow SA, Barres BA (2017) Reactive astrocytes: production, function, and therapeutic potential. Immunity 46(6):957–967. https://doi.org/10.1016/j.immuni.2017.06.006

Liddelow SA, Guttenplan KA, Clarke LE, Bennett FC, Bohlen CJ, Schirmer L, Bennett ML, Münch AE, Chung W-S, Peterson TC et al (2017) Neurotoxic reactive astrocytes are induced by activated microglia. Nature 541(7638):481–487. https://doi.org/10.1038/nature21029

Liu M-L, Zang T, Zou Y, Chang JC, Gibson JR, Huber KM, Zhang C-L (2013) Small molecules enable neurogenin 2 to efficiently convert human fibroblasts into cholinergic neurons. Nat Commun 4:2183. https://doi.org/10.1038/ncomms3183

Liu Y, Yu C, Daley TP, Wang F, Cao WS, Bhate S, Lin X, Still C, Liu H, Zhao D et al (2018) CRISPR activation screens systematically identify factors that drive neuronal fate and reprogramming. Cell Stem Cell 23 (5):758–771.e8. https://doi.org/10.1016/j.stem.2018.09.003

Lujan E, Chanda S, Ahlenius H, Südhof TC, Wernig M (2012) Direct conversion of mouse fibroblasts to self-renewing, tripotent neural precursor cells. Proc Natl Acad Sci 109(7):2527–2532. https://doi.org/10.1073/pnas.1121003109

Ma K, Deng X, Xia X, Fan Z, Qi X, Wang Y, Li Y, Ma Y, Chen Q, Peng H et al (2018) Direct conversion of mouse astrocytes into neural progenitor cells and specific lineages of neurons. Transl Neurodegener 7:29. https://doi.org/10.1186/s40035-018-0132-x

Marques S, Zeisel A, Codeluppi S, van Bruggen D, Mendanha Falcão A, Xiao L, Li H, Häring M, Hochgerner H, Romanov RA et al (2016) Oligodendrocyte heterogeneity in the mouse juvenile and adult central nervous system. Science (New York, NY) 352 (6291):1326–1329. https://doi.org/10.1126/science.aaf6463

Marro S, Pang ZP, Yang N, Tsai M-C, Qu K, Chang HY, Südhof TC, Wernig M (2011) Direct lineage conversion of terminally differentiated hepatocytes to functional neurons. Cell Stem Cell 9(4):374–382. https://doi.org/10.1016/j.stem.2011.09.002

Masserdotti G, Gillotin S, Sutor B, Drechsel D, Irmler M, Jørgensen HF, Sass S, Theis FJ, Beckers J, Berninger B et al (2015) Transcriptional mechanisms of proneural factors and REST in regulating neuronal reprogramming of astrocytes. Cell Stem Cell 17 (1):74–88. https://doi.org/10.1016/j.stem.2015.05.014

Masserdotti G, Gascón S, Götz M (2016) Direct neuronal reprogramming: learning from and for development. Development 143(14):2494–2510. https://doi.org/10.1242/dev.092163

Matsuda T, Irie T, Katsurabayashi S, Hayashi Y, Nagai T, Hamazaki N, Adefuin AMD, Miura F, Ito T, Kimura H et al (2019) Pioneer Factor NeuroD1 rearranges transcriptional and epigenetic profiles to execute microglia-neuron conversion. Neuron 101 (3):472–485.e7. https://doi.org/10.1016/j.neuron.2018.12.010

Matsushima GK, Morell P (2001) The neurotoxicant, cuprizone, as a model to study demyelination and remyelination in the central nervous system. Brain Pathol (Zurich, Switzerland) 11(1):107–116

Mokhtarzadeh Khanghahi A, Satarian L, Deng W, Baharvand H, Javan M (2018) In vivo conversion of astrocytes into oligodendrocyte lineage cells with transcription factor Sox10; Promise for myelin repair in multiple sclerosis. PLoS One 13(9):e0203785. https://doi.org/10.1371/journal.pone.0203785

Morris SA (2016) Direct lineage reprogramming via pioneer factors; a detour through developmental gene regulatory networks. Development 143 (15):2696–2705. https://doi.org/10.1242/dev.138263

Morris SA, Cahan P, Li H, Zhao AM, San Roman AK, Shivdasani RA, Collins JJ, Daley GQ (2014) Dissecting engineered cell types and enhancing cell fate conversion via CellNet. Cell 158(4):889–902. https://doi.org/10.1016/j.cell.2014.07.021

Najm FJ, Lager AM, Zaremba A, Wyatt K, Caprariello AV, Factor DC, Karl RT, Maeda T, Miller RH, Tesar PJ (2013) Transcription factor-mediated reprogramming of fibroblasts to expandable, myelinogenic oligodendrocyte progenitor cells. Nat Biotechnol 31(5):426–433. https://doi.org/10.1038/nbt.2561

Ninkovic J, Götz M (2018) Understanding direct neuronal reprogramming-from pioneer factors to 3D chromatin. Curr Opin Genet Dev 52:65–69. https://doi.org/10.1016/j.gde.2018.05.011

Niu W, Zang T, Zou Y, Fang S, Smith DK, Bachoo R, Zhang C-L (2013) In vivo reprogramming of astrocytes to neuroblasts in the adult brain. Nat Cell Biol 15(10):1164–1175. https://doi.org/10.1038/ncb2843

Niu W, Zang T, Wang L-L, Zou Y, Zhang C-L (2018) Phenotypic reprogramming of striatal neurons into dopaminergic neuron-like cells in the adult mouse brain. Stem Cell Rep 11(5):1156–1170. https://doi.org/10.1016/j.stemcr.2018.09.004

Reich DS, Lucchinetti CF, Calabresi PA (2018) Multiple sclerosis. N Engl J Med 378(2):169–180. https://doi.org/10.1056/NEJMra1401483

Ring KL, Tong LM, Balestra ME, Javier R, Andrews-Zwilling Y, Li G, Walker D, Zhang WR, Kreitzer AC, Huang Y (2012) Direct reprogramming of mouse and human fibroblasts into multipotent neural stem cells with a single factor. Cell Stem Cell 11 (1):100–109. https://doi.org/10.1016/j.stem.2012.05.018

Rivetti di Val Cervo P, Romanov RA, Spigolon G, Masini D, Martín-Montañez E, Toledo EM, La Manno G, Feyder M, Pifl C, Ng Y-H et al (2017) Induction of functional dopamine neurons from human astrocytes in vitro and mouse astrocytes in a Parkinson's disease model. Nat Biotechnol 35 (5):444–452. https://doi.org/10.1038/nbt.3835

Roth AD, Ramírez G, Alarcón R, Von Bernhardi R (2005) Oligodendrocytes damage in Alzheimer's disease: beta amyloid toxicity and inflammation. Biol Res 38 (4):381–387

Rouaux C, Arlotta P (2013) Direct lineage reprogramming of post-mitotic callosal neurons into corticofugal neurons in vivo. Nat Cell Biol 15(2):214–221. https://doi.org/10.1038/ncb2660

Sawcer S, Franklin RJM, Ban M (2014) Multiple sclerosis genetics. Lancet Neurol 13(7):700–709. https://doi.org/10.1016/S1474-4422(14)70041-9

Sheng C, Zheng Q, Wu J, Xu Z, Sang L, Wang L, Guo C, Zhu W, Tong M, Liu L et al (2012) Generation of dopaminergic neurons directly from mouse fibroblasts and fibroblast-derived neural progenitors. Cell Res 22 (4):769–772. https://doi.org/10.1038/cr.2012.32

Sieweke MH (2015) Waddington's valleys and Captain Cook's islands. Cell Stem Cell 16(1):7–8. https://doi.org/10.1016/j.stem.2014.12.009

Son EY, Ichida JK, Wainger BJ, Toma JS, Rafuse VF, Woolf CJ, Eggan K (2011) Conversion of mouse and human fibroblasts into functional spinal motor neurons. Cell Stem Cell 9(3):205–218. https://doi.org/10.1016/j.stem.2011.07.014

Steinbeck JA, Studer L (2015) Moving stem cells to the clinic: potential and limitations for brain repair. Neuron 86(1):187–206. https://doi.org/10.1016/j.neuron.2015.03.002

Su Z, Niu W, Liu M-L, Zou Y, Zhang C-L (2014) In vivo conversion of astrocytes to neurons in the injured adult spinal cord. Nat Commun 5:3338. https://doi.org/10.1038/ncomms4338

Takahashi K, Yamanaka S (2006) Induction of pluripotent stem cells from mouse embryonic and adult fibroblast cultures by defined factors. Cell 126(4):663–676. https://doi.org/10.1016/j.cell.2006.07.024

Tang Y, Le W (2016) Differential roles of M1 and M2 microglia in neurodegenerative diseases. Mol Neurobiol 53(2):1181–1194. https://doi.org/10.1007/s12035-014-9070-5

Thored P, Wood J, Arvidsson A, Cammenga J, Kokaia Z, Lindvall O (2007) Long-term neuroblast migration along blood vessels in an area with transient angiogenesis and increased vascularization after stroke. Stroke 38(11):3032–3039. https://doi.org/10.1161/STROKEAHA.107.488445

Tian E, Sun G, Sun G, Chao J, Ye P, Warden C, Riggs AD, Shi Y (2016) Small-molecule-based lineage reprogramming creates functional astrocytes. Cell Rep 16(3):781–792. https://doi.org/10.1016/j.celrep.2016.06.042

Toft-Hansen H, Füchtbauer L, Owens T (2011) Inhibition of reactive astrocytosis in established experimental autoimmune encephalomyelitis favors infiltration by myeloid cells over T cells and enhances severity of disease. Glia 59(1):166–176. https://doi.org/10.1002/glia.21088

Torper O, Pfisterer U, Wolf DA, Pereira M, Lau S, Jakobsson J, Björklund A, Grealish S, Parmar M (2013) Generation of induced neurons via direct conversion in vivo. Proc Natl Acad Sci 110 (17):7038–7043. https://doi.org/10.1073/pnas.1303829110

Torper O, Ottosson DR, Pereira M, Lau S, Cardoso T, Grealish S, Parmar M (2015) In vivo reprogramming of striatal NG2 glia into functional neurons that integrate into local host circuitry. Cell Rep 12 (3):474–481. https://doi.org/10.1016/j.celrep.2015.06.040

Vadodaria KC, Mertens J, Paquola A, Bardy C, Li X, Jappelli R, Fung L, Marchetto MC, Hamm M, Gorris M et al (2016) Generation of functional human serotonergic neurons from fibroblasts. Mol Psychiatry 21 (1):49–61. https://doi.org/10.1038/mp.2015.161

Victor MB, Richner M, Hermanstyne TO, Ransdell JL, Sobieski C, Deng P-Y, Klyachko VA, Nerbonne JM, Yoo AS (2014) Generation of human striatal neurons by microRNA-dependent direct conversion of fibroblasts. Neuron 84(2):311–323. https://doi.org/10.1016/j.neuron.2014.10.016

Vierbuchen T, Wernig M (2011) Direct lineage conversions: unnatural but useful? Nat Biotechnol 29 (10):892–907. https://doi.org/10.1038/nbt.1946

Vierbuchen T, Ostermeier A, Pang ZP, Kokubu Y, Südhof TC, Wernig M (2010) Direct conversion of fibroblasts to functional neurons by defined factors. Nature 463 (7284):1035–1041. https://doi.org/10.1038/nature08797

Waddington CH (1957) The strategy of the genes; a discussion of some aspects of theoretical biology. Routledge, London. https://doi.org/10.4324/9781315765471

Wang L-L, Zhang C-L (2018) Engineering new neurons: in vivo reprogramming in mammalian brain and spinal cord. Cell Tissue Res 371(1):201–212. https://doi.org/10.1007/s00441-017-2729-2

Wapinski OL, Lee QY, Chen AC, Li R, Corces MR, Ang CE, Treutlein B, Xiang C, Baubet V, Suchy FP et al (2017) Rapid chromatin switch in the direct reprogramming of fibroblasts to neurons. Cell Rep 20 (13):3236–3247. https://doi.org/10.1016/j.celrep.2017.09.011

Xie H, Ye M, Feng R, Graf T (2004) Stepwise reprogramming of B cells into macrophages. Cell 117 (5):663–676

Xu J, Du Y, Deng H (2015) Direct lineage reprogramming: strategies, mechanisms, and applications. Cell Stem Cell 16(2):119–134. https://doi.org/10.1016/j.stem.2015.01.013

Yang N, Ng YH, Pang ZP, Südhof TC, Wernig M (2011) Induced neuronal (iN) cells: how to make and define a neuron. Cell Stem Cell 9(6):517–525. https://doi.org/10.1016/j.stem.2011.11.015

Yang N, Zuchero JB, Ahlenius H, Marro S, Ng YH, Vierbuchen T, Hawkins JS, Geissler R, Barres BA, Wernig M (2013) Generation of oligodendroglial cells by direct lineage conversion. Nat Biotechnol 31 (5):434–439. https://doi.org/10.1038/nbt.2564

Yoo AS, Sun AX, Li L, Shcheglovitov A, Portmann T, Li Y, Lee-Messer C, Dolmetsch RE, Tsien RW, Crabtree GR (2011) MicroRNA-mediated conversion of human fibroblasts to neurons. Nature 476 (7359):228–231. https://doi.org/10.1038/nature10323

Zamanian JL, Xu L, Foo LC, Nouri N, Zhou L, Giffard RG, Barres BA (2012) Genomic analysis of reactive astrogliosis. J Neurosci 32(18):6391–6410

Zhang L, Lei Z, Guo Z, Pei Z, Chen Y, Zhang F, Cai A, Mok YK, Lee G, Swaminathan V et al (2018) Reversing glial scar Back to neural tissue through NeuroD1-mediated astrocyte-to-neuron conversion. bioRxiv 7:261438. https://doi.org/10.1101/261438

Zhu X, Zhou W, Jin H, Li T (2018) Brn2 alone is sufficient to convert astrocytes into neural progenitors and neurons. Stem Cells Dev 27(11):736–744. https://doi.org/10.1089/scd.2017.0250

Adv Exp Med Biol – Cell Biology and Translational Medicine (2019) 6: 49–56
https://doi.org/10.1007/5584_2019_394
© Springer Nature Switzerland AG 2019
Published online: 22 June 2019

Induced Pluripotent Stem Cells for Regenerative Medicine: Quality Control Based on Evaluation of Lipid Composition

Yusuke Nakamura and Yasuo Shimizu

Abstract

Clinical application of induced pluripotent stem cells (iPSCs), which can be differentiated into a wide variety of functional cells, is underway and some clinical trials have already been performed or are ongoing. On the other hand, the risk of carcinogenesis is an issue and the mechanism of cellular reprograming remains unknown. When iPSCs and differentiated cells are used for medical applications, quality control is also important. Here we discuss the possibility of performing quality control of iPSCs by evaluation of phospholipids, which are not just structural components of lipid bilayer membranes, but also have multiple physiological functions. Recently, methods for analysis of lipids have become more widely available and easier to perform. This article reviews the role of iPSCs in regenerative medicine and examines the possibility of using phospholipids for quality control of iPSCs and differentiated cells.

Keywords

Induced pluripotent stem cells · Regenerative medicine · Lipids · Quality control · LC-MS · Imaging mass spectrometry

Abbreviations

ALS	amyotrophic lateral sclerosis
FIH	first-in-human
FOP	Fibrodysplasia ossificans progressive
IMS	imaging mass spectrometry
iPSCs	induced pluripotent stem cells
KLF4	Kruppel like factor 4
LC-MS	liquid chromatography-mass spectrometry
pPE	plasmalogen phosphatidylethanolamine
SM	sphingomyelin
SOX2	SRY-box 2

1 Introduction

1.1 iPSCs and Regenerative Medicine

Human induced pluripotent stem cells (iPSCs) were first established in 2007 by Yamanaka's group (Takahashi et al. 2007). These cells were generated by adding recombinant factors, such as OCT3/4, SRY-box 2 (SOX2), Kruppel like factor 4 (KLF4), or c-Myc, to cultures of somatic cells. iPSCs can be differentiated into various types of cells with potential clinical uses. However, there is a potential risk of carcinogenesis and mechanism of cellular reprograming is still unknown. It

Y. Nakamura and Y. Shimizu (✉)
Department of Pulmonary Medicine and Clinical Immunology, Dokkyo Medical University School of Medicine, Mibu, Tochigi, Japan
e-mail: yasuo-s@dokkyomed.ac.jp

is generally thought that there is a low risk of rejection of iPSCs, but some research has identified rejection of transplanted iPSCs (Zhao et al. 2011). Another group reported the transplantation of undifferentiated iPSCs, but stated that more detailed investigation was needed. There is the possibility of tumorigenic cells emerging from iPSCs or the immune system eliminating them as foreign cells (Okita et al. 2011). Despite these issues, iPSCs have many possible future clinical applications, including drug development and elucidation of the pathophysiology of various diseases.

Cells undergo various changes in the process of differentiation from undifferentiated pluripotent stem cells and also during the development of organs. To maintain the morphology of cells and organs during these processes, a supporting structure is required that preserves the integrity of each cell and/or organ. The cell membrane, which is composed of a lipid bilayer, plays this role and is an essential determinant of the morphology and function of cells and organs. iPSCs show various changes as these cells undergo differentiation depending on the culture conditions. iPSCs are generally cultured with feeder cells. Undifferentiated iPSCs form colonies that can clearly be identified by visual inspection, and the center of each colony shows squamatization after several days. At this time, the iPSC colonies are doughnut-shaped and the border with the surrounding feeder cells gradually becomes unclear. It is considered that these changes of iPSC colony morphology are associated with changes of membrane lipids.

Intracellular vesicular organs, such as the lysosomes and mitochondria, also have membranes composed of lipid bilayers. The lysosomal membrane plays an important role in formation of autophagosomes (which digest unwanted cellular components for recycling), and the role of membrane fusion in the mechanism of autophagy has recently attracted attention.

Thus, lipids are not only important components of cellular architecture, but are also involved in intracellular signaling and metabolism. When iPSCs are used for regenerative medicine, quality control is essential to prevent carcinogenesis and contamination by unwanted cells. Investigation of lipids could be useful for quality control, including maintenance of undifferentiated pluripotent iPSCs and assessment of cells that have been differentiated from iPSCs.

1.2   Clinical Applications

Differentiated iPSC-derived cells have already been employed for several clinical applications. iPSC-derived cells were first applied clinically to the treatment of age-related macular degeneration. In 2014, transplantation of iPSC-derived retinal pigment epithelium was performed, and the safety of this procedure was confirmed as well as improvement of macular degeneration (Mandai et al. 2017). Treatment of Parkinson's disease with iPSC-derived cells has also been studied. In 2017, efficacy of iPSC-derived cells was confirmed in a primate model (Kikuchi et al. 2017). Subsequently, a clinical study was initiated in Japan in August 2018, with transplantation of iPSC-derived dopamine-producing neuronal precursors being performed in November 2018. Furthermore, the efficacy of transplanted iPSC-derived cells for heart failure has been demonstrated in various animal models (Nakane et al. 2017; Ye et al. 2014; Shiba et al. 2016), with the "first-in-human" (FIH) trial commencing in Japan in May 2018. As another potential clinical application, iPSCs have been employed for creation of platelets (Takayama et al. 2010; Ito et al. 2018). In September 2018, the Japanese Ministry of Health, Labour and Welfare approved clinical research on iPSC-derived platelets for patients with aplastic anemia refractory to platelet transfusion, so safety in humans will be evaluated in the future. Moreover, the effect of iPSC-derived neural stem cells/precursor cells on spinal cord injury has been evaluated in an animal model, with good results being obtained (Okubo et al. 2018). In Japan, approval was granted for a clinical investigation of iPSC-derived neural stem cells/precursor cells in patients with spinal cord injury in February 2019, and the results are eagerly anticipated. Finally, studies on the treatment of sickle-cell anemia by genome editing have been conducted and clinical application is under consideration in the U.S.A. (Hanna et al. 2007; Zou et al. 2011).

1.3 Creation of Artificial Organs

The ultimate goal of regenerative medicine is to create functional artificial organs for patients with various diseases. There has been marked progress in the fashioning of mini-livers by self-organization of cultured cells, raising the possibility of this becoming the first iPSC-derived artificial organ. Prolongation of survival has already been confirmed in animal models of hepatic insufficiency after mini-liver transplantation (Takebe et al. 2013, 2017). Creation of organs by self-organization has also been applied to the intestines and kidneys (Spence et al. 2011; Takasato et al. 2015). Pancreatic islets have been created by the blastocyst complementation method, in which the organ is prepared in a heterologous animal host, and pancreatic endocrine function was improved by these artificial islets in an animal model (Yamaguchi et al. 2017). Although the creation of artificial lungs is still at an earlier phase of research, efficient differentiation of type 2 alveolar cells has been achieved and application of these cells is anticipated (Yamamoto et al. 2017). In addition to the self-organization and blastocyst complementation methods, organs can be created by decellularization. In this method, pluripotent stem cell-derived cells are seeded into organs or tissues after decellularization has been performed, leaving the extracellular matrix as a biological scaffold. This procedure has already been investigated for the lungs using iPSC-derived cells (Ren et al. 2015).

1.4 Application of Regenerative Medicine Techniques

Application of iPSC-related techniques has already achieved suggestive results in the treatment of several diseases, as well as in drug development and elucidation of disease pathology. Fibrodysplasia ossificans progressiva (FOP) is a disease in which motor dysfunction occurs due to heterotopic ossification. The import role of mTOR signaling in this disease was confirmed by investigation of the underlying molecular pathology using iPSCs derived from FOP patients, suggesting that an mTOR inhibitor could be employed for treatment (Hino et al. 2017). Based on this finding, a physician-initiated clinical study of the mTOR inhibitor rapamycin for FOP has been underway in Japan since September 2017. With regard to Alzheimer's disease, iPSCs derived from patients have been differentiated into cerebral cortical neurons to investigate treatments that decrease accumulation of amyloid-β, which is considered to be central to the pathogenesis of this disease (Kondo et al. 2017). In the case of amyotrophic lateral sclerosis (ALS), iPSCs derived from patients with SOD1 gene abnormalities were differentiated into motor neurons for comparison with genome-edited control cells to identify molecules suppressing cell death, revealing that phosphorylation of Src/c-Ab1 is involved in motor neuron death (Egawa et al. 2012). Bosutinib (a treatment for chronic myelocytic leukemia) was found to suppress Src/c-Ab1 phosphorylation and a Phase I study assessing its effectiveness for ALS was commenced in March 2019. Moreover, disease-specific iPSCs were created and differentiated into chondrocytes to investigate the pathology of achondroplasia, leading to the possibility of statin treatment being used to improve cartilage formation in this condition (Yamashita et al. 2014). Clinical application of regenerative medicine techniques is also being employed in the treatment of hypertonic cardiomyopathy (Tanaka et al. 2014) and myotonic dystrophy (Ueki et al. 2017). As outlined above, iPSCs are not only useful for direct regenerative medicine applications such as cell transplantation and organ creation, but also for identification of candidate drugs and investigation of disease pathology.

2 Quality Control

It is important to evaluate the quality of cells and tissues used in basic research on iPSCs or clinical applications. When differentiated cells are employed clinically, their properties are evaluated by immunostaining of surface markers and assessment of function, e.g., contractility of the myocardium or secretion of target hormones by endocrine organs. However, quality control has rarely been performed by assessing cellular lipids and the importance of lipids has generally been overlooked. Because lipids have essential embryological and physiological roles, we will next discuss evaluation of lipids for quality control of iPSCs.

3   iPSCs and Lipids

Lipids have been reported to have an important role in differentiation of pluripotent stem cells, normal development, and various diseases, suggesting that evaluation of lipids should be taken into consideration when clinical application of iPSCs is performed. We found that expression of plasmalogen increased during the process of iPSC differentiation into vascular endothelial cells (Nakamura et al. 2017) (Fig. 1). Plasmalogens are glycerophospholipids with a vinyl-ether linkage to a fatty acid at the sn-1 position and an ester linkage at the sn-2 position that are reported to have an antioxidant effect (Braverman and Moser 2012; Zoeller et al.

1988, 2002). Because plasmalogen synthetase deficiency has been reported to cause optic nerve hypoplasia and male sterility in animals, it has been suggested that plasmalogens are involved in the differentiation and maturation of cells (Rodemer et al. 2003). A decrease of plasmalogen has been reported in Alzheimer's disease (Han et al. 2001) and aging (Maeba et al. 2007), as well as in cardiorespiratory diseases such as myocardial infarction (Park et al. 2015) and chronic obstructive pulmonary disease (Wang-Sattler et al. 2008).

Expression of sphingomyelin also changes during the differentiation process. We found a transient decrease of sphingomyelin during the process of iPSC differentiation into vascular endothelial cells and subsequent maturation (Nakamura et al. 2017) (Fig. 1). Sphingosine 1-phosphate is a metabolite of sphingomyelin that is essential for vessel formation in the fetus and its deficiency is thought to be lethal (Liu et al. 2000). Sphingosine 1-phosphate has a role in regulating the permeability of vascular endothelial cells, and is also related to acute respiratory distress syndrome (Sun et al. 2013) and anaphylaxis (Camerer et al. 2009).

As indicated above, lipids have various roles in the differentiation, maturation, and aging of cells and lipid abnormalities are related to various

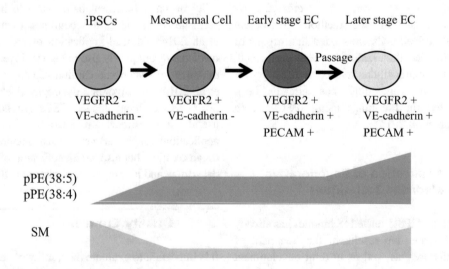

Fig. 1 Changes of lipid components during differentiation of iPSCs into vascular endothelial cells. *pPE* Plasmalogen phosphatidylethanolamine, *SM* Sphingomyelin

diseases. Accordingly, lipid components may determine the physiological characteristics of differentiated cells, suggesting that further investigation of lipids is warranted in relation to regenerative medicine using iPSCs.

4   Evaluation of Lipids

Some lipids can be evaluated indirectly by labeling, such as detection of phosphatidylserine to identify cells undergoing apoptosis. However, lipids cannot be easily labeled with antibodies, unlike proteins and nucleic acids. Recently developed analytical methods have made it possible to analyze lipids in more detail. Lipid components are generally extracted from cells or tissues for evaluation by chromatography-mass spectrometry or imaging mass spectrometry (IMS), which can directly assess two-dimensional structures.

Liquid chromatography-mass spectrometry (LC-MS) is a common method for analysis of lipids, which requires extraction of the target lipids before measurement. Among various lipid extraction methods, the Bligh-Dyer method employing methanol and chloroform can be used to easily extract phospholipids for analysis (Bligh and Dyer 1959). When LC-MS is performed, the mass-to-charge ratio (m/z) of an ionized substance is specified using a mass detector, after which the ionized substance is fragmented with argon gas and identified by analysis of the fragment ions. According to reports concerning the methods of ionization and chromatography, identification of a target substance is simple if m/z is specified with reference to previous data (Shui et al. 2011). We have used LC-MS to analyze the changes of lipids during the process of iPSC differentiation into vascular endothelial cells (Fig. 1).

IMS can be used to evaluate the secondary structure of lipids in tissues or cultured cells. After a tissue or cell specimen is placed on an electrically conductive slide, lipid analysis is performed by mass spectrometry (Fig. 2), with matrix-assisted   laser   desorption/ionization

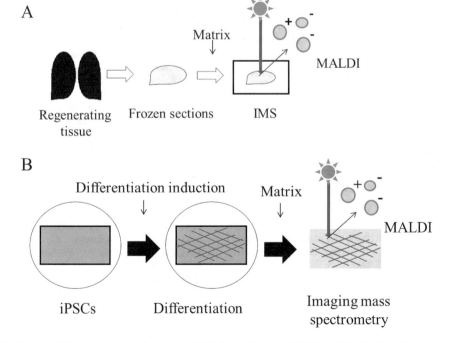

Fig. 2 Application of imaging mass spectrometry (IMS) for evaluation of iPSCs. (**a**) Evaluation of tissue created from differentiated cells. (**b**) Inducing the differentiation of iPSCs on an electrically conductive slide

generally being employed for molecular ionization (Tanaka et al. 1988). We have confirmed that IMS is applicable to cultured iPSCs and have used this method to show that the distribution of phosphatidylcholine differs between the differentiated and undifferentiated parts of iPSC colonies (Shimizu et al. 2017).

5   Conclusions

iPSCs are associated with a low risk of rejection and are currently being used clinically for various regenerative medicine applications. Although some problems still remain to be solved, such as the potential risk of cancer and the mechanisms for initialization of differentiation, iPSCs seem to be a promising tool that will be increasingly used for regenerative medicine. If iPSC-derived cells are widely employed for clinical applications in the future, quality control will become more important. iPSC-derived cells have previously been evaluated on the basis of function and surface marker expression, with little attention being paid to assessment of cellular lipids. However, lipids are not only important structural components of membranes (lipid bilayer), but are also required for differentiation and development as well as physiological cellular functions. In this review, we outlined the clinical applications of iPSCs and suggested that evaluation of lipids could be employed for quality control. It has recently become possible to perform detailed analysis of lipids by methods such as LC-MS and IMS. These methods may become useful tools for quality control based on evaluation of lipids when iPSC-derived cells and tissues are used for clinical applications in the future.

Acknowledgements This work was supported by a Japan Society for the Promotion of Science (JSPS) KAKENHI Grant, Japan to Y.S (16 K09553).

Author Contributions Y.N. wrote the manuscript. Y.S. designed and wrote the manuscript. Both authors reviewed the manuscript.

References

Bligh EG, Dyer WJ (1959) A rapid method of total lipid extraction and purification. Can J Biochem Physiol 37 (8):911–917

Braverman NE, Moser AB (2012) Functions of plasmalogen lipids in health and disease. Biochim Biophys Acta 1822(9):1442–1452. https://doi.org/10.1016/j.bbadis.2012.05.008

Camerer E, Regard JB, Cornelissen I, Srinivasan Y, Duong DN, Palmer D, Pham TH, Wong JS, Pappu R, Coughlin SR (2009) Sphingosine-1-phosphate in the plasma compartment regulates basal and inflammation-induced vascular leak in mice. J Clin Invest 119 (7):1871–1879

Egawa N, Kitaoka S, Tsukita K, Naitoh M, Takahashi K, Yamamoto T, Adachi F, Kondo T, Okita K, Asaka I, Aoi T, Watanabe A, Yamada Y, Morizane A, Takahashi J, Ayaki T, Ito H, Yoshikawa K, Yamawaki S, Suzuki S, Watanabe D, Hioki H, Kaneko T, Makioka K, Okamoto K, Takuma H, Tamaoka A, Hasegawa K, Nonaka T, Hasegawa M, Kawata A, Yoshida M, Nakahata T, Takahashi R, Marchetto MC, Gage FH, Yamanaka S, Inoue H (2012) Drug screening for ALS using patient-specific induced pluripotent stem cells. Sci Transl Med 4 (145):145ra104. https://doi.org/10.1126/scitranslmed.3004052

Han X, Holtzman DM, McKeel DW Jr (2001) Plasmalogen deficiency in early Alzheimer's disease subjects and in animal models: molecular characterization using electrospray ionization mass spectrometry. J Neurochem 77(4):1168–1180

Hanna J, Wernig M, Markoulaki S, Sun CW, Meissner A, Cassady JP, Beard C, Brambrink T, Wu LC, Townes TM, Jaenisch R (2007) Treatment of sickle cell anemia mouse model with iPS cells generated from autologous skin. Science 318(5858):1920–1923. https://doi.org/10.1126/science.1152092

Hino K, Horigome K, Nishio M, Komura S, Nagata S, Zhao C, Jin Y, Kawakami K, Yamada Y, Ohta A, Toguchida J, Ikeya M (2017) Activin-a enhances mTOR signaling to promote aberrant chondrogenesis in fibrodysplasia ossificans progressiva. J Clin Invest 127(9):3339–3352. https://doi.org/10.1172/JCI93521

Ito Y, Nakamura S, Sugimori N, Shigemori T, Kato Y, Ohno M, Sakuma S, Ito K, Kumon H, Hirose H, Okamoto H, Nogawa M, Iwasaki M, Kihara S, Fujio K, Matsumoto T, Higashi N, Hashimoto K, Sawaguchi A, Harimoto KI, Nakagawa M, Yamamoto T, Handa M, Watanabe N, Nishi E, Arai F, Nishimura S, Eto K (2018) Turbulence activates platelet biogenesis to enable clinical scale ex vivo production. Cell 174(3):636–648.e18. https://doi.org/10.1016/j.cell.2018.06.011

Kikuchi T, Morizane A, Doi D, Magotani H, Onoe H, Hayashi T, Mizuma H, Takara S, Takahashi R, Inoue H, Morita S, Yamamoto M, Okita K, Nakagawa M, Parmar M, Takahashi J (2017) Human iPS cell-derived dopaminergic neurons function in a primate Parkinson's disease model. Nature 548 (7669):592–596. https://doi.org/10.1038/nature23664

Kondo T, Imamura K, Funayama M, Tsukita K, Miyake M, Ohta A, Woltjen K, Nakagawa M, Asada T, Arai T, Kawakatsu S, Izumi Y, Kaji R, Iwata N, Inoue H (2017) iPSC-based compound screening and in vitro trials identify a synergistic anti-amyloid β combination for Alzheimer's disease. Cell Rep 21(8):2304–2312. https://doi.org/10.1016/j.celrep.2017.10.109

Liu Y, Wada R, Yamashita T, Mi Y, Deng CX, Hobson JP, Rosenfeldt HM, Nava VE, Chae SS, Lee MJ, Liu CH, Hla T, Spiegel S, Proia RL (2000) Edg-1, the G protein-coupled receptor for sphingosine-1-phosphate, is essential for vascular maturation. J Clin Invest 106 (8):951–961. https://doi.org/10.1172/JCI10905

Maeba R, Maeda T, Kinoshita M, Takao K, Takenaka H, Kusano J, Yoshimura N, Takeoka Y, Yasuda D, Okazaki T, Teramoto T (2007) Plasmalogens in human serum positively correlate with high- density lipoprotein and decrease with aging. J Atheroscler Thromb 14(1):12–18

Mandai M, Watanabe A, Kurimoto Y, Hirami Y, Morinaga C, Daimon T, Fujihara M, Akimaru H, Sakai N, Shibata Y, Terada M, Nomiya Y, Tanishima S, Nakamura M, Kamao H, Sugita S, Onishi A, Ito T, Fujita K, Kawamata S, Go MJ, Shinohara C, Hata KI, Sawada M, Yamamoto M, Ohta S, Ohara Y, Yoshida K, Kuwahara J, Kitano Y, Amano N, Umekage M, Kitaoka F, Tanaka A, Okada C, Takasu N, Ogawa S, Yamanaka S, Takahashi M (2017) Autologous induced stem-cell-derived retinal cells for macular degeneration. N Engl J Med 376(11):1038–1046. https://doi.org/10.1056/NEJMoa1608368

Nakamura Y, Shimizu Y, Horibata Y, Tei R, Koike R, Masawa M, Watanabe T, Shiobara T, Arai R, Chibana K, Takemasa A, Sugimoto H, Ishii Y (2017) Changes of plasmalogen phospholipid levels during differentiation of induced pluripotent stem cells 409B2 to endothelial phenotype cells. Sci Rep 7 (1):9377. https://doi.org/10.1038/s41598-017-09980-x

Nakane T, Masumoto H, Tinney JP, Yuan F, Kowalski WJ, Ye F, LeBlanc AJ, Sakata R, Yamashita JK, Keller BB (2017) Impact of cell composition and geometry on human induced pluripotent stem cells-derived engineered cardiac tissue. Sci Rep 7:45641. https://doi.org/10.1038/srep45641

Okita K, Nagata N, Yamanaka S (2011) Immunogenicity of induced pluripotent stem cells. Circ Res 109 (7):720–721. https://doi.org/10.1161/RES.0b013e318232e187

Okubo T, Nagoshi N, Kohyama J, Tsuji O, Shinozaki M, Shibata S, Kase Y, Matsumoto M, Nakamura M, Okano H (2018) Treatment with a gamma-secretase inhibitor promotes functional recovery in human iPSC- derived transplants for chronic spinal cord injury. Stem Cell Rep 11(6):1416–1432. https://doi.org/10.1016/j.stemcr.2018.10.022

Park JY, Lee SH, Shin MJ, Hwang GS (2015) Alteration in metabolic signature and lipid metabolism in patients with angina pectoris and myocardial infarction. PLoS One 10(8):e0135228. https://doi.org/10.1371/journal.pone.0135228

Ren X, Moser PT, Gilpin SE, Okamoto T, Wu T, Tapias LF, Mercier FE, Xiong L, Ghawi R, Scadden DT, Mathisen DJ, Ott HC (2015) Engineering pulmonary vasculature in decellularized rat and human lungs. Nat Biotechnol 33(10):1097–1102. https://doi.org/10.1038/nbt.3354

Rodemer C, Thai TP, Brugger B, Kaercher T, Werner H, Nave KA, Wieland F, Gorgas K, Just WW (2003) Inactivation of ether lipid biosynthesis causes male infertility, defects in eye development and optic nerve hypoplasia in mice. Hum Mol Genet 12 (15):1881–1195

Shiba Y, Gomibuchi T, Seto T, Wada Y, Ichimura H, Tanaka Y, Ogasawara T, Okada K, Shiba N, Sakamoto K, Ido D, Shiina T, Ohkura M, Nakai J, Uno N, Kazuki Y, Oshimura M, Minami I, Ikeda U (2016) Allogeneic transplantation of iPS cell-derived cardiomyocytes regenerates primate hearts. Nature 538 (7625):388–391. https://doi.org/10.1038/nature19815

Shimizu Y, Satou M, Hayashi K, Nakamura Y, Fujimaki M, Horibata Y, Ando H, Watanabe T, Shiobara T, Chibana K, Takemasa A, Sugimoto H, Anzai N, Ishii Y (2017) Matrix-assisted laser desorption/ionization imaging mass spectrometry reveals changes of phospholipid distribution in induced pluripotent stem cell colony differentiation. Anal Bioanal Chem 409(4):1007–1016. https://doi.org/10.1007/s00216-016-0015-x

Shui G, Stebbins JW, Lam BD, Cheong WF, Lam SM, Gregoire F, Kusonoki J, Wenk MR (2011) Comparative plasma lipidome between human and cynomolgus monkey: are plasma polar lipids good biomarkers for diabetic monkeys? PLoS One 6(5):e19731. https://doi.org/10.1371/journal.pone.0019731

Spence JR, Mayhew CN, Rankin SA, Kuhar MF, Vallance JE, Tolle K, Hoskins EE, Kalinichenko VV, Wells SI, Zorn AM, Shroyer NF, Wells JM (2011) Directed differentiation of human pluripotent stem cells into intestinal tissue in vitro. Nature 470(7332):105–109. https://doi.org/10.1038/nature09691

Sun X, Ma SF, Wade MS, Acosta-Herrera M, Villar J, Pino-Yanes M, Zhou T, Liu B, Belvitch P, Moitra J, Han YJ, Machado R, Noth I, Natarajan V, Dudek SM, Jacobson JR, Flores C, Garcia JG (2013) Functional promoter variants in sphingosine 1-phosphate receptor 3 associate with susceptibility to sepsis-associated acute respiratory distress syndrome. Am J Physiol Lung Cell Mol Physiol 305(7):L467–L477. https://doi.org/10.1152/ajplung.00010.2013

Takahashi K, Tanabe K, Ohnuki M, Narita M, Ichisaka T, Tomoda K, Yamanaka S (2007) Induction of pluripotent stem cells from adult human fibroblasts by defined factors. Cell 131(5):861–872. https://doi.org/10.1016/j.cell.2007.11.019

Takasato M, Er PX, Chiu HS, Maier B, Baillie GJ, Ferguson C, Parton RG, Wolvetang EJ, Roost MS, Chuva de Sousa Lopes SM, Little MH (2015) Kidney organoids from human iPS cells contain multiple lineages and model human nephrogenesis. Nature 526 (7574):564–568. https://doi.org/10.1038/nature15695

Takayama N, Nishimura S, Nakamura S, Shimizu T, Ohnishi R, Endo H, Yamaguchi T, Otsu M, Nishimura K, Nakanishi M, Sawaguchi A, Nagai R, Takahashi K, Yamanaka S, Nakauchi H, Eto K (2010) Transient activation of c-MYC expression is critical for efficient platelet generation from human induced pluripotent stem cells. J Exp Med 207(13):2817–2830. https://doi.org/10.1084/jem.20100844

Takebe T, Sekine K, Enomura M, Koike H, Kimura M, Ogaeri T, Zhang RR, Ueno Y, Zheng YW, Koike N, Aoyama S, Adachi Y, Taniguchi H (2013) Vascularized and functional human liver from an iPSC-derived organ bud transplant. Nature 499 (7459):481–484. https://doi.org/10.1038/nature12271

Takebe T, Sekine K, Kimura M, Yoshizawa E, Ayano S, Koido M, Funayama S, Nakanishi N, Hisai T, Kobayashi T, Kasai T, Kitada R, Mori A, Ayabe H, Ejiri Y, Amimoto N, Yamazaki Y, Ogawa S, Ishikawa M, Kiyota Y, Sato Y, Nozawa K, Okamoto S, Ueno Y, Taniguchi H (2017) Massive and reproducible production of liver buds entirely from human pluripotent stem cells. Cell Rep 21 (10):2661–2670. https://doi.org/10.1016/j.celrep.2017.11.005

Tanaka K, Waki H, Ido Y, Akita S, Yoshida Y, Yoshida T (1988) Protein and polymer analyses up to m/z 100 000 by laser ionization time-of-flight mass spectrometry. Rapid Commun Mass Spectrom 2:151–153

Tanaka A, Yuasa S, Mearini G, Egashira T, Seki T, Kodaira M, Kusumoto D, Kuroda Y, Okata S, Suzuki T, Inohara T, Arimura T, Makino S, Kimura K, Kimura A, Furukawa T, Carrier L, Node K, Fukuda K (2014) Endothelin-1 induces myofibrillar disarray and contractile vector variability in hypertrophic cardiomyopathy-induced pluripotent stem cell-derived cardiomyocytes. J Am Heart Assoc 3(6):e001263. https://doi.org/10.1161/JAHA.114.001263

Ueki J, Nakamori M, Nakamura M, Nishikawa M, Yoshida Y, Tanaka A, Morizane A, Kamon M, Araki T, Takahashi MP, Watanabe A, Inagaki N, Sakurai H (2017) Myotonic dystrophy type 1 patient-derived iPSCs for the investigation of CTG repeat instability. Sci Rep 7:42522. https://doi.org/10.1038/srep42522

Wang-Sattler R, Yu Y, Mittelstrass K, Lattka E, Altmaier E, Gieger C, Ladwig KH, Dahmen N, Weinberger KM, Hao P, Liu L, Li Y, Wichmann HE, Adamski J, Suhre K, Illig T (2008) Metabolic profiling reveals distinct variations linked to nicotine consumption in humans--first results from the KORA study. PLoS One 3(12):e3863. https://doi.org/10.1371/journal.pone.0003863

Yamaguchi T, Sato H, Kato-Itoh M, Goto T, Hara H, Sanbo M, Mizuno N, Kobayashi T, Yanagida A, Umino A, Ota Y, Hamanaka S, Masaki H, Rashid ST, Hirabayashi M, Nakauchi H (2017) Interspecies organogenesis generates autologous functional islets. Nature 542(7640):191–196. https://doi.org/10.1038/nature21070

Yamamoto Y, Gotoh S, Korogi Y, Seki M, Konishi S, Ikeo S, Sone N, Nagasaki T, Matsumoto H, Muro S, Ito I, Hirai T, Kohno T, Suzuki Y, Mishima M (2017) Long-term expansion of alveolar stem cells derived from human iPS cells in organoids. Nat Methods 14 (11):1097–1106. https://doi.org/10.1038/nmeth.4448

Yamashita A, Morioka M, Kishi H, Kimura T, Yahara Y, Okada M, Fujita K, Sawai H, Ikegawa S, Tsumaki N (2014) Statin treatment rescues FGFR3 skeletal dysplasia phenotypes. Nature 513(7519):507–511. https://doi.org/10.1038/nature13775

Ye L, Chang YH, Xiong Q, Zhang P, Zhang L, Somasundaram P, Lepley M, Swingen C, Su L, Wendel JS, Guo J, Jang A, Rosenbush D, Greder L, Dutton JR, Zhang J, Kamp TJ, Kaufman DS, Ge Y, Zhang J (2014) Cardiac repair in a porcine model of acute myocardial infarction with human induced pluripotent stem cell-derived cardiovascular cells. Cell Stem Cell 15(6):750–761. https://doi.org/10.1016/j.stem.2014.11.009

Zhao T, Zhang ZN, Rong Z, Xu Y (2011) Immunogenicity of induced pluripotent stem cells. Nature 474 (7350):212–215. https://doi.org/10.1038/nature10135

Zoeller RA, Morand OH, Raetz CR (1988) A possible role for plasmalogens in protecting animal cells against photosensitized killing. J Biol Chem 263 (23):11590–11596

Zoeller RA, Grazia TJ, LaCamera P, Park J, Gaposchkin DP, Farber HW (2002) Increasing plasmalogen levels protects human endothelial cells during hypoxia. Am J Physiol Heart Circ Physiol 283(2):H671–H679. https://doi.org/10.1152/ajpheart.00524.2001

Zou J, Mali P, Huang X, Dowey SN, Cheng L (2011) Site-specific gene correction of a point mutation in human iPS cells derived from an adult patient with sickle cell disease. Blood 118(17):4599–4608. https://doi.org/10.1182/blood-2011-02-335554

Adv Exp Med Biol – Cell Biology and Translational Medicine (2019) 6: 57–70
https://doi.org/10.1007/5584_2019_371
© Springer Nature Switzerland AG 2019
Published online: 16 April 2019

Decellularized Adipose Tissue: Biochemical Composition, *in vivo* Analysis and Potential Clinical Applications

Omair A. Mohiuddin, Brett Campbell, J. Nicholas Poche, Caasy Thomas-Porch, Daniel A. Hayes, Bruce A. Bunnell, and Jeffrey M. Gimble

Abstract

Decellularized tissues are gaining popularity as scaffolds for tissue engineering; they allow cell attachment, proliferation, differentiation, and are non-immunogenic. Adipose tissue is an abundant resource that can be decellularized and converted in to a bio-scaffold. Several methods have been developed for adipose tissue decellularization, typically starting with freeze thaw cycles, followed by washes with hypotonic/hypertonic sodium chloride solution, isopropanol, detergent (SDS, SDC and Triton X-100) and trypsin digestion. After decellularization, decellularized adipose tissue (DAT) can be converted into a powder, solution, foam, or sheet to allow for convenient subcutaneous implantation or to repair external injuries. Additionally, DAT bio-ink can be used to 3D print structures that closely resemble physiological tissues and organs. Proteomic analysis of DAT reveals that it is composed of collagens (I, III, IV, VI and VII), glycosaminoglycans, laminin, elastin, and fibronectin. It has also been found to retain growth factors like VEGF and bFGF after decellularization. DAT inherently promotes adipogenesis when seeded with adipose stem cells *in vitro*, and when DAT is implanted subcutaneously it is capable of recruiting host stem cells and forming adipose tissue in rodents. Furthermore, DAT has promoted healing in rat models of full-thickness skin wounds and peripheral nerve injury. These findings suggest that DAT is a promising candidate for repair of

O. A. Mohiuddin and B. A. Bunnell
Center for Stem Cell Research & Regenerative Medicine, Tulane University School of Medicine, New Orleans, LA, USA

B. Campbell
School of Medicine, Tulane University, New Orleans, LA, USA

J. N. Poche
School of Medicine, Louisiana State University, New Orleans, LA, USA

C. Thomas-Porch
Department of Biology and Chemistry, Southern University and A&M College, Baton Rouge, LA, USA

D. A. Hayes
Department of Biomedical Engineering, Pennsylvania State University, State College, PA, USA

J. M. Gimble (✉)
Center for Stem Cell Research & Regenerative Medicine, Tulane University School of Medicine, New Orleans, LA, USA

LaCell LLC, New Orleans, LA, USA
e-mail: jgimble@tulane.edu

soft tissue defects, and is suitable for breast reconstruction post-mastectomy, wound healing, and adipose tissue regeneration. Moreover, since DAT's form and stiffness can be altered by physicochemical manipulation, it may prove suitable for engineering of additional soft and hard tissues.

Keywords

Biochemical composition · Biological scaffold · Clinical applications · Decellularized adipose tissue · Tissue engineering

Abbreviations

ASC	adipose stem cell
bFGF	basic fibroblast growth factor
DAT	decellularized adipose tissue
ECM	extra cellular matrix
EDC	1-Ethyl-3-(-3-dimethylaminopropyl) carbodiimide hydrochloride
ELISA	enzyme-linked immunosorbent assay
GAG	glycosaminoglycan
GPDH	glycerol-3-pohosphate dehydrogenase
hDAT	human decellularized adipose tissue
IHC	immuno-histochemistry
MCS	methacrylated chondroitin sulfate
MGC	methacrylated glycol chitosan
MRI	magnetic resonance imaging
MSC	mesenchymal stem cell
NHS	N-hydroxysuccinimide
OEhMSC	Osteogenically enhanced human mesenchymal stem cell
PEG	Polyethylene glycol
SDC	Sodium deoxycholate
SDS	Sodium dodecyl sulfate
VEGF	vascular endothelial growth factor

1   Introduction

The persistent increase in demand for donor tissues for transplants over the last few decades has stimulated the development of *in vitro* tissue engineering and regenerative strategies (Parmaksiz et al. 2016). Tissue engineering requires a minimum of two primary components; viable cells and a minimally immunogenic scaffold that can support cell proliferation and local delivery of growth factors and cytokines. The scaffolds established thus far are categorized as synthetic scaffolds, extracellular matrix (ECM) protein (e.g. collagen, laminin etc.) derived scaffolds, decellularized tissue-based scaffolds or a combination thereof (Costa et al. 2017). Decellularized tissue-based scaffolds are useful tools for tissue engineering, primarily because they can mimic the native tissue microenvironment better than alternative synthetic scaffolds.

While autologous, allogeneic, or xenogeneic grafts have been used previously as transplants for heart valves, skin and coronary arteries, the presence of foreign antigens in these grafts often has caused adverse immune reactions (Costa et al. 2017). Decellularized tissue is processed to remove immunogenic epitopes including cell surface antigens, cytoplasmic components and nucleic acids (Crapo et al. 2011; Gilbert et al. 2006). The retained ECM, is generally conserved between species and does not elicit an immune response (Gilbert et al. 2006). ECM consists of collagens, glycosaminoglycans (GAGs), laminins, growth factors and other biochemical components (Gilpin and Yang 2017). These components serve as important survival, proliferation and migration cues for stem cells and provide signaling that leads to differentiation along specific lineages (Agmon and Christman 2016; Gilbert et al. 2006). Decellularized tissue-based scaffolds actively communicate with cells seeded on them, as evidenced by the scaffolds ability to influence cell differentiation and tissue remodeling, features not typically associated with synthetic scaffolds (Costa et al. 2017). These characteristics make decellularized tissue-based scaffolds superior biomaterials for tissue engineering.

Decellularized tissue-based scaffolds have been generated from dermis, intestinal mucosa, urinary bladder, pericardium, and adipose, among other tissues (Costa et al. 2017; Parmaksiz et al. 2016). The potential clinical uses of these scaffolds are under investigation for a wide range of applications. Various 'off-the-shelf' tissue-engineered skin substitutes for dermal and epidermal regeneration have become available (Costa et al. 2017; Parmaksiz et al. 2016); however, there remains a void in composites for subdermal adipose tissue-engineering. Although skin grafts and tissue-engineered epidermal substitutes are useful for regenerating the superficial tissue they do not adequately restore the subdermal adipose tissue (Chung et al. 2017). Thus, decellularized adipose tissue (DAT) a promising scaffold for plastic surgery applications, it can serve as a non-immunogenic substitute to fat tissue grafting for augmentation or reconstruction of soft tissues (Han et al. 2015).

Human adipose tissue obtained after lipectomy, abdominoplasty or breast reduction is discarded as medical waste (Schneider et al. 2017), therefore unlike other human tissues, it is an abundant resource for bio-scaffolds (Flynn 2010; Song et al. 2018). Consequently, recent research has focused on preparing DAT that can be utilized for tissue engineering. Several studies have shown that DAT can be used for *in vitro* adipose tissue engineering and *in vivo* adipose tissue regeneration (Han et al. 2015; Wang et al. 2013). The role of DAT is not limited to adipose tissue engineering since it has potential application to heal wounds, regenerate cartilage, support breast tissue reconstruction and promote nerve repair (Choi et al. 2012; Haddad et al. 2016; Lin et al. 2011; Woo et al. 2015).

Decellularization of adipose tissue was first described in 2010 (Flynn 2010). Subsequently, several published articles have reported alternative methods for efficiently decellularizing adipose tissue (Brown et al. 2011; Choi et al. 2011; Song et al. 2018). The majority of publications have focused on converting DAT into diverse physicochemical forms. These modifications are aimed at making a scaffold that allows for easier encapsulation of cells as well as better cell proliferation and differentiation. Attempts have also been made to transform DAT into an injectable scaffold which can be implanted in a non-invasive manner (Young et al. 2011).

This review will summarize the methods utilized for adipose tissue decellularization and the biochemical composition of DAT. Furthermore, physical forms of DAT developed thus far will be discussed, with emphasis on pre-clinical studies and potential clinical applications.

2 Adipose Tissue Decellularization and Physicochemical Manipulation

The ECM of the human body is a gel-like, fibrous network that provides mechanical support and biochemical signals to cells that make up the body's tissues. ECM includes structural proteins, growth factors, proteoglycans, GAGs, and proteolytic enzymes (Costa et al. 2017). It is currently being harvested, processed, and utilized in a clinical capacity for a variety of tissue regeneration applications (Flowers et al. 2017). Removal of parenchymal and stromal cells from the whole tissue by various decellularization techniques minimizes the potential for *in vivo* rejection of the resultant ECM (Crapo et al. 2011). Adipose tissue can be harvested from both xenogeneic and allogeneic sources. Porcine DAT is being explored as a xenograft, while DAT from allogeneic human donors is routinely harvested via abdominoplasty, liposuction, or breast reduction procedures (Flynn 2010).

Once acquired, adipose tissue is physically, chemically and/or enzymatically processed to produce DAT (Fig. 1). Methods for tissue decellularization differ based on the intended application of the DAT. The initial stages for adipose tissue decellularization often involve methods to lyse cells within the tissue, including repeated freeze-thaw cycles or multiple washes in hypotonic or hypertonic solutions (Flynn 2010; Morissette Martin et al. 2018; Roehm et al. 2016). Adipose tissue can be minced by mechanical means or frozen and ground into a powder in order to increase its surface area for further disruption (Choi et al. 2012; Pati et al. 2015). To remove lipids from the proteinaceous ECM, polar extraction with isopropanol is frequently

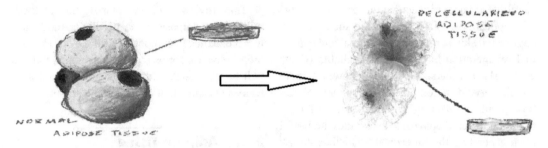

Fig. 1 Native adipose tissue vs decellularized adipose tissue

employed (Flynn 2010; Zhang et al. 2016a). Centrifugation or enzymatic digestion using lipase are also employed to accomplish the same goal (Brown et al. 2011; Choi et al. 2011; Thomas-Porch et al. 2018). Ionic detergents, such as sodium dodecyl sulfate (SDS) or sodium deoxycholate (SDC), can be utilized to disrupt cellular membrane components, while non-ionic detergents, such as Triton X-100, act to remove residual lipids by disturbing lipid-lipid and lipid-protein interactions (Costa et al. 2017; Giatsidis et al. 2018; Tan et al. 2017; Young et al. 2011; Yu et al. 2013). The highly specific proteolytic enzyme, trypsin, is commonly used in conjunction with the chelating agent, EDTA, for further breakdown of cell membrane proteins (Dunne et al. 2014; Wang et al. 2013); however, overexposure to trypsin can disrupt the ECM (Costa et al. 2017). Since the resultant DAT is intended for use for *in vivo* grafting, it is routinely treated with nucleases, such as RNAse and DNAse, to limit the scope of potential immunogenic responses (Lin et al. 2011; Yu et al. 2017). The culmination of a successful decellularization procedure yields a stable, biologically active polymer with little immunogenicity.

While some studies have characterized DAT without further physicochemical manipulation, several groups have improved its mechanical property's tunability by conversion to a hydrogel (Young et al. 2011). After digestion in a protease solution, such as pepsin or alpha-amylase (dissolved in acetic or hydrochloric acid), solubilized DAT can self-polymerize under physiological pH and temperature (Tan et al. 2017; Young et al. 2011; Yu et al. 2013). Furthermore, the DAT hydrogel displays thermosensitive

gelation properties, remaining in a liquid state at cooler temperatures. It can, therefore, be used in applications that require biomaterial polymerization *in vivo*. A DAT-hydrogel can also be used as bio-ink to print 3D constructs for *in vitro* or *in vivo* tissue engineering. While the mechanisms involved in the polymerization of collagens, the principle components of DAT-hydrogels, are poorly understood, it is believed that hydrogen bonding, Vander Waals' forces, and ionic interactions are primarily responsible (Tan et al. 2017). The mechanical properties of a hydrogel, such as stiffness, porosity, storage modulus, and gelation time, can be altered based on ECM concentration (Ghuman et al. 2016).

Further improvement in mechanical tunability of DAT-derived hydrogels can be achieved by functionalization of the DAT components prior to polymerization through a variety of chemical reactions. Combination of DAT with biologic or synthetic polymers can yield greater mechanical strength (Cheung et al. 2014; Dong et al. 2012). For example, DAT can be cross-linked using 1-Ethyl-3-(-3-dimethylaminopropyl) carbodiimide hydrochloride (EDC) and *N*-hydroxysuccinimide (NHS), resulting in increased resistance to enzymatic degradation *in vivo* while maintaining cytocompatibility (Wu et al. 2012). A composite hydrogel made from DAT and polyethylene glycol (PEG) displayed tunable mechanical and degradation properties based on varying DAT concentrations (Li et al. 2018). Another method of synthesizing cross-linked hydrogels includes the combination of DAT with either methacrylated glycol chitosan (MGC) or methacrylated chondroitin sulfate (MCS) which polymerizes through a photocatalyzed reaction induced by low-intensity UV light (Cheung et al. 2014). Increasing concentrations of the methacrylated polymers can lead to greater hydrogel stiffness (Cheung et al. 2014). Hypothetically, incorporation of these cross-linking strategies, copolymers, and polymerization methods DAT-hydrogel fabrication will increase the final product's applicability to an expanded range of tissue engineering challenges.

3   Biochemical Composition of DAT

4   Mechanical Properties of DAT

Decellularized tissue needs to possess a particular mechanical strength to retain biological activity. There is evidence to suggest that tissue stiffness is one of the factors that drives the differentiation of stem cells to specific cell lineages (Breuls et al. 2008; Gilpin and Yang 2017; Levy-Mishali et al. 2009). Therefore, creation of a tunable scaffold that can be tailored in terms of stiffness to support the formation of different tissue types is likely to prove pivotal.

Chemical component	Tissue source and form	Assay method	Result	References
Total collagen	C57BL/6 mouse adipose tissue	Sircol collagen assay kit Biocolor	1211.1 ± 223.7 µg/100 mg hydrated DAT	Lu et al. (2014)
	Porcine adipose tissue		332.9 ± 12.1 µg/mg ECM dry weight	Choi et al. (2012)
GAGs	C57BL/6 mouse adipose tissue	Blyscan GAG assay kit	15.6 ± 1.4 µg/100 mg hydrated DAT	Lu et al. (2014)
	Porcine adipose tissue		85 ± 0.7 µg/mg ECM dry weight	Choi et al. (2012)
			768.3 ± 52.2–11.09 ± 43.1 µg/g ECM dry weight	Brown et al. (2011)
	Human adipose tissue		39.67 ± 2.31 µg/g ECM dry weight	Song et al. (2018)
	Human adipose tissue derived hydrogel		2.18 ± 0.32 µg/mg ECM dry weight	Young et al. (2011)
	Human adipose tissue microparticles	Alcian blue colorimetric assay kit	1.72 ± 0.64 µg/mg ECM dry weight	Wang et al. (2013)
	Fischer 344 rat adipose tissue		0.54 ± 0.58 µg/mg ECM dry weight	Zhang et al. (2016a)
Collagen I	Human adipose tissue	IHC	Present	He et al. (2018)
	Porcine adipose tissue		Present	Brown et al. (2011 and Zhao et al. (2018)
	Human adipose tissue derived hydrogel		Present	Young et al. (2011)
Collagen III	Human adipose tissue derived hydrogel	IHC	Present	Young et al. (2011)
	Porcine adipose tissue		Present	Brown et al. (2011)

(continued)

Chemical component	Tissue source and form	Assay method	Result	References
Collagen IV	C57BL/6 mouse adipose tissue	IHC	Present	Lu et al. (2014)
	Human adipose tissue		Present	Flynn (2010), Giatsidis et al. (2018), He et al. (2018), Song et al. (2018) and Zhao et al. (2018)
	Porcine adipose tissue		Present	Brown et al. (2011)
	Human adipose tissue derived hydrogel		Present	Young et al. (2011)
Collagen VI	Human adipose tissue	IHC	Present	Thomas-Porch et al. (2018)
Collagen VII	Porcine adipose tissue	IHC	Present	Brown et al. (2011)
Laminin	C57BL/6 mouse adipose tissue	IHC	Present	Lu et al. (2014)
	Human adipose tissue		Present	Flynn (2010), Giatsidis et al. (2018), He et al. (2018), Song et al. (2018) and Zhao et al. (2018)
	Porcine adipose tissue		Present	Brown et al. (2011)
	Fischer 344 rat adipose tissue		Present	Zhang et al. (2016a)
	Human adipose tissue derived hydrogel		Present	Young et al. (2011)
Elastin	Porcine adipose tissue	Fastin elastin assay kit	152.6 ± 4.5 µg/mg ECM dry weight	Choi et al. (2012)
Fibronectin	Human adipose tissue	IHC	Present	Giatsidis et al. (2018 and Zhao et al. (2018)
Vascular endothelial growth factor (VEGF)	C57BL/6 mouse adipose tissue	ELISA	42.9 ± 25.2 pg/100 mg hydrated DAT	Lu et al. (2014)
	Porcine adipose tissue		15.2 ± 130–27.6 ± 1.2 pg/g ECM dry weight	Brown et al. (2011)
	Human adipose tissue microparticles	IHC	Present	Wang et al. (2013)
	Fischer 344 rat adipose tissue		Present	Zhang et al. (2016a)
Basic fibroblast growth factor (bFGF)	C57BL/6 mouse adipose tissue	ELISA	360.7 ± 120.5 pg/100 mg hydrated DATs.	Lu et al. (2014)
	Porcine adipose tissue		1840.5 ± 92.3–2551.8 ± 148.1 pg/g ECM dry weight	Brown et al. (2011)
	Fischer 344 rat adipose tissue	IHC	Present	Zhang et al. (2016a)

Common tests conducted to determine DAT mechanical strength measure Young's modulus, elastic modulus, and tensile strength. Independent studies have reported Young's modulus of human DAT (hDAT) as 8.312 MPa (Choi et al. 2011) and 65.7 ± 5.97 MPa (Song et al. 2018), whereas DAT microcarriers were found to have Young's modulus of 0.73 ± 0.23 KPa (Yu et al. 2017). Human DAT-MGC and DAT-MCS composite hydrogels exhibited Young's modulus of 30.1 ± 4.0 KPa and 37.1 ± 5.0 KPa respectively (Cheung et al. 2014). Young's modulus of hDAT derived porous foams formulated with different concentrations of DAT was found to range from 2.42 ± 0.65 to 4.01 ± 0.46 KPa (Yu et al. 2013). Tensile strength was determined to be 87.4 ± 23.1 KPa (Roehm et al. 2016) and 128.57 ± 13.15 KPa (Choi et al. 2012) for porcine DAT and 220 ± 50 KPa for hDAT (Song et al. 2018).

Omidi et al. compared the mechanical properties of DAT sourced from different fat depots. Young's modulus calculated for several different source tissues including breast, abdomen, pericardium, omentum and thymic remnant, was found to range from 2.109 ± 0.685 to 3.46 ± 1.21 KPa (Omidi et al. 2014).

5 Applications of DAT

5.1 Modified DAT Forms for *in vivo* Applications

5.1.1 Powder

A powdered form of DAT has been developed for subcutaneous injection. DAT was freeze-dried, and manually milled into a powder and suspended in cell culture medium. The suspension permitted cell invasion, proliferation, and survival *in vitro* (Choi et al. 2009). Suspensions with and without adipose stem cells (ASC) were injected subcutaneously into nude mice. Histology and RT-qPCR for adipogenic genes showed both experimental groups induced adipogenesis with organized intracellular lipid droplets and angiogenesis; however, the changes were more prominent in the suspensions seeded with ASC (Choi et al. 2009).

5.1.2 Injectable Liquid

Likewise, an injectable form of DAT has been established and tested for soft tissue defect applications. After being lyophilized, DAT can be digested with pepsin to form a viscous

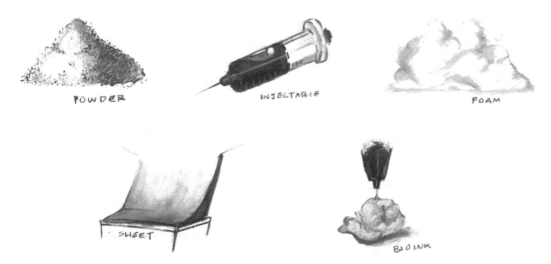

Fig. 2 Modified forms of DAT for *in vivo* applications; powder, injectable liquid, foam, scaffold sheet, bio-ink

suspension. When injected into rats or mice, the scaffold has displayed the potential to promote angiogenesis (Giatsidis et al. 2019) and adipogenesis (Zhao et al. 2018). It has also been shown that injectable DAT alone can instruct host cells to remodel the tissue and thus, the scaffold has the potential for stimulating adipose tissue regeneration (Choi et al. 2012; Tan et al. 2017; Wu et al. 2012; Young et al. 2011; Zhang et al. 2016b; Zhao et al. 2018). Injectable DAT undergoes rapid volume loss after *in vivo* implantation, whereas crosslinking liquid DAT with synthetic polymers reduces degradation of the scaffold while retaining its capability to allow cell attachment and adipo-induction (Cheung et al. 2014; Li et al. 2018; Wu et al. 2012). However, cross-linking introduces a trade-off between biocompatibility and volume persistence.

5.1.3   Foam

DAT was converted in to a porous foam by solubilization using α-amylase followed by freeze-drying. The α-amylase treatment is milder than the pepsin treatment, thus retaining the structural integrity of collagens, and removing the need for a chemical crosslinker (Yu et al. 2013). The porous property of the foam improves cellular infiltration, viability, and intracellular signaling. Another procedure was established to construct "bead foams" which are fused interconnected networks of porous DAT beads. This was accomplished by electrospraying DAT solution followed by freezing and lyophilization. Glycerol-3-phosphate dehydrogenase (GPDH) enzyme activity, RT-qPCR analysis of adipogenic gene expression, and intracellular lipid accumulation studies served as strong evidence that the porous foam and bead foams supported adipogenesis (Yu et al. 2013). Subcutaneous implants in a rat model further demonstrated that the porous foam and bead foams were inherently adipogenic and biocompatible (Yu et al. 2013), suggesting that the composites are a viable soft defect filler.

5.1.4   Scaffold Sheet

While injections and foams have several potential applications for subcutaneous tissue regeneration, a bandage-like sheet could be useful for the treatment of topical injuries. A thin scaffold sheet was fabricated from DAT by casting the acellular tissue in a shallow mold and lyophilizing it. Based on mechanical tests the sheets were determined to have sufficient mechanical integrity for easy handling (Kim et al. 2012). The scaffold sheets were seeded with different cell types to establish their potential to support cell adhesion and proliferation. Live/dead assays confirmed that several different human cells types successfully integrated into the sheet. (Kim et al. 2012). The potential applications of this sheet are considerable due to its ability to support a wide variety of cell types. Furthermore, a DAT sheet could be applied non-invasively and topically (Fig. 2).

5.1.5   Bio-ink

Apart from the biochemical composition of the scaffold and the inherent presence of growth factors, another key regulator of cell differentiation and tissue remodeling is its three-dimensional architecture. The advent of 3-D printing has allowed custom designed fabrication of scaffolds in specific shapes and structures to promote the formation of specific tissue types (Pati et al. 2014). DAT can be enzymatically digested and converted into liquid bio-ink for 3D printing. Pati et al. developed dome-shaped tissue constructs using DAT bio-ink and ASC. *In vitro* experiments showed that the cells remained highly viable in the constructs and displayed adipogenic differentiation. The tissue constructs also exhibited remodeling and adipose tissue formation when implanted subcutaneously in mice (Pati et al. 2014, 2015).

5.2   *In vivo* Applications

5.2.1   Adipose Tissue Engineering

DAT's most obvious clinical use is as a filler or implant in situations where soft tissue damage has occurred due to traumatic injury or tumor resection. By providing important microenvironmental signals that promote adipogenesis and angiogenesis, DAT can resolve soft tissue defects.

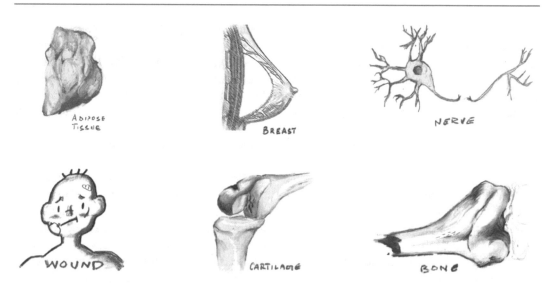

Fig. 3 *In vivo* applications of DAT; adipose tissue engineering, wound healing, breast reconstruction, cartilage engineering, nerve repair, bone regeneration

The capability of DAT to form adipose tissue after subcutaneous implantation has been demonstrated in different animal models. Additionally, DAT in its native form or any of the modified forms (discussed in Sect. 5.1) has been found to be adipo-inductive *in vivo* and recruits host stem cells (Han et al. 2015). DAT microparticles seeded with ASC formed adipose tissue within 30 days of implantation in Fischer rats (Wang et al. 2013). DAT-MCS composite hydrogels and DAT foam, seeded with ASC, displayed significant adipogenesis in Wistar rats after 12 weeks (Cheung et al. 2014; Han et al. 2015; Yu et al. 2013). ASC seeded on 3D printed polycaprolactone-DAT composite scaffolds underwent adipogenic differentiation in nude mice (Pati et al. 2015). Viscous DAT suspension with ASC caused adipose tissue formation 4–12 weeks after implantation in mice (Choi et al. 2012; Young et al. 2011; Zhang et al. 2016b). There is also evidence that DAT alone in the absence of cells possesses the potential to promote adipogenesis through the recruitment of host stromal/stem cells (Tan et al. 2017).

5.2.2 Wound Healing

Lacerations that penetrate to the depth of the subdermal tissue and full-thickness third-degree burns may require surgical intervention and skin grafting. Without an intervention to repair the defect, the patient is at risk of infection, unable to regulate body temperature, and likely to develop excess fluid loss and electrolyte imbalance. DAT fabrications have the potential to limit these complications as well as eliminate the disadvantages of conventional treatment methods.

To address this issue, a bilayer sheet was developed for wound dressing. The top layer was composed of a chitosan membrane integrated with titanium dioxide prepared by electrospinning chitosan solution with titanium dioxide solution. This top sheet provided an external bacterial barrier while a DAT bottom layer served to promote subdermal tissue regeneration (Woo et al. 2015). *In vivo* studies using a full-thickness wound in rats showed that the bilayer sheets induced faster regeneration of granulation tissue and reduced epidermal scar formation compared to controls (Woo et al. 2015). This bilayer sheet, as well as the bandage-like sheet (Kim et al. 2012), are promising wound dressings, especially in clinical settings with damaged subdermal tissue and increased risk of bacterial penetration.

5.2.3 Breast Reconstruction

With 1 in 8 women in the United States facing a lifetime risk of breast cancer development

(Schneider et al. 2014) and an increasing annual number of mastectomies being performed, there is a clear need for advances in breast reconstruction biomaterials. Haddad et al. evaluated the biomechanical efficacy of DAT as a reconstructive material from breast defects using a computational approach. The study used numerical models to compare deformation that a normal breast endures due to gravitational loading conditions versus deformation in DAT under the same conditions. These numerical models were developed using magnetic resonance imaging (MRI) data of breasts from female volunteers. DAT from several different physiological depots were analyzed and the results demonstrated that breasts reconstructed with these DAT materials did not show contour defects. Furthermore, the breasts reconstructed with DAT sourced from breast and abdominal sites demonstrated more similar deformation compared to native breast tissue than those breasts reconstructed with DAT sourced from other depots. (Haddad et al. 2016). These results are encouraging for expanded use of DAT in plastic surgery applications. DAT may prove to be a practical biomaterial for mammoplasty in which the goal is to reshape or modify the appearance of the breast. Theoretically, a DAT scaffold will prove superior to a synthetic implant by allowing the breast to retain its natural texture.

5.2.4 Cartilage Tissue Engineering

Cartilage acts as a cushion to protect the bones from impact and allows joints to easily glide and bend without pain. Damage to cartilage can occur from trauma such as a meniscus tear in the knee or from degenerative changes due to osteoarthritis (Beaufils et al. 2017). Joint surface defects are ubiquitous and difficult to renew, especially full-thickness defects. These injuries have limited capability of healing due to the avascular nature of cartilage and declining chondrocytes function with age and injury (Sophia Fox et al. 2009). Current treatment strategies such as microfracture lead to the replacement of the original hyaline cartilage with a less durable, less resilient, and sub-optimal fibrocartilage (Goldberg et al. 2017). Autologous chondrocyte implantation is an alternative therapy involving harvesting, culturing, and re-implanting a patient's cells into the defect (Goldberg et al. 2017). The use of cartilage-derived ECM for cartilage regeneration does not have promising clinical application because of limited cartilage sources. While investigators have conjectured that a wider application of DAT includes cartilage regeneration, research currently remains limited to *in vitro* studies. Choi et al. investigated whether adipose stem cells seeded on DAT could be induced to undergo chondrogenic differentiation. ASC and DAT scaffolds cultured in chondrogenic medium with 10 ng/mL TGF-β1 for 45 days successfully formed cartilaginous tissue with cartilage-specific collagens II and XI, proteoglycans, and GAGs (Choi et al. 2012). These experiments indicate that a DAT can support chondrogenic differentiation of human ASC and has potential utility in synthesizing a cartilage-like tissue *in vitro*. Further studies will be necessary to determine whether a DAT scaffold with chondrogenic-induced ASC can generate a cartilage-like tissue *in vivo* with volume persistence or can promote healing in a cartilage defect model.

5.2.5 Nerve Repair

The utility of DAT for nerve repair was examined in a rat model of peripheral nerve injury. A 5 mm long nerve section was resected from the cavernous nerves, which mediate erectile function. DAT seeded with ASC was grafted into the site of nerve injury and erectile function was assessed 3 months later. DAT treated rats displayed better erectile function than controls, the results were not statistically significant (Lin et al. 2011).

5.2.6 Bone Regeneration

Although bone has a remarkable capacity for regeneration, critical-size fractures are unable to heal (Clough et al. 2015a). These defects generally require an invasive surgical procedure or autologous bone grafting. Bone grafting involves the harvesting of bone, which often leads to prolonged donor site pain, hospitalization, and risk of infection (Clough et al. 2015a; Strong et al. 2016). Scaffolds that have successfully promoted cell adhesion, proliferation, and osteogenic differentiation of ASC include natural polymers such as chitosan/gelatin, decellularized bovine tendon and decellularized bovine trabecular bone (Clough et al. 2015b; Elgali et al. 2017). Often, synthetic bone scaffolds are not biologically compatible and do not effectively support angiogenesis or proliferation (Amini et al. 2012). Additionally, the source of tendon and bone-derived scaffolds is not abundant enough for wide orthopedic use. Consequently, there remains an orthopedic demand for a biocompatible composite with translatable clinical relevancy.

Mesenchymal stem cells (MSC) derived ECM scaffolds have displayed improved bone healing in mouse model of calvarial defect (Zeitouni et al. 2012). Additionally, osteogenically enhanced MSC (OEhMSC) seeded on gelatin foam promoted bone regeneration in mouse model of femoral defect (Clough et al. 2015a). While there are presently no *in vivo* studies that have analyzed the use of DAT for bone regeneration, ASC seeded on DAT have displayed osteogenic differentiation when cultured in osteogenic media (Guneta et al. 2017). Future research needs to be performed to demonstrate osteogenically enhanced ASC adhere and proliferate in DAT. Furthermore, experiments validating DATs efficacy as a promoter of bone growth in a bone defect model would be essential (Fig. 3).

5.3   3-Dimensional Cell Culture

In addition to clinical *in vivo* applications, DAT has potential use for *in vitro* research. DAT's biomimetic properties make it an ideal 3D cell culturing system because of its ability to very closely simulate the *in vivo* microenvironment.

It has been shown by several studies that ASC can maintain high viability and proliferation when seeded on native DAT (Choi et al. 2012; Fan et al. 2014; Flynn 2010; Kim et al. 2012; Song et al. 2018; Young et al. 2011; Yu et al. 2017; Zhang et al. 2016a) or chemically modified DAT (Brown et al. 2015; Cheung et al. 2014; Li et al. 2018; Pati et al. 2015; Turner et al. 2012; Wu et al. 2012). DAT powder displayed lower attachment of ASC in comparison to 2D culture

plate, but exhibited increased proliferation rates (Choi et al. 2009). Apart from ASC, DAT also allows attachment and proliferation of human aortic smooth muscle cells, human chondrocytes, human umbilical vein endothelial cells (Kim et al. 2012; Zhang et al. 2016a), human dermal fibroblasts (Choi et al. 2012) and neuroblasts (Roehm et al. 2016). DAT has been found to be adipo-inductive *in vitro* and ASC seeded on DAT are able to differentiate in to adipocytes without adipogenic induction (Brown et al. 2015; Cheung et al. 2014; Flynn 2010; Lin et al. 2016; Pati et al. 2015; Tan et al. 2017; Turner et al. 2012; Yu et al. 2013).

5.3.1   In Vitro Cancer Research

DAT has been used as a 3D cell culturing system to study breast cancer growth and drug response (Dunne et al. 2014). DAT scaffold was compared to standard 2D cell culture and 3D Matrigel™ scaffolds. The proliferation profile of the *in vivo* xenografts was found to be similar to that of the cells in DAT scaffolds *in vitro*, with a long lag phase and a significantly slower proliferation compared to that of the 2D and Matrigel™ culture proliferation profiles (Dunne et al. 2014). These results demonstrated the superiority of DAT as a cell culturing system to mimic *in vivo* conditions.

6   Conclusions

The easy accessibility of source tissue makes DAT an attractive candidate as a regenerative medical bio-scaffold and *in vitro* research tool. Decellularization of adipose tissue has been optimized through repeated studies and it has been shown that DAT retains most of its structural proteins as well as functional molecules such as VEGF and bFGF. These biochemical components make DAT a bioactive system that is capable of recruiting and housing cells, providing them with adipogenic signals and promoting neovascularization. DAT is an adaptable scaffold, as it can be modified in to an array of different chemical compositions and physical forms. This theoretically means that a wide variety of tissue types could be engineered in DAT.

The obvious clinical application of DAT rests on its homologous use for regeneration of soft tissue defects. However, based on pre-clinical animal studies and *in vitro* work, DAT has the potential for wider ranging applications including peripheral nerve, cartilage, and bone regeneration.

References

Agmon G, Christman KL (2016) Controlling stem cell behavior with decellularized extracellular matrix scaffolds. Curr Opin Solid State Mater Sci 20:193–201. https://doi.org/10.1016/j.cossms.2016.02.001

Amini AR, Laurencin CT, Nukavarapu SP (2012) Bone tissue engineering: recent advances and challenges. Crit Rev Biomed Eng 40:363–408

Beaufils P, Becker R, Kopf S, Matthieu O, Pujol N (2017) The knee meniscus: management of traumatic tears and degenerative lesions. EFORT Open Rev 2:195–203

Breuls RGM, Jiya TU, Smit TH (2008) Scaffold stiffness influences cell behavior: opportunities for skeletal tissue engineering. Open Orthop J 2:103–109

Brown BN et al (2011) Comparison of three methods for the derivation of a biologic scaffold composed of adipose tissue extracellular matrix. Tissue Eng Part C Methods 17:411–421. https://doi.org/10.1089/ten.tec.2010.0342

Brown CF, Yan J, Han TT, Marecak DM, Amsden BG, Flynn LE (2015) Effect of decellularized adipose tissue particle size and cell density on adipose-derived stem cell proliferation and adipogenic differentiation in composite methacrylated chondroitin sulphate hydrogels. Biomed Mater (Bristol, England) 10:045010. https://doi.org/10.1088/1748-6041/10/4/045010

Cheung HK, Han TT, Marecak DM, Watkins JF, Amsden BG, Flynn LE (2014) Composite hydrogel scaffolds incorporating decellularized adipose tissue for soft tissue engineering with adipose-derived stem cells. Biomaterials 35:1914–1923. https://doi.org/10.1016/j.biomaterials.2013.11.067

Choi JS et al (2009) Human extracellular matrix (ECM) powders for injectable cell delivery and adipose tissue engineering. J Control Release 139:2–7. https://doi.org/10.1016/j.jconrel.2009.05.034

Choi JS et al (2011) Decellularized extracellular matrix derived from human adipose tissue as a potential scaffold for allograft tissue engineering. J Biomed Mater Res A 97:292–299. https://doi.org/10.1002/jbm.a.33056

Choi YC, Choi JS, Kim BS, Kim JD, Yoon HI, Cho YW (2012) Decellularized extracellular matrix derived from porcine adipose tissue as a xenogeneic biomaterial for tissue engineering. Tissue Eng Part C Methods 18:866–876. https://doi.org/10.1089/ten.TEC.2012.0009

Chung B, O'Mahony GD, Lam G, Chiu DTW (2017) Adipose tissue-preserved skin graft: applicability and long-term results. Plast Reconstr Surg 140:593–598. https://doi.org/10.1097/prs.0000000000003623

Clough BH, McCarley MR, Gregory CA (2015a) A simple critical-sized femoral defect model in mice. J Vis Exp. https://doi.org/10.3791/52368

Clough BH et al (2015b) Bone regeneration with osteogenically enhanced mesenchymal stem cells and their extracellular matrix proteins. J Bone Miner Res 30:83–94

Costa A, Naranjo JD, Londono R, Badylak SF (2017) Biologic scaffolds. Cold Spring Harb Perspect Med 7:a025676. https://doi.org/10.1101/cshperspect.a025676

Crapo PM, Gilbert TW, Badylak SF (2011) An overview of tissue and whole organ decellularization processes. Biomaterials 32:3233–3243

Dong Y et al (2012) "One-step" preparation of thiol-ene clickable PEG-based thermoresponsive hyperbranched copolymer for in situ crosslinking hybrid hydrogel. Macromol Rapid Commun 33:120–126. https://doi.org/10.1002/marc.201100534

Dunne LW, Huang Z, Meng W, Fan X, Zhang N, Zhang Q, An Z (2014) Human decellularized adipose tissue scaffold as a model for breast cancer cell growth and drug treatments. Biomaterials 35:4940–4949. https://doi.org/10.1016/j.biomaterials.2014.03.003

Elgali I, Omar O, Dahlin C, Thomsen P (2017) Guided bone regeneration: materials and biological mechanisms revisited. Eur J Oral Sci 125:315–337

Fan X, Tian C, Fu Y, Li X, Deng L, Lu Q (2014) Preparation and characterization of acellular adipose tissue matrix. Zhongguo Xiu Fu Chong Jian Wai Ke Za Zhi 28:377–383

Flowers SA et al (2017) Lubricin binds cartilage proteins, cartilage oligomeric matrix protein, fibronectin and collagen II at the cartilage surface. Sci Rep 7:13149. https://doi.org/10.1038/s41598-017-13558-y

Flynn LE (2010) The use of decellularized adipose tissue to provide an inductive microenvironment for the adipogenic differentiation of human adipose-derived stem cells. Biomaterials 31:4715–4724. https://doi.org/10.1016/j.biomaterials.2010.02.046

Ghuman H, Massensini AR, Donnelly J, Kim SM, Medberry CJ, Badylak SF, Modo M (2016) ECM hydrogel for the treatment of stroke: characterization of the host cell infiltrate. Biomaterials 91:166–181. https://doi.org/10.1016/j.biomaterials.2016.03.014

Giatsidis G, Guyette JP, Ott HC, Orgill DP (2018) Development of a large-volume human-derived adipose acellular allogenic flap by perfusion decellularization. Wound Repair Regen 26(2):245–250. https://doi.org/10.1111/wrr.12631

Giatsidis G et al (2019) Preclinical optimization of a shelf-ready, injectable, human-derived, Decellularized allograft adipose matrix. Tissue Eng A 25:271–287. https://doi.org/10.1089/ten.TEA.2018.0052

Gilbert TW, Sellaro TL, Badylak SF (2006) Decellularization of tissues and organs. Biomaterials 27:3675–3683. https://doi.org/10.1016/j.biomaterials.2006.02.014

Gilpin A, Yang Y (2017) Decellularization strategies for regenerative medicine: from processing techniques to applications. Biomed Res Int 2017:9831534. https://doi.org/10.1155/2017/9831534

Goldberg A, Mitchell K, Soans J, Kim L, Zaidi R (2017) The use of mesenchymal stem cells for cartilage repair and regeneration: a systematic review. J Orthop Surg Res 12:39. https://doi.org/10.1186/s13018-017-0534-y

Guneta V, Zhou Z, Tan NS, Sugii S, Wong MTC, Choong C (2017) Recellularization of decellularized adipose tissue-derived stem cells: role of the cell-secreted extracellular matrix in cellular differentiation. Biomater Sci 6:168–178. https://doi.org/10.1039/c7bm00695k

Haddad SM, Omidi E, Flynn LE, Samani A (2016) Comparative biomechanical study of using decellularized human adipose tissues for post-mastectomy and post-lumpectomy breast reconstruction. J Mech Behav Biomed Mater 57:235–245. https://doi.org/10.1016/j.jmbbm.2015.12.005

Han TT, Toutounji S, Amsden BG, Flynn LE (2015) Adipose-derived stromal cells mediate in vivo adipogenesis, angiogenesis and inflammation in decellularized adipose tissue bioscaffolds. Biomaterials 72:125–137. https://doi.org/10.1016/j.biomaterials.2015.08.053

He Y et al (2018) Optimized adipose tissue engineering strategy based on a neo-mechanical processing method. Wound Repair Regen 26(2):163–171. https://doi.org/10.1111/wrr.12640

Kim BS, Choi JS, Kim JD, Choi YC, Cho YW (2012) Recellularization of decellularized human adipose-tissue-derived extracellular matrix sheets with other human cell types. Cell Tissue Res 348:559–567. https://doi.org/10.1007/s00441-012-1391-y

Levy-Mishali M, Zoldan J, Levenberg S (2009) Effect of scaffold stiffness on myoblast differentiation. Tissue Eng Part A 15:935–944. https://doi.org/10.1089/ten.tea.2008.0111

Li S et al (2018) Hybrid synthetic-biological hydrogel system for adipose tissue regeneration. Macromol Biosci 18(11):e1800122. https://doi.org/10.1002/mabi.201800122

Lin G et al (2011) Cavernous nerve repair with allogenic adipose matrix and autologous adipose-derived stem cells. Urology 77:1509–e1501–1508. https://doi.org/10.1016/j.urology.2010.12.076

Lin CY, Liu TY, Chen MH, Sun JS (2016) An injectable extracellular matrix for the reconstruction of epidural fat and the prevention of epidural fibrosis. Biomed Mat (Bristol, England) 11:035010. https://doi.org/10.1088/1748-6041/11/3/035010

Lu Q, Li M, Zou Y, Cao T (2014) Delivery of basic fibroblast growth factors from heparinized decellularized adipose tissue stimulates potent de novo adipogenesis. J Control Release 174:43–50. https://doi.org/10.1016/j.jconrel.2013.11.007

Morissette Martin P, Shridhar A, Yu C, Brown C, Flynn LE (2018) Decellularized adipose tissue scaffolds for soft tissue regeneration and adipose-derived stem/stromal cell delivery. Methods Mol Biol (Clifton, NJ) 1773:53–71. https://doi.org/10.1007/978-1-4939-7799-4_6

Omidi E, Fuetterer L, Reza Mousavi S, Armstrong RC, Flynn LE, Samani A (2014) Characterization and assessment of hyperelastic and elastic properties of decellularized human adipose tissues. J Biomech 47:3657–3663. https://doi.org/10.1016/j.jbiomech.2014.09.035

Parmaksiz M, Dogan A, Odabas S, Elcin AE, Elcin YM (2016) Clinical applications of decellularized extracellular matrices for tissue engineering and regenerative medicine. Biomed Mater (Bristol, England) 11:022003. https://doi.org/10.1088/1748-6041/11/2/022003

Pati F et al (2014) Printing three-dimensional tissue analogues with decellularized extracellular matrix bioink. Nat Commun 5:3935. https://doi.org/10.1038/ncomms4935

Pati F, Ha DH, Jang J, Han HH, Rhie JW, Cho DW (2015) Biomimetic 3D tissue printing for soft tissue regeneration. Biomaterials 62:164–175. https://doi.org/10.1016/j.biomaterials.2015.05.043

Roehm KD, Hornberger J, Madihally SV (2016) In vitro characterization of acelluar porcine adipose tissue matrix for use as a tissue regenerative scaffold. J Biomed Mater Res A 104:3127–3136. https://doi.org/10.1002/jbm.a.35844

Schneider AP, Zainer CM, Kubat CK, Mullen NK, Windisch AK (2014) The breast cancer epidemic: 10 facts. Linacre Q 81:244–277. https://doi.org/10.1179/2050854914y.0000000027

Schneider S, Unger M, van Griensven M, Balmayor ER (2017) Adipose-derived mesenchymal stem cells from liposuction and resected fat are feasible sources for regenerative medicine. Eur J Med Res 22:17

Song M, Liu Y, Hui L (2018) Preparation and characterization of acellular adipose tissue matrix using a combination of physical and chemical treatments. Mol Med Rep 17:138–146. https://doi.org/10.3892/mmr.2017.7857

Sophia Fox AJ, Bedi A, Rodeo SA (2009) The basic science of articular cartilage: structure, composition, and function. Sports Health 1:461–468

Strong AL et al (2016) Obesity inhibits the osteogenic differentiation of human adipose-derived stem cells. J Transl Med 14:27. https://doi.org/10.1186/s12967-016-0776-1

Tan QW et al (2017) Hydrogel derived from decellularized porcine adipose tissue as a promising biomaterial for soft tissue augmentation. J Biomed Mater Res A 105:1756–1764. https://doi.org/10.1002/jbm.a.36025

Thomas-Porch C et al (2018) Comparative proteomic analyses of human adipose extracellular matrices decellularized using alternative procedures. J Biomed Mater Res A 106(9):2481–2493. https://doi.org/10.1002/jbm.a.36444

Turner AE, Yu C, Bianco J, Watkins JF, Flynn LE (2012) The performance of decellularized adipose tissue microcarriers as an inductive substrate for human adipose-derived stem cells. Biomaterials 33:4490–4499. https://doi.org/10.1016/j.biomaterials.2012.03.026

Wang L, Johnson JA, Zhang Q, Beahm EK (2013) Combining decellularized human adipose tissue extracellular matrix and adipose-derived stem cells for adipose tissue engineering. Acta Biomater 9:8921–8931. https://doi.org/10.1016/j.actbio.2013.06.035

Woo CH, Choi YC, Choi JS, Lee HY, Cho YW (2015) A bilayer composite composed of TiO2-incorporated electrospun chitosan membrane and human extracellular matrix sheet as a wound dressing. J Biomater Sci Polym Ed 26:841–854. https://doi.org/10.1080/09205063.2015.1061349

Wu I, Nahas Z, Kimmerling KA, Rosson GD, Elisseeff JH (2012) An injectable adipose matrix for soft-tissue reconstruction. Plast Reconstr Surg 129:1247–1257. https://doi.org/10.1097/PRS.0b013e31824ec3dc

Young DA, Ibrahim DO, Hu D, Christman KL (2011) Injectable hydrogel scaffold from decellularized human lipoaspirate. Acta Biomater 7:1040–1049. https://doi.org/10.1016/j.actbio.2010.09.035

Yu C, Bianco J, Brown C, Fuetterer L, Watkins JF, Samani A, Flynn LE (2013) Porous decellularized adipose tissue foams for soft tissue regeneration. Biomaterials 34:3290–3302. https://doi.org/10.1016/j.biomaterials.2013.01.056

Yu C, Kornmuller A, Brown C, Hoare T, Flynn LE (2017) Decellularized adipose tissue microcarriers as a dynamic culture platform for human adipose-derived stem/stromal cell expansion. Biomaterials 120:66–80. https://doi.org/10.1016/j.biomaterials.2016.12.017

Zeitouni S et al (2012) Human mesenchymal stem cell-derived matrices for enhanced osteoregeneration. Sci Transl Med 4:132ra155. https://doi.org/10.1126/scitranslmed.3003396

Zhang Q et al (2016a) Decellularized skin/adipose tissue flap matrix for engineering vascularized composite soft tissue flaps. Acta Biomater 35:166–184. https://doi.org/10.1016/j.actbio.2016.02.017

Zhang S, Lu Q, Cao T, Toh WS (2016b) Adipose tissue and extracellular matrix development by injectable Decellularized adipose matrix loaded with basic fibroblast growth factor. Plast Reconstr Surg 137:1171–1180. https://doi.org/10.1097/prs.0000000000002019

Zhao Y, Fan J, Bai S (2018) Biocompatibility of injectable hydrogel from decellularized human adipose tissue in vitro and in vivo. J Biomed Mater Res B Appl Biomater. https://doi.org/10.1002/jbm.b.34261

Adv Exp Med Biol – Cell Biology and Translational Medicine (2019) 6: 71–85
https://doi.org/10.1007/5584_2019_338
© Springer Nature Switzerland AG 2019
Published online: 2 February 2019

Decellularization Concept in Regenerative Medicine

Özge Sezin Somuncu

Abstract

Decellularized organs and tissues are effectively utilized in a diversity of regenerative medicine purposes, and the decellularization approaches employed differ as broadly as the tissues/organs of concern. Biological scaffold substances formed by extracellular matrix (ECM) are mostly produced with methods that include decellularization of tissues. Conservation of the multifaceted arrangement and three-dimensional (3D) construction of the ECM is very wanted but it is documented that almost every approach of decellularization cause disturbance of the organization and possible forfeiture of surface organization and conformation. The competence of cell elimination from a tissue is reliant on the basis of the tissue and the precise physical, chemical, and enzymatic approaches that are utilized. Here, the most frequently applied and newly developed decellularization techniques are designated, organ engineering with decellularized scaffolds for different organs, recent knowledge in the field are explained.

Keywords

Decellularization · Regenerative medicine · Recellularization · Tissue engineering

Ö. S. Somuncu (✉)
Department of Medical Biology, Bahçeşehir University
Faculty of Medicine, İstanbul, Turkey
e-mail: ozgesezin.somuncu@vsh.bau.edu.tr

Abbreviation

3D	Three dimensional
ADSCs	Adipose-derived stem cells
CC10	Secretoglobin Family 1A
CCSP	Clara cell secretory protein
CD 31	Cluster of differentiation 31
CK18	Keratin 18
ECM	Extracellular matrix
ESCs	Embryonic stem cells
FOXJ1	Forkhead box protein J1
GAG	Glycosaminoglycans
Mg	Miligram
MIN-6	Mouse insulinoma 6
Ng	Nano gram
Nkx2.1	NK2 Homeobox 1
PBS	Phosphate buffered saline
PDGFR	Platelet derived growth factor receptor
ProSPC	Alveolar type 2 cell marker
SDS	Sodium dodecyl sulfate
SPC	Pulmonary-associated surfactant protein C
TNF	Tumor necrosis factor
TTF-1	Transcription termination factor 1

1   Introduction

The antiquity of medical expedient manufacturing is oversupplied with power-driven materials designed for organ utility replacement;

nevertheless, these non-biological tools have been mainly ineffective, since they were proposed to imitate one or more occupations of a specific organ, yet not the organ in its wholeness. Beings, dissimilar to medicinal tools, are not produced from metal and nylon. Medicinal tools have aimed the management of an illness, whereas regenerative medicine has the prospective to compromise a treatment. If a person really desires to substitute the occupation of a human organ, why not begin with a completely practical, vigorous organ afterwards (Seetapun and Ross 2017)?

Regenerative medicine embraces the potential to substitute or restore human cells, tissue or organs so as to reestablish the healthy function lost because of disease and injury. Through the arrangement of innovative biomaterials with cells, one of the purposes of regenerative medicine is to generate autologous tissue grafts for future replacement therapies (Garreta et al. 2017). Innately reproduced constituents have demonstrated to be greater to artificial polymers as matrix scaffolds, which may preserve the hierarchical complication of natural tissues. Indeed, these constituents can be utilized to form environments of enlarged complexity: from micro-tissues to organ scaffolds shaped through decellularization of complete tissues (Taylor et al. 2018). Studies show the existence of significant bioactive constituents covering collagen types, basic fibroblastic growth factor, transforming growth factor-β, vascular endothelial growth factor upon decellularization of dissimilar tissues. Preserving greater quantities of active molecules upsurges the chance of the medical usage of that decellularized tissue (Parmaksiz et al. 2016). In this present work, we assess the newest progressions in the practice of decellularization and recellularization technologies for the production of autologous tissue grafts and organ replacement therapies.

2   Concept of Decellularization

Decellularization is a technique which comprises the elimination of cellular apparatuses from the tissue to obtain extracellular matrix (ECM) patterns with a multifaceted combination of

physical and practical proteins for maintaining architecturally systematized entities to work as bio-derivative scaffold (Rana et al. 2017). Decellularization may be custom-made with the specific properties of an idyllic scaffold fulfilling bio-comformity, non-immunogenicity and the capacity to deliver physical, mechanical, chemical and biotic signals for cell connection, propagation, relocation, differentiation and sustained utility (Rana et al. 2017).

It has been stated that decellularized scaffolds display their organ-precise attitude pertaining to the foundation of the original tissue ECM for organ renovation. Decellularized scaffolds have been presented to be convenient applicants for consignment undertaking practicalities through procuring the mechanical force of collagen filaments, the springiness of elastin backbones and the hydration-attachment occupations of proteoglycans from their non-natural ECM matrices (Rana et al. 2017). The usage of innate vasculature and fast partition of debris and excess liquid from tissue meaningfully expands the superiority of scaffold for reconstruction of novel organ/tissue (Khan et al. 2014).

By way of expending biological matrices which recapitulate innate tissue to variable levels, it is likely to boost the chemical, mechanical and vascular environment of injured tissues possibly both to enhance cell preservation or endurance and to review natural signals for cell conductivity, improving the efficiency of cell-level overhaul (Taylor et al. 2018). Decellularized ECM has been effectively utilized to restructure numerous varieties of tissues and organs covering lung, heart valve, blood vessel, esophagus, urinary bladder, kidney, cornea, trachea (Gilpin and Yang 2017). Concept of decellularization is stated in Fig. 1.

3   Decelllularization Methods

The method of decellularization necessitates the separation of the ECM from any particular tissue with negligible forfeiture, injury or disturbance, whereas exploiting the elimination of cellular substance. This can be accomplished through

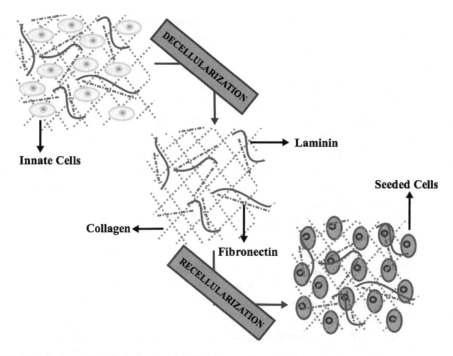

Fig. 1 Model of tissue decellularization. Artificial tissues may be produced by decellularization of native tissue. ECM then may be reseeded with cells that "maturate" the substance (e.g., to improve its renovation or vascularization) or "get matured" in the direction of a precise utility (e.g., to multiply or differentiate)

the joint implementation of physical, chemical, enzymatic and apoptosis-inducing approaches (Tapias and Ott 2014). Cell and genetic constituent exclusion is serious in avoiding immune rejection of the structure to seeded cells. The standards for evaluating the efficiency of elimination of these constituents is proposed like this: the decellularized ECM must carry fewer than 50 ng double stranded DNA per mg ECM dry heaviness, a smaller amount of 200 base pair DNA fragment size, no detectable nuclear material (Gilpin and Yang 2017).

While effective decellularization was attained for different organs, still much energy should be focused on the description of normalized decellularization procedures alongside the ultimate aim to develop well matched and tailored organ scaffolds for medical purposes. In order to achieve that, subjects covering degradation, bioconformity, pathogenicity and immunogenic responses should be further deliberated (Garreta et al. 2017).

3.1 Chemical and Enzymatic Methods

Different kinds of chemicals have been utilized in decellularization, containing surfactants, acids and bases. Surfactants, the most usual decellularizing materials, characteristically employ via lysing cells over disturbing the phospholipid cell membrane. These agents are categorized attribute to their charge, as they are ionic, nonionic, or zwitterionic. Acids such as peracetic acid and bases covering sodium hydroxide degrade the cell membrane and nuclear material by using their intrinsically charged features (Gilpin and Yang 2017).

Sodium dodecyl sulfate (SDS) has been effective in a quantity of submissions in decellularization according to its capability to proficiently eliminate cells and genetic content. SDS therapies have encountered the optimal necessities of whole cell elimination and removal of 90% of host DNA in numerous categories of tissues and organs, comprising rat arm, porcine cornea, procine myocardium, porcine heart valve, porcine small intestine, porcine kidney, human vein, rat-porcine-human lungs and human heart (Gilpin and Yang 2017).

While SDS can effectively eradicate unsolicited innate components of the tissue, it may be harmful to the organizational and signaling proteins. Destructive structural proteins and factors not only may avert the cells from occupying the tissue as earlier but similarly stops the full maintenance of its mechanical features. This consequence is mainly established in tinnier tissues and cell sheets. Although most surfactant-treated tissues must classically be rinsed with solutions like phosphate buffered saline (PBS), SDS is more challenging to be detached because of its ionic nature. Triton-X is a non-ionic less harsh and less damaging surfactant. It is oftentimes utilized to remove the remnant SDS. Not only is Triton-X advantageous in rinsing, but also it is regularly employed as a decellularizing agent unaccompanied (Gilpin and Yang 2017).

Enzymes can be employed to take aim at the leftover nucleic acids upon cell fraction or the peptide bonds that bind the cells to the ECM. They incline to endure in noteworthy numbers in the tissue and may inflame a further immune retort (Bourgine et al. 2013). Trypsin is an enzyme frequently utilized with EDTA that functions by pleating the cell-matrix linkages. Utilized in the treatment of porcine pulmonary valves, whole cell and genetic material elimination was detected after 24 h (Gilpin and Yang 2017).

Acids and bases counter with and denature proteins, solubilize cell constituents and change nucleic acids, therefore rupture the cells. They are not choosy and so modify also ECM apparatuses, particularly collagens, glycosaminoglycans (GAG) and growth factors (Bourgine et al. 2013).

3.2   Mechanical Methods

Approaches that can physically/mechanically decellularize the ECMs contain temperature and compression actions that operate to remove cells over mishmash of lysing the cells and abolishing cell-matrix adhesive proteins. Especially, physical therapies comprise the utilization of freeze-thaw, high hydrostatic pressure, or supercritical carbon dioxide to entirely eradicate constituent cells and genetic materials (Gilpin and Yang 2017).

Freeze-thaw operation effactually lyses cells inside tissues and organs. However, the consequential membranous and intracellular substances endure if not detached via consequent dispensation. A solitary freeze-thaw cycle may decrease opposing immune reactions covering leukocyte permeation in vascular ECM scaffolds. Repetitive freeze-thaw cycles can be utilized throughout decellularization and do not suggestively upsurge the forfeiture of ECM proteins from tissue (Crapo et al. 2011).

Hydrostatic force necessitates moderately slight time period and may be more efficient than detergents or enzymes for eradicating cells from vasculature and corneal tissues, while the baric construction of ice crystals can disturb ECM structure. Augmented temperature through pressure decellularization averts ice crystal construction but then again disturb ECM regarding to the related upsurge in entropy (Crapo et al. 2011).

3.3   Decellularization by Programmed Cell Death

Current decellularization approaches necessitate a compromise amongst effective cellular elimination and ECM conservation, as enhancing the action harshness for a more comprehensive decellularization seriously fallouts in a heightened ECM disturbance. The apoptotic paths may be intentionally stimulated over a clean and coordinated procedure via the conveyance of suitable signals. Throughout the entire apoptotic progression, the cellular substance is retained precisely inside the plasma membrane and the apoptotic bodies. The immunogenic cellular components do not outflow into the nearby environment, therefore averting a redundant inflammatory response. This is contrary to present decellularization methods, encouraging necrosis, cell exploding and the discharge of immunogenic material in the adjacent surroundings (Bourgine et al. 2013).

Kiss-of-death approach completely depends on the extrinsic pathway initiation by the distribution of precise ligands that attach their analogous death receptors of the TNF (Tumor necrosis factor) superfamily. Lethal-environmental-conditioning method is persuaded by moderating environmental influences such as temperature, pH in addition to carbon dioxide/oxygen, nitric oxide and hydrogen peroxide content. Contrasting to the physical freeze&thaw method triggering necrosis, induction of apoptosis through temperature variations necessitates low differences, in either hyperthermic or hypothermic ranges. Death-engineering approach depends on the instigation of any of the two apoptotic pathways by the routine of a genetic approach. Apoptosis stimulation could theoretically be attained by controlling the expression level of crucial genes elaborated in the pathway (Bourgine et al. 2013).

4 Pros and Cons of Decellularization

Decellularization may harm the ECM and membrane material of the organ. Most effective technique for the exact tissue form ought to be assessed for overpowering the difficulty. One other key problem with decellularized scaffold is to cultivate the practical anticipated cell types inside this and further to transplant *in vivo* to acquire purposeful organs. Therefore, there is a requirement for cells suitable to be transplanted, apposite medium and skill (Khan et al. 2014). The effects of each and individual decellularization method and their operation of actions are shown in Fig. 2.

Utilizing xenogenic scaffolds have a misgiving about antigen that can trigger the refusal. Although the decellularization is completed

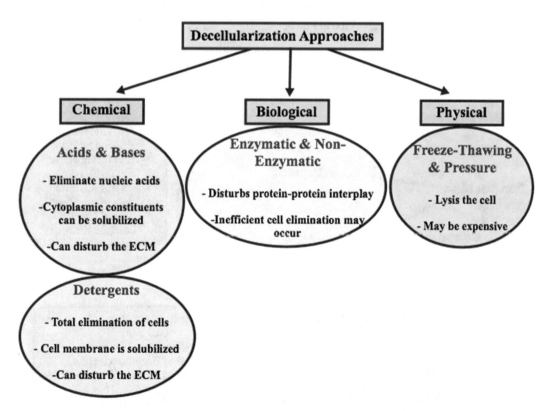

Fig. 2 Methods of decellularization. Their operation of actions, along with the advantages and disadvantages of each method

faultlessly, a specific surface antigen stays at nominal level called galactosyl α (1,3) galactose in xenogenic tissues and is inattentive in human that may activate severe refusal (Khan et al. 2014). Decellularized scaffold maintain the vasculature and reinstating blood stream precisely is less puzzling, nonetheless the problem stays to be replied of how long it can be hold. Since any partly re-endothelialized vasculature carries danger of thrombosis prominent to confined organ failure (Khan et al. 2014).

Even though the decellularized tissue and organ-derived matrices are like the in vivo ECM, they cannot imitate the expansion progression of tissue or body. In order to simulate the modelling of the ECM through expansion, the utilization of stem cells which are regulated at diverse periods of growth is essential. Hypothetically, the provision of ECM imitating the entire growth is conceivable if embryonic stem cells or induced pluripotent cells are expended (Hoshiba et al. 2010). The probable outcomes of decellularization is specified in Fig. 3, respectively.

The main pro of decellularized ECM is maintenance of the organizational entirety at the macro and micro measure, such that the subsequent scaffolds carry parallel tensile forte to innate tissues and preserve maximum of the connatural natural vasculature (Taylor et al. 2018). Decellularized ECM may moderate cell attitude: adharence, relocation and differentiation. Conservation of ultrastructure and conformation persuade advantageous tissue organization and remodeling (Garreta et al. 2017).

5   Recellularization

Recellularization is a progression of producing practical cells inside the decellularized organ scaffold to generate entirely operative simulated organ. Seeding of precise sort of cells with high propagation and regulated differentiation capacity is essential to recellularize the decellularized organ scaffold. Every organ has dissimilar compositions with divergent utilities. Thus,

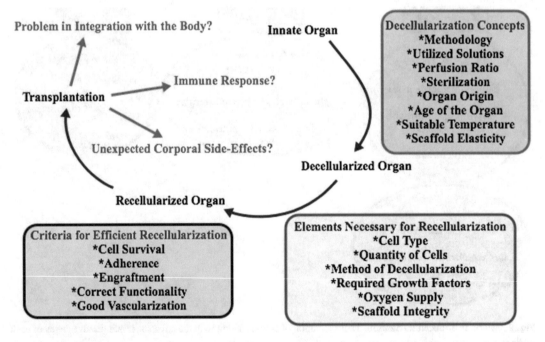

Fig. 3 Schematic demonstration indicating bioartificial organ expansion utilizing decellularization and recellularization methods and their probable application

utilization of more suitable cells stays as an obstacle to restore a whole organ (Khan et al. 2014).

To accurately form an organ or tissue necessitates producing the particular assemblies of that tissue together with any vascular or ductal constituents and secondary assemblies supplied with cells, counting inhabitants of occupant stem or progenitor cells for constant organ conservation (Fig. 4). Rejuvenation of an organ needs both parenchymal and non-parenchymal cell foundations. Preferably both would be originating from non-immunogenic bases and would bring matured functioning tissue. Therefore, a naive foundation of cells to reconstruct any particular organ would appear to be that real organ. Conversely, most adult organs either do not cover cells which are adequately proliferative to produce the quantities of cells required for reconstructing the entire structure or do not comprise cells that may be simply collected. For each organ type dissimilar approaches and practical features are required to consider before emerging a healthier solution for the existing difficulties (Khan et al. 2014; Badylak et al. 2011). The factors that should be estimated during recellularization is indicated in Fig. 5.

6 Stem Cells Deliver an Innovative Aspect to Decellularized Scaffold-Related Tissue Engineering

Stem cells have turned into an essential segment of tissue engineering and deliver a novel aspect to decellularized scaffold-related tissue engineering especially, either as a cell foundation for recellularization or as a basis for producing ECM matrix (Badylak et al. 2011). Stem cells are stated as undifferentiated cells which can turn into precise lineages (potency) and can split by themselves to generate more stem cells (self-regeneration) (Clevers 2015). Fortified with their innate features, decellularized scaffolds coated with stem cells are inviting more courtesy in essential research and medical studies and are now being operated for interpreting practical organs. Nevertheless, the appliances underlying the interface of stem cells with decellularized scaffolds have still not been fully understood (Rana et al. 2017; Agmon and Christman 2016). The demonstration of stem cell seeding on decellularized scaffolds is presented in Fig. 6.

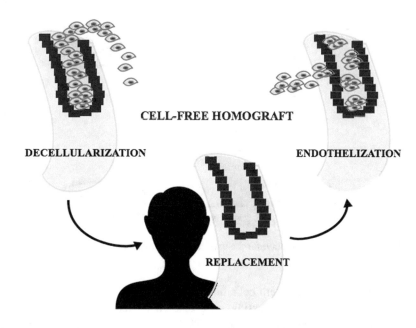

Fig. 4 Concepts of endothelialized biomaterials.
Decellularized scaffolds may be implanted to generate an elementary internal vasculature. Upon implantation, the embedded endothelial cells are expected to connect with the host tissue to generate a perusable construct

CELL-FREE HOMOGRAFT

DECELLULARIZATION

ENDOTHELIZATION

REPLACEMENT

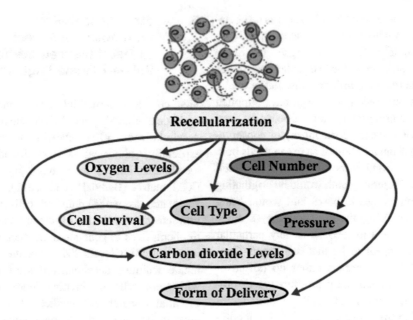

Fig. 5 Recellularization factors that are important for suitable integration with the decellularized scaffold

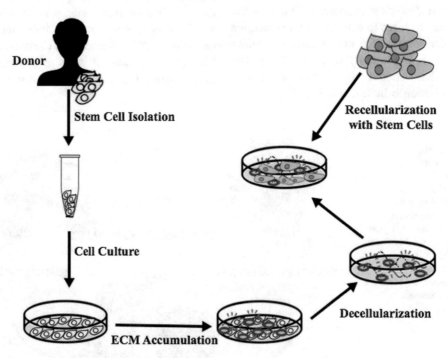

Fig. 6 Diagram demonstration for the expansion of practical tissues by decellularization and recellularization through stem cell seeding

Murine embryonic stem cells have been persuaded to positively expressing lung epithelial phenotype comprising alveolar (TTF-1 and SPC) and airway (CCSP) epithelium later to differentiation of definitive endoderm utilizing Activin A to simulate the Nodal signaling followed through adherent cell culture in small airways growth medium or use of more discerning differentiation medium. Cellularizing decellularized rat scaffolds with mouse ESCs led to superior survival compared to cells seeded onto non-lung matrices and also deceptive differentiation in the direction of various lineages that showed zone-precise distribution covering club cells (CC10+), Alveolar epithelial type II (AT II) cells (CK18+, proSPC+), endothelial cells (CD31+) and mesenchymal cells (PDGFRa +). Therefore, it is hypothetically conceivable to detect differentiation of the most nascent of stem cells along the lines of growth only by seeding them on acellular scaffolds. Correspondingly, mouse ESCs differentiated to Nkx2.1+, proSPC + ATII-like cells seeded onto mouse decellularized scaffolds that were injected subcutaneously distributed to airways (FoxJ1^{pos}) and alveolar regions (proSPC+ or PDGFRa+ cells) and conserved phenotype expression for 14 days. Separately, host-derived endothelial cells infiltrated the scaffolds signifying that practical vascularization could happen (Wagner et al. 2013).

Studies established the viability of decellularized scaffolds seeded with autologous adipose-derived stem cells (ADSCs) for cartilage fault overhaul in rabbits. The results indicated that the ADSC-seeded decellularized scaffolds prompted cartilage tissue restoration similar to natural cartilage in point of physical features and chemical constituents. Hence, the expansion of decellularized scaffolds, together with stem cells, is an enthusing field of research that unlocks a novel area in stem cell-related tissue engineering (Kang et al. 2014).

Several cell types including bone marrow stem cells, umbilical stem cells, embryonic stem cells, fetal cardiomyocytes, smooth muscle cells, and dermal fibroblasts, were examined as cell foundations for spot recellularization with the purpose of fabrication of porcine cardiac patches. Between them, bone marrow stem cells were discovered to have a capability to renew myocardium, persuade angiogenesis, and be permitted from ethical issues (Wang et al. 2010).

7 Regenerative Medicine

Encouraging preclinical and experimental data thus far upkeep the opportunity for handling both chronic acute illnesses, and for regenerative medicine to abet diseases happening through an extensive group of organ systems and contexts covering skin wounds, cardiovascular diseases and traumas, managements for specific forms of cancer, etc. The existing treatment of transplantation of intact organs and tissues to cure organ and tissue insufficiency and forfeiture agonizes from restricted donor quantity and frequently serious immune difficulties, but these problems may possibly be avoided by the utilization of regenerative medicine approaches (Mao and Mooney 2015). Over the last 20 years, regenerative medicine has developed with a goal of directly restoring/renewing damaged or unhealthy tissue instead of treating disease indications (Taylor et al. 2018; Rijal 2017). Decellularized ECM coming from tissues or complete organs has significant regenerative benefits. Regardless of numerous stated compensations and the accomplishment in premature clinical studies, this know-how has not entirely been interpreted into standard medical practice so far (Parmaksiz et al. 2016).

8 Recent Research on Decellularized Tissue in Regenerative Medicine

8.1 Heart

Heart failure is one of the foremost reasons of hospitalization and mortality, effecting almost 15 million people in Europe (Heidenreich et al. 2011). A progressive phase of heart failure, also recognized as end-stage heart failure, is rising and this propensity grants itself as a difficulty to both medical doctors and healthcare organizations. Heart failure is described as having cardiac output that is incompetent for living, and diagnosis is predominantly poor, with around 25–50% of patients dying in 5 years after diagnosis (Zia et al. 2016).

Ott et al. have indicated an original technique of perfusion decellularization that can produce complete organ scaffolds. The extension of a cannula through the ascending aorta permitted backward coronary perfusion by detergents. This type of process attained the elimination of the cellular compartment of an entire heart. That was then recellularized with newborn rat cardiomyocytes. The construct was distributed inside of the heart scaffold over transmural injection, whereas endothelial cells were inserted through the aorta. The final structure was capable of contract up to 2% of the usual contractile function (Ott et al. 2008).

Ng et al. stated a parallel method where they seeded the decellularized heart with embryonic stem cells that botched to produce practical heart regarding to poor cardiac myocyte differentiation competence (Ng et al. 2011). Then, Lu et al. established the manufacturing efficient human heart tissues via recellularization of decellularized mouse hearts by human induced pluripotent stem cell-derived cardiac myocyte progenitors. Cells were transported over coronary vessels of the heart which also showed cardiomyocyte propagation, differentiation, myofilament development, impulsive contraction, production of mechanical strength and response to medications (Lu et al. 2013).

8.2 Liver

Liver is the biggest interior organ carrying vital physiological utilities, comprising vitamin storing, protein synthesis and detoxification. On behalf of people with end-stage liver failures, that are frequently linked with hepatitis, bile duct illnesses or liver cancers, liver transplantation has turned into first-rate management. It delivers improved excellence of life and affordability. The 5 year survival ratio after liver replacement is over 70%. Nonetheless, the number of patients who require transplantation surpasses suitable donors. In only United States of America, it was stated that over 16,000 patients demand transplantation every year but just 6000 liver replacements are accomplished because of lack of apposite donors. The lack of suitable donors consequences in eighteen deceases daily. There is a necessity for inventive methods to develop practical liver substitutes for end-stage liver failure patients (Meng et al. 2017).

Uygun et al. have indicated an exhaustive perspective of decellularizing rat livers and seeding them with rat primary hepatocytes, presenting encouraging hepatic utility and the capacity of heterotopically transplant these engineered livers into animals for almost 8 h (Uygun et al. 2010).

Hassanein et al. have showed recellularization by the bile duct supported practical allogenic and xenogenic cell proliferation on a decellularized rat liver scaffold. Replacement of the recellularized paradigm has resulted in fast implant thrombosis, likely secondary to exposed collagen in the deendothelialized vasculature. Development in the field is restricted by the incapability to produce a completely endothelialized structure covering different cell lines similar to an innate organ (Hassanein et al. 2017).

8.3 Pancreas

More than 1500 islet replacement practices have been implemented globally since 2000. However only about 7.5% of the transferred patients persist insulin independent 5 years after procedure. Existing transplantation methods cover the infusion of islets into the liver portal vein. The main difficulties of intraportal islet infusion are hemorrhage and thrombosis whereas other factors like anoikis, hypoxia, and inflammation related immune response also consequence in damage of islet utility and graft loss (Salvatori et al. 2014).

Claudius et al. have presented the perfusion decellularization of cadaveric rat and human pancreases for production of ECM scaffolds aiming to maintain the innate islet cell niche to enhance islet endurance and utility after replacement. Recellularization with human islets created practical endocrine tissue *in vitro* and retreated the diabetic phase after transplantation in rat models (Conrad et al. 2010).

Wu et al. have tried to create a microenvironment mimicking the innate pancreas that is suitable for not only cell development but also cellular utility exertion. They utilized a decellularized mouse pancreas as 3D scaffold in the experimentation. MIN-6 (Mouse Insulinoma 6) cells were seeded on the bioscaffold. Moreover, *in vivo* plantation of the recellularized bioscaffold presented its capacity of regulating blood glucose. Nevertheless, they detected the progressively swelling blood glucose, which suggested the seeded cells might perish because of the absence of incessant source of nourishment (Wu et al. 2015).

8.4 Trachea

Idyllic approaches for renovating the tracheal structure and reestablishing tracheal utility upon tracheal injury or subtraction have not been settled. Artificial constituents have been utilized as unconventional alternates. Studies stated good outcomes for different tracheal engineering techniques with countless compatible materials. Conversely, the restorations have motivated on preserving the unity of the tubular organization, and difficulties covering body reactions and infections have not been completely addressed, with the long-term healing effect of this approach lasting unclear (Hung et al. 2016).

Gray et al. composed a tissue-engineered scaffold from xenologous decellularized leporine tracheal segment seeded with autologous amniotic mesenchymal stem cells and equated it with decellularized scaffolds. The consequences of the animal study established supreme endurance with full epithelization for the engineered scaffolds with amplified elastin plane later to transplantation (Gray et al. 2012).

Zang et al. have collected from Brown Norway rats (donor) and Lewis rats (receiver) were decellularized with frequent detergent/enzymatic methods. Decellularized Brown Norway tracheal matrix scaffolds were recellularized with Lewis rat stem cell–derived chondrocytes from outside and tracheal epithelial cells from inside to produce a layered tracheal structure. Decellularized tracheal matrix scaffold did not provoke noteworthy allograft refusal or remote body answer *in vivo*. While the structure sustained reepithelialization, stem

cell–induced chondrocytes unsuccessful for engrafting in the heterotopic background (Zang et al. 2013).

8.5 Lung

In only United Stated, more than 24 million patients have indication of compromised lung utility, triggering extensive physical and financial weight. Presently, lung transplantation is the solitary absolute cure for people with end-stage lung disease. Nevertheless, vigorous organ unavailability averts transplantation from being a useful answer for the mainstream of these patients. The present donor lung unavailability is also worsened by the delicateness of lung itself. Lung tissues are simply injured and frequently conceded throughout the progression of transplantation. Along with these restrictions, patients are obligatory to be on permanent immunosuppressive drugs after lung transplantation, causing intense lessening in standard of life and an improved susceptibility to pulmonary contaminations. The formation of a disinfected, autologously cell sourced artificial lung would reduce the disease allied with immunosuppression and the lack of donor organ, and it would also deliver tailor lung substitutes for patients with an extensive variety of functional requirements (Balestrini et al. 2015).

Pulmonary tissue engineering has concentrated on renewal endorsed by decellularized scaffold *in vivo* and *in vitro*. Throughout decellularization, the organizational proteins and pertinent cytokines of extracellular matrix were preserved, while cellular constituents were detached. Epithelial and endothelial cells were seeded onto trachea and vessels, two groups discovered that efficient gas conversation may be produced 6 h later in rats with recellularized lungs (Yu et al. 2016).

8.6 Bone

Treatment options to serious bone deficiencies because of trauma, corruptions, tumors or hereditary disorders are based on autologous or allogeneic bone grafts. Conversely, severe donor-site indisposition, elevated risk of contagions and location inadequacies hamper them as maintainable alternatives. Unconventional tissue engineering methods, in spite of the important developments of the field, are not yet capable of delivering reliably effective clarifications for repetitive therapeutic use. Evidently, there is the necessity to establish original or superior approaches able to encounter patients' requests (Papadimitropoulos et al. 2015).

Fröhlich et al. established an *in vitro* 0.5 cm sized bone structure utilizing human adipose derived stem cells, decellularized bone scaffolds and perfusion bioreactors. Later to 5 weeks of refinement, introduction of osteogenic supplements to the culture medium meaningfully improved the construct cellularity and the aggregates of bone matrix constituents covering collagen, bone sialoprotein and bone osteopontin (Frohlich et al. 2010).

8.7 Kidney

Regarding the universal upsurge of patients with renal failure, the expansion of practical renal replacement treatments have gained substantial attention and new tools are promptly advancing. Currently expended renal replacement therapies inefficiently eliminate accumulating waste products, resulting in the uremic syndrome. A more desired management choice is kidney trans plantation, but the lack of donor organs and the rising amount of patients remaining for a transplant permit the expansion of original technologies (Jansen et al. 2014).

Porcine kidneys were effectively decellularized, suggesting the probability of utilizing these transplantable scaffolds to build engineered kidney clinically pertinent. Entire porcine kidneys were decellularized and orthotopically *in vivo* relocated, then prophylaxis was controlled with anticoagulant therapy. Immune cells in the pericapsular area and thrombosis happened owing to the absence of endothelial cells (Yu et al. 2016).

9 Workbench to Bedside Interpretation of Repopulated Decellularized Scaffolds

Conveying scaffolds to medical excellence and measure is only one of numerous phases headed for the renewal of feasible and operative organs. At the modern level of expertise, an effort to review the whole progression of embryogenesis from a single cell to organogenesis in a laboratory looks barely convincing. This kind of an accomplishment would necessitate broad culture intervals to produce tissues of quantifiable dimension and a diversity of tissues to allow interplay, and would cause obvious ethical problems (Song and Ott 2011).

There have been couple of attempts for the production of innate ECM with practical retrieval. Ott et al. and Petersen et al. established transplant endurance for limited hours of recellularized lung scaffold into rats with passable oxygen and carbon dioxide interchange with suitable pressure/volume interactions. The rat passed away owing to pulmonary edema subsequent to respiratory failure (Khan et al. 2014).

Song et al. produced an engineered rat kidney by decellularization and recellularization method by human umbilical cord blood-derived endothelial cells and presented urine making accompanied by minor macromolecular sieving and reabsorption capacity (Khan et al. 2014). Laronda et al. instigated sexual maturity in mice following decellularized ovary transplant. They established that ovarian transplant seeded with primary ovarian cells on a decellularized matrix may deliver a niche for steroid and peptide hormone assembly that can start adolescence in ovariectomized mice (Laronda et al. 2015).

Hung et al. performed segmental organ decellularization on trachea and employed autotransplantation on rabbit models. Even though the respirational epithelium renewal on the internal surface seemed to be pleasing, the tubular assemblies were not able to be preserved later to replacement that eventually led to the decease of the animals (Hung et al. 2016).

Hopkins, et al. evaluated in a baboon model the hemodynamics and human leukocyte antigen immunogenicity of recurrently inserted engineered human and baboon heart valve scaffolds. Valve impairment interrelated with markers for more powerful inflammatory incitement. The verified bioengineering techniques condensed antigenicity of both human and baboon valves. Replacement valves from both species were found hemodynamically equal to innate valves (Hopkins et al. 2013).

Duisit et al. formerly stated the use of the perfusion-decellularization method in porcine ear and human face models, with effective implant decellularization, conservation and convenience of the vascular system, appropriate for seeding with fresh cells and conformity. They then theorized that this knowledge may be pertained to human ear grafts. They found that vascularized and multifaceted auricular scaffolds may be acquired from human basis to deliver a platform for additional useful auricular tissue engineered creates, thus offering a perfect path to the vascularized complex tissue engineering method (Duisit et al. 2018).

10 Conclusion

Regenerative medicine embraces the potential of renewing tissues and organs by either encouraging formerly irremediable tissues to overhaul themselves or engineering them ex vivo. Decellularized constituents have the capacity to grasp the exclusive principles presented by autografts that elucidates the many dissimilar areas of utilization for these knowledges. According to the tissue form, allogeneic or xenogeneic tissues maintain their tissue precise features and bioactive composites following decellularization. Consequently, decellularized ECM materials have the capability to enable reformative renovation of the injured tissue. Regulation of construction, organization of donor tissues and organs, efficacy and welfare assessment, superiority control, application proprieties and strategies, calibration and consciences are serious subjects for the expansion and commercialization of decellularized ECM scaffolds. The suitable approach for producing a 3D structure is displayed in Fig. 7.

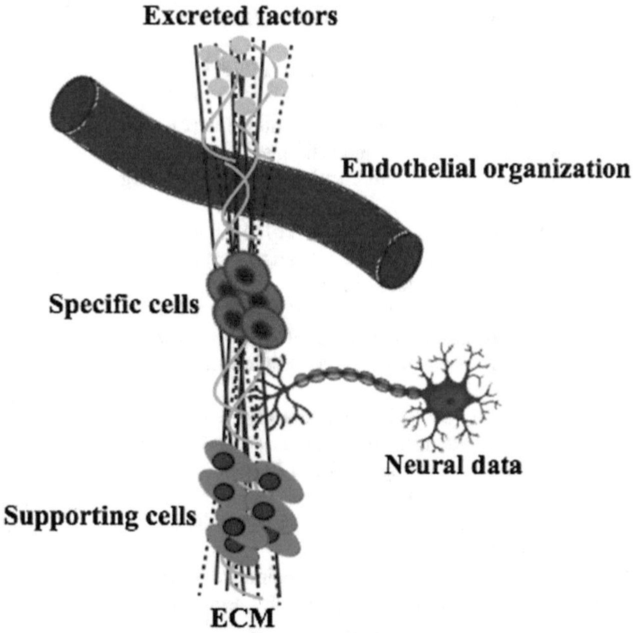

Fig. 7 Common approach for creating a suitable 3D structure. The precise range of ECM features, emblematic for designated normal tissue can be established. A decellularized scaffold with toning factors and appropriate cells can be identified and united in vitro for growth of specific, custom-made 3D system

Many of the existing efforts target emerging approaches for growth and stemness might convey innovative solutions relevant to organ engineering in the anticipatable future. Solid organ redevelopment founded on perfusion-decellularized innate ECM scaffolds carries excessive aptitude for patients distressing end-organ failure, nonetheless evidently remains an aspiring aim. Effort in the direction of that target will include numerous fields and construct intermediate remedial products and highpoints that expand our knowledge on stem and progenitor cell fate in organ expansion and disease.

References

Agmon G, Christman KL (2016) Controlling stem cell behavior with decellularized extracellular matrix scaffolds. Curr Opin Solid State Mater Sci 20 (4):193–201

Badylak SF, Taylor D, Uygun K (2011) Whole-organ tissue engineering: decellularization and recellularization of three-dimensional matrix scaffolds. Annu Rev Biomed Eng 13:27–53

Balestrini JL et al (2015) Production of decellularized porcine lung scaffolds for use in tissue engineering. Integr Biol 7(12):1598–1610

Bourgine PE et al (2013) Tissue decellularization by activation of programmed cell death. Biomaterials 34 (26):6099–6108

Clevers H (2015) What is an adult stem cell? Science 350 (6266):1319–1320

Conrad C et al (2010) Bio-engineered endocrine pancreas based on decellularized pancreatic matrix and mesenchymal stem cell/islet cell coculture. J Am Coll Surg 211(3):S62–S62

Crapo PM, Gilbert TW, Badylak SF (2011) An overview of tissue and whole organ decellularization processes. Biomaterials 32(12):3233–3243

Duisit J et al (2018) Perfusion-decellularization of human ear grafts enables ECM-based scaffolds for auricular vascularized composite tissue engineering. Acta Biomater 73:339–354

Frohlich M et al (2010) Bone grafts engineered from human adipose-derived stem cells in perfusion bioreactor culture. Tissue Eng A 16(1):179–189

Garreta E et al (2017) Tissue engineering by decellularization and 3D bioprinting. Mater Today 20 (4):166–178

Gilpin A, Yang Y (2017) Decellularization strategies for regenerative medicine: from processing techniques to applications. Biomed Res Int 2017:1–13

Gray FL et al (2012) Prenatal tracheal reconstruction with a hybrid amniotic mesenchymal stem cells-engineered construct derived from decellularized airway. J Pediatr Surg 47(6):1072–1079

Hassanein W et al (2017) Recellularization via the bile duct supports functional allogenic and xenogenic cell growth on a decellularized rat liver scaffold. Organogenesis 13(1):16–27

Heidenreich PA et al (2011) Forecasting the future of cardiovascular disease in the United States a policy statement from the American Heart Association. Circulation 123(8):933–944

Hopkins RA et al (2013) Bioengineered human and allogeneic pulmonary valve conduits chronically implanted orthotopically in baboons: hemodynamic performance and immunologic consequences. J Thorac Cardiovasc Surg 145(4):1098

Hoshiba T et al (2010) Decellularized matrices for tissue engineering. Expert Opin Biol Ther 10(12):1717–1728

Hung SH et al (2016) Preliminary experiences in trachea scaffold tissue engineering with segmental organ decellularization. Laryngoscope 126(11):2520–2527

Jansen J et al (2014) Biotechnological challenges of bioartificial kidney engineering. Biotechnol Adv 32 (7):1317–1327

Kang HJ et al (2014) In vivo cartilage repair using adipose-derived stem cell-loaded decellularized cartilage ECM scaffolds. J Tissue Eng Regen Med 8 (6):442–453

Khan AA et al (2014) Repopulation of decellularized whole organ scaffold using stem cells: an emerging technology for the development of neo-organ. J Artif Organs 17(4):291–300

Laronda MM et al (2015) Initiation of puberty in mice following decellularized ovary transplant. Biomaterials 50:20–29

Lu TY et al (2013) Repopulation of decellularized mouse heart with human induced pluripotent stem cell-derived cardiovascular progenitor cells. Nat Commun 4:2307

Mao AS, Mooney DJ (2015) Regenerative medicine: current therapies and future directions. Proc Natl Acad Sci U S A 112(47):14452–14459

Meng FW et al (2017) Whole liver engineering: a promising approach to develop functional liver surrogates. Liver Int 37(12):1759–1772

Ng SLJ et al (2011) Lineage restricted progenitors for the repopulation of decellularized heart. Biomaterials 32 (30):7571–7580

Ott HC et al (2008) Perfusion-decellularized matrix: using nature's platform to engineer a bioartificial heart. Nat Med 14(2):213–221

Papadimitropoulos A et al (2015) Engineered decellularized matrices to instruct bone regeneration processes. Bone 70:66–72

Parmaksiz M et al (2016) Clinical applications of decellularized extracellular matrices for tissue engineering and regenerative medicine. Biomed Mater 11 (2):022003

Rana D et al (2017) Development of decellularized scaffolds for stem cell-driven tissue engineering. J Tissue Eng Regen Med 11(4):942–965

Rijal G (2017) The decellularized extracellular matrix in regenerative medicine. Regen Med 12(5):475–477

Salvatori M et al (2014) Extracellular matrix scaffold technology for bioartificial pancreas engineering: state of the art and future challenges. J Diabetes Sci Technol 8(1):159–169

Seetapun D, Ross JJ (2017) Eliminating the organ transplant waiting list: the future with perfusion decellularized organs. Surgery 161(6):1474–1478

Song JJ, Ott HC (2011) Organ engineering based on decellularized matrix scaffolds. Trends Mol Med 17(8):424–432

Tapias LF, Ott HC (2014) Decellularized scaffolds as a platform for bioengineered organs. Curr Opin Organ Transplant 19(2):145–152

Taylor DA et al (2018) Decellularized matrices in regenerative medicine. Acta Biomater 74:74–89

Uygun BE et al (2010) Organ reengineering through development of a transplantable recellularized liver graft using decellularized liver matrix. Nat Med 16(7):814–U120

Wagner DE et al (2013) Can stem cells be used to generate new lungs? Ex vivo lung bioengineering with decellularized whole lung scaffolds. Respirology 18(6):895–911

Wang B et al (2010) Fabrication of cardiac patch with decellularized porcine myocardial scaffold and bone marrow mononuclear cells. J Biomed Mater Res A 94a(4):1100–1110

Wu D et al (2015) 3D culture of min-6 cells on decellularized pancreatic scaffold: in vitro and in vivo study. Biomed Res Int 2015:432645

Yu YL et al (2016) Decellularized scaffolds in regenerative medicine. Oncotarget 7(36):58671–58683

Zang MQ et al (2013) Decellularized tracheal matrix scaffold for tracheal tissue engineering: in vivo host response. Plast Reconstr Surg 132(4):549e–559e

Zia S et al (2016) Hearts beating through decellularized scaffolds: whole-organ engineering for cardiac regeneration and transplantation. Crit Rev Biotechnol 36(4):705–715

Adv Exp Med Biol – Cell Biology and Translational Medicine (2019) 6: 87–106
https://doi.org/10.1007/5584_2019_381
© Springer Nature Switzerland AG 2019
Published online: 9 May 2019

Synovium-Derived Mesenchymal Stem/Stromal Cells and their Promise for Cartilage Regeneration

Janja Zupan, Matej Drobnič, and Klemen Stražar

Abstract

Adult tissues are reservoirs of rare populations of cells known as mesenchymal stem/stromal cells (MSCs) that have tissue-regenerating features retained from embryonic development. As well as building up the musculoskeletal system in early life, MSCs also replenish and repair tissues in adult life, such as bone, cartilage, muscle, and adipose tissue. Cells that show regenerative features at least *in vitro* have been identified from several connective tissues. Bone marrow and adipose tissue are the most well recognized sources of MSCs that are already used widely in clinical practice. Regenerative medicine aims to exploit MSCs and their tissue regeneration even though the underlying mechanisms for their beneficial effects are largely unknown. Despite many studies that have used various tissue-derived MSCs, the most effective tissue source for orthopedic procedures still remains to be identified. Another question that needs to be addressed is how to evaluate autologous MSCs (i.e., patient derived). Previous studies have suggested the features of bone-marrow-derived MSCs can differ widely between

individuals, and can be changed in particular in patients suffering from some forms of degenerative disorder, such as osteoarthritis. The synovium is a thin membrane that protects the synovial joints, and it is a rich source of MSCs that show great potential for regenerative medicine. Here, we review synovium-derived MSCs from reports on basic and clinical studies. We discuss their potential to treat cartilage defects caused by either degeneration or trauma, and what needs to be done in further research toward their better exploitation for joint regeneration.

Keywords

Animal studies · Clinical studies · *In-vitro* studies · Mesenchymal stem/stromal cells · Synovium

1 Introduction

Mesenchymal stem/stromal cells (MSCs) are heterogeneous populations of stem cells (Sacchetti et al. 2016) that reside in adult tissues that have the unique ability for multilineage differentiation into bone, cartilage, muscle, and adipose tissues (Fellows et al. 2016). MSCs were first described for bone marrow by Friedenstein and colleagues (Friedenstein et al. 1970), and have since been found in many other adult tissues, such as bone, muscle, adipose tissue, synovium, skin, and other

J. Zupan
Faculty of Pharmacy, Department of Clinical Biochemistry, University of Ljubljana, Ljubljana, Slovenia

M. Drobnič and K. Stražar (✉)
Department of Orthopaedic Surgery, University Medical Centre Ljubljana, Ljubljana, Slovenia
e-mail: klemen.strazar@kclj.si

connective tissues. The scientific pursuit of these cells has been long hampered by their rare frequency in adult tissues and their lack of specific markers *in vivo*. Their propensity for plastic adherence allowed a series of *in-vitro* studies to evaluate cultures of MSCs and their potency for cartilage, bone and adipose tissue differentiation. It is likely that *in-vitro* expanded MSCs represent a population comprised of multiple types of distinct stem cells, as well as mature cells (Chan et al. 2018).

Transgenic animal models that can trace MSCs from early development through adult life and through health and disease (Fuchs and Horsley 2011) have led the *in-vivo* scientific research of MSCs. Use of transgenic animal models in combination with tissue injury models has provided priceless information about the identities and regenerative abilities of MSCs. Using these approaches, specific MSC populations have been identified in bone marrow (Zhou et al. 2014; Worthley et al. 2015; Chan et al. 2018) and synovium (Roelofs et al. 2017). It has also been clearly shown that these MSCs replenish bone and cartilage in adult life, and can contribute to tissue repair following injury; some can even form rudimentary joints *de novo* (Roelofs et al. 2017).

As MSCs are endogenously present in several tissues and contribute to tissue repair in adult life, they show great promise for regenerative medicine in degenerative joint disorders. Osteoarthritis, in particular, is the most common joint disorder, and it results from a combination of the breakdown of a joint and the attempt by the body to repair the damage. In humans, endogenous cartilage repair is ineffective, and the poor healing capacity of cartilage after injury can lead to osteoarthritis. The risk of developing osteoarthritis also increases with age (Arthritis Research UK 2018). Stem-cell exhaustion and their decreased regeneration potential have been proposed as hallmarks of aging in humans (Partridge et al. 2018). It has been shown that bone-marrow-derived MSCs in patients with hip osteoarthritis have low proliferative potential, and they are less active in chondrogenic and adipogenic differentiation (Murphy et al. 2002). Moreover, another

study reported that MSCs in subchondral bone from patients with late-stage hip osteoarthritis increase in number in areas of damage, but show perturbations that can lead to damage escalation (Campbell et al. 2016).

Controversy remains over whether articular cartilage itself has MSCs and which tissue-derived MSCs, if any, contribute to cartilage repair in humans. Recently, increasing evidence has indicated that chondrogenic abilities and cartilage repair features can be attributed to synovium-derived MSCs (Mak et al. 2016; Roelofs et al. 2017; Yao et al. 2018; De Bari and Roelofs 2018; Zayed et al. 2018; Enomoto et al. 2018; Shimomura et al. 2018; Murata et al. 2018; Jia et al. 2018b). Synovium is the soft tissue that lines the spaces of diarthrodial joints, tendon sheaths, and bursae. It includes the continuous surface layer of cells (i.e., the intima) and the underlying tissue (i.e., the subintima). The intima consists of macrophages and fibroblasts, while the subintima includes blood and lymphatic vessels, where both resident fibroblasts and infiltrating cells are found in a collagenous extracellular matrix. Between the intimal surfaces there is a small amount of fluid, which is usually rich in hyaluronan (Smith 2011).

Synovial hyperplasia is a very common phenomenon following joint injury. However, it is not only inflammation in the joint that sustains synovial hyperplasia. It has been shown that hyperplasia is also underpinned by proliferation of MSCs that respond upon joint injury (Kurth et al. 2011). Recently, a subpopulation of synovial MSCs (defined as a Gdf5-lineage derived from the interzone of the early embryonic joint structure) was shown to contribute to cartilage repair postnatally (Roelofs et al. 2017).

There is gathering evidence that this tiny membrane wrapped around the synovial joints is a rich source of MSCs, which has been little recognized in comparison to bone marrow and adipose tissue. Here, we provide an overview of MSCs derived from synovium. We start with basic and clinical studies, with a focus on their potential for regeneration of cartilage defects caused by either degeneration or trauma. We also discuss the best lines for further research toward better exploitation of synovium MSCs in regenerative medicine.

2 Basic Studies

Here we provide an overview of the basic studies that have investigated synovium-derived MSCs. *In vitro*, human and animal synovium tissues have been used to characterize MSCs using laboratory methods. *In vivo* in animal studies, synovium-derived MSCs have been implanted in different animal models of osteoarthritis, to define their regenerative features.

2.1 In-Vitro Studies

The *in-vitro* studies to date are summarized in Table 1. Minimal criteria to define cells cultured *in vitro* as MSCs were suggested by the International Society for Cell Therapy (ISCT) in 2006 (Dominici et al. 2006). These criteria included plastic adherence, more than 95% positivity for CD73, CD90, and CD105, and less than 2% positivity for CD45, CD34, CD11b/14, CD79α/19, and human leukocyte antigen (HLA)-DR, combined with trilineage differentiation potential; i.e., chondrogenic, adipogenic, and osteogenic. In investigations into these three criteria, the majority of *in-vitro* studies into synovium and synovial-fluid-derived MSCs have been performed in humans. Despite the minimal criteria set by the ISCT, the cell isolation and cultivation methods used still vary widely, with a general lack of standard identification (Dominici et al. 2006; Lv et al. 2014). Therefore, it is difficult to compare the data across these studies.

Most of these human studies have investigated synovium from the knee joint, with synovium from the hip joint less well recognized (Hermida-Gómez et al. 2011; Murata et al. 2018). The sites of synovium in the hip joint that have been shown to be sources of MSCs are given in Fig. 1.

De Bari et al. (2001) were the first to demonstrate that MSC-like cells can be isolated from synovium of knee joints. Sakaguchi et al. (2005) showed this synovium to be a superior reservoir of MSCs in comparison with bone marrow, adipose tissue, periosteum, and muscle in patients with anterior cruciate ligament injury. The advantage of their study was that five different tissue sources of MSCs were compared within the same patient (Sakaguchi et al. 2005). Using this approach, the influence of concomitant conditions and other variables such as age, sex, and body mass index, among others, can be eliminated. Similar findings were reported in mice, where infrapatellar fat-pad-derived MSCs were shown to have higher proliferative potential and similar or higher multilineage potential in comparison with bone-marrow and muscle-derived MSCs. A recent study compared two different synovium sources of MSCs in patients with femoracetabular impingement syndrome (Murata et al. 2018). They observed that MSCs from the cotyloid fossa synovium have higher proliferation and differentiation potential than those from the paralabral synovium. Hence, they suggested the use of cotyloid-fossa-synovium-derived MSCs for stem-cell therapy.

Efforts have also been made to identify the optimal tissue source of MSCs in particular joint disorders. Kohno et al. (2017) compared synovium-derived MSCs in patients with osteoarthritis and rheumatoid arthritis. They showed that cell yields, surface markers, and the chondrogenic potentials of synovial MSCs from both of these patient groups were comparable, and so they indicated that synovium derived from patients with rheumatoid arthritis represents a promising source of MSCs for cartilage and meniscus regeneration. As mentioned above, in such comparisons of two groups of patients, it is imperative to exclude other sources of donor-to-donor variations that might influence the features of the MSCs, such as age, sex, concomitant conditions, and disorders.

Some studies have compared synovium-derived MSCs from patients with osteoarthritis to those from healthy controls. The aim was to determine whether osteoarthritis affects MSCs in synovium in the same way as it affects, for instance, MSCs in bone-marrow (Murphy et al. 2002; Campbell et al. 2016). Generally, the

Table 1 Summary of *in-vitro* studies using synovium-derived MSCs

Species	MSC source	Methods used	Outcomes	Reference
Human	Paralabral and cotyloid fossa synovium from 18 patients with femoroacetabular impingement syndrome during hip arthroscopy	CFU-F, trilineage differentiation	Higher CFU-F, adipogenesis, osteogenesis and chondrogenesis in cotyloid fossa compared to paralabral synovium	Murata et al. (2018)
Rabbit	Knee synovium and bone marrow from tibia	Macroscopic analysis, histology	More MSCs observed at site of defective ligament in the knee	Morito et al. (2008)
Human	Synovial fluid from 22 patients with meniscus injury and eight controls with no history of knee injury	*In-vitro* expansion, CFU-F, trilineage differentiation, surface epitopes	Higher CFU-F of MSCs from patients. CFU-F correlated with post-injury period. No difference in trilineage potential and surface epitopes	Matsukura et al. (2014)
Human	Fibrous synovium from 28 patients and fibrous and adipose synovium from six patients during total-knee arthroplasty	Suspended synovium culture model: CFU-F, trilineage differentiation, surface epitopes	MSCs from suspended synovium culture model formed multipotent colonies and expressed CD73, CD90, CD105, CD44. Higher numbers of MSCs from fibrous synovium in comparison with adipose synovium	Katagiri et al. (2017)
Human	Synovial fluid from 17 patients and synovium from eight patients with temporomandibular joint osteoarthritis, during surgical debridement or joint disk perforation	*In-vitro* expansion, CFU-F, trilineage differentiation, surface epitopes	Similar proliferative and differentiation potentials between synovial fluid and synovium-derived MSCs. Higher expression of CD73, CD90, CD105, CD44 in synovial fluid MSCs	Yao et al. (2018)
Human	Surface, stromal and perivascular regions of synovium from 10 patients during total knee arthroplasty	Immunostaining for 19 markers, cell sorting, proliferation, surface epitopes, trilineage differentiation	Surface MSCs, CD55+ CD271−; stromal MSCs, CD55− CD271−; perivascular MSCs, CD55− CD271+. Highest proliferation, chondrogenic, osteogenic potential of perivascular MSCs	Mizuno et al. (2018)
Human	Synovium collected by direct biopsy and arthroscopic trocar shaver-blade filtrate, and synovial fluid MSCs from 19 patients with traumatic injuries, traumatic inflammation, osteoarthritis	Isolation efficacy, growth kinetics, surface epitopes, culture expansion under hypoxic conditions, trilineage differentiation	Isolation efficacy >75% for all three groups. No difference in surface markers CD73, CD90, CD105. No difference in trilineage potential. Higher growth kinetics of arthroscopic shaver-blade-derived MSCs, also under hypoxic conditions	Ferro et al. (2019)

(continued)

Table 1 (continued)

Species	MSC source	Methods used	Outcomes	Reference
Mouse	Synovium from the infrapatellar fat pad, bone marrow flushed from femur and tibia, and muscle from quadriceps	*In-vitro* expansion, CFU-F, trilineage differentiation, surface epitopes	Higher proliferative potential, growth kinetics, CFU-F, PDGFRα expression of synovium MSCs compared to muscle and bone-marrow MSCs. Similar or higher osteogenic, adipogenic, and chondrogenic potential of synovium MSCs compared to muscle and bone marrow MSCs	Futami et al. (2012)
Human	Synovium from 20 osteoarthritis patients during hip arthroplasty and six healthy donors during organ donation	Immunostaining of tissue sections, trilineage differentiation, surface epitopes	Higher expression of CD44, CD90, CD105 in intimal lining in healthy donors, and diffused expression in osteoarthritic patients. Higher expression of CD44, CD90, CD105 in MSCs from patients with osteoarthritis. No comparison of trilineage potential	Hermida-Gómez et al. (2011)
Human	Synovium from eught patients with rheumatoid arthritis and eight patients with osteoarthritis, during total knee arthroplasty	Cell yields, surface markers, trilineage differentiation	No difference in cell yields, surface markers CD73, CD90, CD105, CD44 expression, and trilineage potential	Kohno et al. (2017)
Human	Synovium, bone marrow, adipose tissue, periosteum and muscle from eight donors during anterior cruciate ligament reconstruction surgery for ligament injury	Cell yields, culture expansion, trilineage differentiation, surface epitopes	Highest colony numbers and cell numbers per colony in bone marrow compared to other MSCs. MSCs from all tissues retained proliferation ability even at passage 10. Higher chondrogenic potential in synovium MSCs. No difference in surface epitopes	Sakaguchi et al. (2005)
Human	Synovium from six donors post mortem and from patients during knee arthroplasty	*In-vitro* expansion, CFU-F, senescence staining, telomerase activity, trilineage differentiation	Synovium MSCs expanded intensively in monolayers, with limited senescence, showed trilineage differentiation even at single clone level. Donor age, cell passaging, and cryopreservation did not affect multilineage potential of MSCs	De Bari et al. (2001)
Human	Infrapatellar fat pad and synovial fluid from six patients undergoing total knee replacement or anterior cruciate ligament surgery	Tissue histology, growth kinetics, trilineage differentiation, surface epitopes, stimulation with interferon-γ	MSCs from both tissues positive for CD73, CD90, CD105, and showed multipotency. Synovium-derived MSCs had significantly faster proliferation rates. MSCs from both tissues increased production of human leukocyte antigen-DR (HLA-DR) following interferon-γ stimulation	Garcia et al. (2016)

(continued)

Table 1 (continued)

Species	MSC source	Methods used	Outcomes	Reference
Human	Synovium from patients undergoing knee replacement for degenerative arthritis	CFU-F, surface epitopes, chondrogenesis	CD105, CD166, CD10, CD13, CD44, CD49a, CD73 positive and chondrogenic	Jo et al. (2007)
Human	Synovial MSCs collected during knee arthroscopy as irrigation fluid at inception, after initial inspection of the joint, after agitation of the synovium (mobilized MSCs)	CFU-F, surface epitopes, trilineage differentiation, adherence to blood clots and fibrin scaffolds	Mobilized synovial MSCs showed the highest CFU-F trilineage differentiation, standard MSC phenotype, and adhered to various fibrin scaffolds	Baboolal et al. (2018)
Rabbit	Synovial fluid from articular cavity of knee	*In-vitro* expansion, CD90 sorting, trilineage differentiation	CD90 sorted *in vitro* expanded MSCs showed expression of CD44, CD90, and trilineage potential	Jia et al. (2018a)
Human	Synovium and bone marrow from same patients undergoing total knee arthroplasty	Low-affinity nerve growth factor receptor and THY-1 sorting, CFU-F, trilineage differentation	Low-affinity nerve growth factor receptor and THY-1 sorted synovial MSCs showed higher CFU-F and enhanced adipogenic and chondrogenic differentiation in comparison with bone-marrow-derived MSCs	Ogata et al. (2015)

CFU-F, colony forming unit in fibroblast assay; PDGFRα, platelet-derived growth factor receptor α.

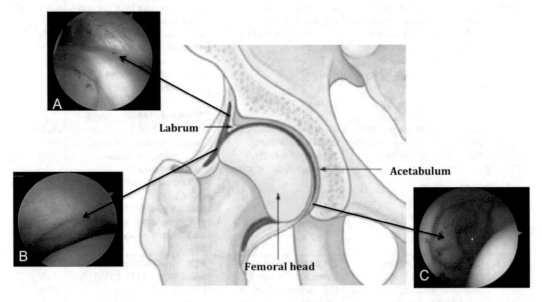

Fig. 1 Sites of synovium in the hip joint shown to be reservoirs of MSCs, accessible by arthroscopy
A – periacetabular sulcus
B – inner surface of the joint capsule
C – cotyloid fossa

healthy controls were post-mortem donors with no evidence of joint disorders (De Bari et al. 2001; Hermida-Gómez et al. 2011). Interestingly, Hermida-Gómez et al. (2011) showed that synovium in patients with osteoarthritis who were undergoing total hip replacement contained more cells that expressed MSC markers than synovium from joints with no cartilage damage. These latter cells were obtained from healthy donors who were undergoing organ donation (Hermida-Gómez et al. 2011). No differences in multilineage potential was seen between these groups, which would suggest that MSCs from patients with osteoarthritis have functional deficiencies. Ferro et al. (2019) investigated whether different methods of synovium biopsy affect MSCs, here as direct biopsy *versus* arthroscopic trocar shaver-blade filtrate. They compared MSCs derived from synovium using these two methods of synovium collection with MSCs isolated from synovial fluid. All of their samples showed MSC-like characteristics, while arthroscopic shaver-blade-derived MSCs also showed higher proliferation (Ferro et al. 2019). Apart from synovial tissue, MSCs can also be isolated from synovial fluid from the knee (Morito et al. 2008; Matsukura et al. 2014; Garcia et al. 2016) and temporomandibular joint (Yao et al. 2018). Similar to synovium, synovial fluid in patients with osteoarthritis has also been shown to contain greater numbers of MSCs than for healthy volunteers, although with no difference in their multipotency (Morito et al. 2008; Matsukura et al. 2014).

In addition to the MSC markers CD73, CD90, and CD105 that were suggested by the ISCT (Dominici et al. 2006), CD44 (i.e., the hyaluronan receptor) is frequently analyzed in *in-vitro* studies. Hyaluronan and hyaluronic acid bind to CD44, which appears to be the main factor responsible for the constant volume of the synovial fluid that serves as a cushion for synovial tissue and as a reservoir of lubricant for cartilage (Smith 2011). Synovial MSCs express high levels of CD44 (Hermida-Gómez et al. 2011; Kohno et al. 2017; Katagiri et al. 2017; Yao et al. 2018). CD44 has an important role during synovial joint development, prior to cavitation. It is

expressed in the interzone and the articular surfaces. After cavitation, hyaluronic acid binds to CD44 on synovium and articular surfaces. This facilitates tissue separation and helps to create a functional joint cavity (de Sousa et al. 2014). On this basis, CD44 might be a true marker of synovial MSCs, which are retained from early embryonic development through adult life. However, as with other MSC markers, CD44 is not exclusive to synovial MSCs.

Mizuno et al. (2018) compared MSCs derived from different regions of synovium; i.e., from the surface, stromal, and perivascular regions. They reported that combination of CD55 and CD271 as markers can differentiate between synovial MSCs from these three regions. Moreover, they also showed that MSCs from the perivascular region that were CD55 negative and CD271 positive had the highest proliferative and chondrogenic potentials. Garcia et al. (2016) also challenged MSCs from infrapatellar fad pads and synovial fluid with interferon γ in an *in-vitro* setting. Both MSCs responded with increased levels of HLA-DR, thus showing an immunomodulatory capacity that confers further therapeutic value to these cells with regard to treatment of the inflammatory aspect of osteoarthritis.

Synovium-derived MSCs have also been isolated from knees of rabbits (Jia et al. 2018a) and mice (Futami et al. 2012). The most challenging step here was the dissection of synovium from the knees of these small animal species. Indeed, larger animals are more commonly used for *in-vivo* studies of synovium-derived MSCs.

In summary, most of the *in-vitro* studies to demonstrate that MSC-like cells can be isolated from adult synovium have investigated the proliferation and trilineage potential, and the limited set of standard markers of CD73, CD90, and CD105, plus the additional marker CD44, of synovium or synovial fluid-derived MSCs in human knee joints. There is a lack of studies that have compared synovium of patients with osteoarthritis and healthy donors to identify changes in these MSCs that would provide better understanding of the mechanisms underlying joint degeneration in osteoarthritis. Future research on synovial MSCs requires verification of the MSCs markers identified in animal studies (Zhou et al. 2014; Worthley et al. 2015; Roelofs et al. 2017) and in tissue sources other than synovium (Chan et al. 2018). Immunomodulation of synovial MSCs, and in particular their immunosuppression, needs to be evaluated using appropriate *in-vitro* tests. It might also be of interest to investigate the potential of synovial MSCs for tendinocyte differentiation, due to the increasing prevalence of tendon and ligament degeneration, and for myocyte differentiation, due to their potential for muscle repair.

2.2 In-Vivo Animal Studies

The *in-vivo* animal studies carried out to date using synovium-derived MSCs are summarized in Table 2. Synovium-derived cell therapies have most commonly been tested in mouse, rat, rabbit, pig, and equine models of osteoarthritis. To define the regenerative features of MSCs *in vivo*, joint defects must first be created in these animals. Osteoarthritis in animal models is simulated by creating full-thickness cartilage defects or ligament injuries (Kuyinu et al. 2016). Due to the ease of access for surgical induction of osteoarthritis and implantation of MSCs, the knee is the preferred joint in all of these animal species.

In addition to determination of the *in-vivo* effects of implanted MSCs on joint regeneration, the major advantage of these studies is that they also provide some *in-vitro* assessment of the implanted cells, to demonstrate that they were indeed MSCs. This is feasible because the majority of the *in-vivo* studies with animals have used *in-vitro* expanded cells. Due to the experimental nature of these studies, no strict standards are required, such as good manufacturing practice, unlike for clinical studies using *in-vitro* expanded MSCs. The majority of these expanded MSCs are used for implantation, while the standard MSC tests are performed on a small aliquot (i.e., for surface MSC markers, trilineage differentiation, chondrogenesis).

In most of these studies, pure preparations of *in-vitro* expanded MSCs have been implanted, while studies that have embedded MSCs in some kind of a scaffold or have preconditioning them in chondrogenic media have been infrequent. Jia et al. (2018a, b) showed that prechondrogenesis of synovial-fluid-derived MSCs did not improve cartilage repair in a rabbit chondral-defect model. This observation might be explained by the loss of paracrine function of the MSCs by pre-differentiation.

There is some evidence that rather than contributing to tissue repair themselves, MSCs have an immunomodulatory action and stimulate the endogenous cells of the recipients toward tissue repair (for review, see Čamernik et al. 2018). However, apart from rare *in-vitro* studies (e.g., Garcia et al. 2016), little is known about the immunomodulation of synovial MSCs and their potential to supress inflammation.

To evaluate the *in-vivo* effects of implanted MSCs on cartilage regeneration following joint injury, the preferable methods of assessment are arthroscopy, histology, and magnetic resonance imaging. Most studies have investigated these effects at 4–12 weeks post injury or post MSC implantation. Koga et al. (2008) compared the *in-vivo* cartilage regeneration of MSCs isolated from bone marrow, synovium, adipose tissue, and muscle of adult rabbits. They embedded these MSCs in collagen gel and transplanted them into full-thickness cartilage defects of rabbit knees, where synovium-derived and bone marrow-derived MSCs formed a more abundant cartilage matrix than adipose-derived and muscle-derived MSCs. When implanting synovium-derived MSCs into osteochondral defects of pigs, Nakamura et al. (2012) reported that leaving a suspension of synovial MSCs on the cartilage defect for 10 min before closure of the wound allowed the cells to adhere within the defect first, which promoted improved cartilage repair.

In the light of the dilemma of whether to use autologous or allogeneic MSCs for tissue regeneration, there have also been studies where xenogenic cell therapies were used (Ozeki et al. 2016; Zayed et al. 2018; Neybecker et al. 2018). None of these studies observed adverse events associated with xenogenic cell therapies. Ozeki et al. (2016) and Zayed et al. (2018) also observed some degree of chondroprotection in their animal

Table 2 Summary of the *in-vivo* animal studies

Species	Number (n)	Model of osteoarthritis	Cell therapy used	Methods used	Duration	Outcomes	Reference
Rabbit	NA	Full thickness osteochondral defects in trochlear groove	Bone marrow, synovium, adipose tissue and muscle-derived MSCs embedded in collagen gel	*In vitro*: Proliferation and chondrogenesis	4, 12 weeks	*In vitro*: Higher proliferation of synovium and muscle-derived MSCs, more cartilage formed by bone marrow and synovium-derived MSCs	Koga et al. (2008)
				In vivo: Histology of the cartilage matrix		*In vivo*: More cartilage produced by synovium and bone marrow-derived MSCs	
Pig	16	Full thickness osteochondral defects in weight-bearing area of medial femoral condyles	Bone marrow, periosteum, muscle, adipose tissue, suprapatellar pouch synovium-derived MSCs	*In vitro*: Chondrogenesis	4, 12 weeks	*In vitro*: Higher chondrogenic potential of synovial MSCs	Nakamura et al. (2012)
				In vivo: Arthroscopy, histology, MRI		*In vivo*: Oswestry arthroscopy, ICRS and modified Wakitani score higher with synovial MSCs. Placing suspension of synovial MSCs on cartilage defect 10 min before wound closure promoted cartilage repair	
Rabbit	NA	Patellar groove cartilage defects	Autologous synovial fluid MSCs cultured for 3 weeks *in vitro* in either chondrogenic or normal medium, intra-articular injection once a week for 4 weeks	*In vitro*: Surface epitopes, trilineage differentiation	4 weeks	*In vitro*: Expression of CD73, CD90, CD44, and trilineage differentiation	Jia et al. (2018b)
				In vivo: Arthroscopy, histology, MRI		*In vivo*: Hyaline-like cartilage was observed in defects treated with synovium-fluid-derived MSCs in comparison with fibrocartilage formed in the defects treated with MSCs in chondrogenic medium	

(continued)

Table 2 (continued)

Species	Number (n)	Model of osteoarthritis	Cell therapy used	Methods used	Duration	Outcomes	Reference
Mouse	NA	Joint surface knee injury (medial parapatellar arthrotomy)	C57BL/6 mice received two different nucleosides in drinking water to label MSCs in synovium *in vivo*	Histology and immunohistochemistry	4, 8, 12 days post-injury	First evidence of resident MSCs in knee joint synovium that undergo proliferation and chondrogenic differentiation following injury *in vivo*	Kurth et al. (2011)
Equine, rat	NA	Full thickness articular cartilage defects in trochlear grooves of distal femur in rats	Xenogenic implantation of equine-derived bone marrow and synovial fluid MSCs encapsulated in neutral agarose scaffold	*In vitro*: Proliferation, viability, chondrogenesis	1, 12 weeks	*In vitro*: Higher chondrogenic potential of synovial fluid MSCs. MSCs seeded on agarose construct were metabolically active and viable	Zayed et al. (2018)
				In vivo: Macroscopic and histology		*In vivo*: Better macroscopic and histological result of articular cartilage in knee treated with MSCs than in control	
Rats	40	Partial thickness cartilage defect on the medial femoral condyle	Intra-articular injection of allogeneic synovium-derived *in-vitro* expanded MSCs at three different times (injury, 1, 2 weeks post injury)	*In vitro*: Surface markers, trilineage differentiation	6 weeks post injury	*In vitro*: MSCs expressed CD90 and demonstrated trilineage differentiation potential	Enomoto et al. (2018)
				In vivo: Histology		*In vivo*: Significantly higher histological score in group with MSCs at time of injury. MSCs distributed in synovium, not in cartilage surrounding defective area	
Mouse	52	Focal, full-thickness knee cartilage defect	Sca-1+ synovial MSCs from MRL/MpJ "super-healer" and C57BL6 mice	*In vitro*: Trilineage differentiation	At injury, 2, 4 weeks post injury	*In vitro*: Similar trilineage potential of synovial MSCs from MRL and C57BL6 mice	Mak et al. (2016)
				In vivo: Histology, MRI		*In vivo*: Increased cartilage repair 4 weeks post injury with MSCs from both mice. C57BL6 mice injected with MRL-derived MSCs showed greatest change in MRI signal intensity in defect site, in comparison to imaging directly after injury	

Species	Number	Model	Method	Analysis	Timepoint	Results	Reference
Rat	76	Anterior cruciate ligament transection in Lewis rats	Xenogenic implantation of human synovial MSCs (10^6); periodic *versus* single injection	*In vivo*: Histology and flow cytometry	12 weeks	Periodic injections maintained MSCs in knees and showed higher chondroprotective effects in comparison with single MSCs migrated mainly into synovium and retained undifferentiated	Ozeki et al. (2016)
Mouse	12	Joint surface knee injury (medial parapatellar arthrotomy)	Gdf5-lineage of synovial MSCs (endogenous and after allogeneic transplantation)	*In vitro*: Trilineage differentiation, synoviogenesis / *In vivo*: Histology	4, 8 weeks	Gdf5-lineage cells contribute to cartilage repair via yes-associated protein (yap)	Roelofs et al. (2017)
Rat	16	Anterior cruciate ligament transection	Xenogenic injections of human synovial fluid-derived MSCs obtained from donors with advanced knee osteoarthritis in rat knee at day 7 and 14 following anterior cruciate ligament transection	*In vitro*: Immunophenotype, trilineage differentiation, chondrogenic induction in collagen sponges / *In vivo*: Macroscopic, histology	4, 8 weeks post anterior cruciate ligament transection	*In vitro*: CD73+, CD90+, CD105+, CD34−, CD45− immunophenotype and multilineage. Potency, chondrogenic induction (TGF-β1 ± BMP-2) in collagen sponges induced expression of chondrogenic and extracellular matrix genes / *In vivo*: No chondroprotection nor inflammation in rat knees injected with MSCs	Neybecker et al. (2018)

NA, not assigned; ICRS, International Cartilage Regeneration & Joint Preservation Society; MRI, magnetic resonance imaging; TGF-β1, transforming growth factor β1; BMP-2, bone morphogenetic factor 2.

models of osteoarthritis, while in contrast to chondroprotection, Neybecker et al. (2018) observed lack of inflammation, which suggested immunosuppressive effects of this cell therapy. Mak et al. (2016) investigated Sca-1 positive, chondrogenesis-capable mouse synovial MSCs that were derived from an MRL/MpJ 'super-healer' mouse strain, to determine whether these might regenerate cartilage injury better than those derived from 'nonhealer' C57BL6 mice. However, as no differences were seen, they suggested that regardless of strain background, synovial MSCs have beneficial effects when injected into an injured joint.

Enomoto et al. (2018) evaluated the effects of timing of intra-articular injection of MSCs on healing of a partial-thickness cartilage defect in rat. Synovium of infrapatella fat-pad-derived MSCs were implanted at three different times: time of injury, and 1 week and 2 weeks post injury. In their hands, the early intra-articular injection of MSCs enhanced cartilage healing. They also traced fluorescently labeled MSCs 1 day after implantation, and interestingly, they found them distributed in synovium, not in the cartilage surrounding the defect. Kurth et al. (2011) were the first to evaluate endogenous MSCs, and they showed that adult synovium is a site of functional MSCs that contributes to cartilage repair following joint surface injury (Kurth et al. 2011).

It has also been shown that the subpopulation of synovial MSCs with a Gdf5 lineage underpins the synovial hyperplasia and contributes to cartilage repair in a mouse model of join surface injury (Roelofs et al. 2017). The transcriptional co-factor known as Yes associated protein (Yap) was up-regulated after injury, and its conditional ablation in the Gdf5-lineage cells prevented synovial lining hyperplasia and decreased the contribution of the Gdf5-lineage cells to cartilage repair.

Together, these studies indicated the pivotal role of the resident MSCs in the joint, and in particular in synovium, which become activated upon a stimulus such as articular cartilage injury. These then respond by either contributing to cartilage repair themselves or through their paracrine functions, such as immunosuppression, to promote joint regeneration. Immunomodulatory potential of human synovium-derived MSCs has been shown in collagen-induced arthritis in mice (Yan et al. 2017). Human synovium-derived MSCs injected into inflamed joints of mice suppressed immune responses via immunoregulatory cell expansion. Further animal studies are awaited to demonstrate the same for osteoarthritis.

Another study similar to that of Enomoto et al. (2018) investigated the effects of single and repetitive intra-articular injections of MSCs in a rat osteoarthritis model (Ozeki et al. 2016). Histological analysis of the femoral and tibial cartilage showed that a single injection of the MSCs was ineffective, while weekly injections for 12 weeks had significant chondroprotective effects.

In summary, most of the *in-vivo* studies have investigated the effects of *in-vitro* expanded synovial MSCs on regeneration of cartilage injuries in models from small animals, such as mice and rats. Due to the experimental nature of these studies, these can provide the evidence that the implanted cells are indeed MSC-like cells. Moreover, they allow detailed analysis of the regenerated tissue, such as the presence of collagen type II. They can also provide the proof of concept that xenogeneic cell implantation also works for cartilage regeneration. However, most of the implanted MSCs were collected from 'healthy' animal donors. It would be particularly interesting to determine whether synovial MSCs derived from mouse models of osteoarthritis or from 'aged' mice can still regenerate cartilage. Further animal studies that investigate the cartilage regeneration capabilities of recently identified specific MSC subpopulations (Chan et al. 2018) are also awaited.

Finally, what needs to be learnt from the preclinical studies is not only the optimal tissue source(s) of MSCs, but also the timing and frequency of the implanted cell therapies. High quality experimental studies and efforts for effective translation from preclinical studies to clinical trials are still required (Xing et al. 2018).

3   Clinical Studies

Several clinical studies have shown the promising potential of MSC-based therapies in the treatment of osteoarthritis and traumatic cartilage lesions, with minimal adverse effects seen (Harrell et al. 2019; Ha et al. 2019; Reissis et al. 2016). Most of these studies have been conducted on knees using MSCs derived from bone marrow or adipose tissue. Unfortunately, the present critical systematic reviews do not provided sufficient level of evidence that intra-articular–derived MSCs are indeed effective for hyaline cartilage regeneration or repair, for symptoms attenuation, and for functional restoration (Ha et al. 2019).

It has been estimated that the great majority of the available clinical studies have some risk of reporting bias. Further, the optimum source and concentration of MSCs to fulfill clinical expectations remain to be identified. Also, subpopulations of MSCs differ in their chondrogenic differentiation potential and immunomodulatory capability (Im et al. 2005; Waldner et al. 2018).

These clinical studies are also difficult to compare due to the inconsistency in their delivery methods into the affected joints (Ha et al. 2019). MSCs applied directly to the site of a lesion using three-dimensional (3D) scaffolds are believed to have better healing potential, compared to MSCs injected percutaneously (Coelho et al. 2012). Furthermore, percutaneous injections of MSCs might represent a risk for undesired dissemination of the cells into noncartilage tissue, although this remains to be shown (Roffi et al. 2018).

Some clinical studies include adjuvant procedures that can have significant influence on the outcome of the treatment, and might also increase the risk of bias; e.g., concomitant injections of platelet-rich plasma or hyaluronic acid, microfracture of the subchondral bone, or corrective osteotomies (Ha et al. 2019). Similarly, to evaluate the outcomes of treatments, inconsistent, and on several occasions irregular, tools have been used (Ha et al. 2019). Some studies have provided information about the postoperative conditions of the cartilage using MRI or even

histological evaluation of biopsies obtained during second-look arthroscopy, although the long-term fate of the treated cartilage remains unclear.

Autologous MSCs have also been used in elderly patients, which again results in outcome inconsistencies between trials due to the age-related decrease in MSC proliferation. Unlike pharmaceutical drugs with defined chemical structures and functions, identification and functional characterisation of MSCs have not yet been standardized, thus making it difficult to produce MSCs with consistent biological activities on a large scale for clinical trials (Lee and Wang 2017).

Clinical research on synovium-derived MSCs is still in its infancy. Review of the currently available literature revealed only three reports of their clinical use in the knee (Table 3). Only two of these reports are clinical studies on case series, while the other is a case report.

Sekiya et al. (2015) reported on the injection of synovium-derived MSCs that had been expanded in 10% autologous human serum. This study included 10 adult patients with International Cartilage Repair Society grade 3 or 4 focal lesions of the femoral condyle, but no control group (Sekiya et al. 2015). Suspensions of synovium-derived MSCs were applied arthroscopically directly into the lesion, which was positioned facing upward to allow 10 min of static exposure. At the final follow-up (median, 48 months), all 10 of these patients reported significant symptomatic improvements. Their MRI scores also significantly increased after the treatments. Biopsy specimens of cartilage were obtained arthroscopically in only four of these patients, and histological evaluation revealed hyaline cartilage in three of them. In five patients, concomitant anterior cruciate ligament reconstruction was carried out, and in two, meniscal repair, and these might have had a significant influence on the final outcome, thus presenting high risk of bias. Here, for implantation, the authors prepared passage 0 MSCs that were expanded with autologous human serum over 14 days. In agreement with their previous study, these synovium MSCs showed better potential for expansion, compared to bone-derived MSCs (Nimura et al. 2008).

Table 3 Summary of clinical studies using synovium-derived MSCs

Study design	Diagnosis	Number of patients (n)	Age (years)	Cell therapy	Route of transplantation	Follow-up	Treatment outcome	Reference
Clinical study	Trauma-induced femoral condyle defects	10	20–43	Synovial MSCs expanded with autologous serum	Arthroscopic implantation of cell suspension for 10 min using syringe	37–80 months	MRI score: Significantly increased after cell therapy. Arthroscopy: Improved quality of cartilage defect. Histology: Hyaline cartilage in three and fibrous in one patient. Lysholm score: Significantly increased. Tegner activity level scale: Did not decrease	Sekiya et al. (2015)
Case series (level of evidence, 4)	Isolated full-thickness cartilage defects of knee (<5cm²; ICRS grade III, IV)	5	28–46	Autologous synovial membrane-derived MSCs, cultured, scaffold free	Implantation without use of sutures or fixation glue	24 months	Safety: No adverse events recorded. Self-assessed clinical scores: Significantly improved. Arthroscopy and MOCART: Secure defect filling confirmed. Tissue biopsy: Repair tissue with composition and structure of hyaline cartilage formed	Shimomura et al. (2018)
Case report, controlled laboratory study	Isolated full-thickness cartilage defects of knee (<5cm²; ICRS grade III or IV)	1	34	Autologous synovial membrane-derived MSCs, cultured, scaffold free, and high dose steroid therapy	Implantation without sutures or fixation glue	3, 7 weeks	MSCs at 3 weeks post high-dose steroid therapy failed to generate functional construct *in vitro*, and recovered 7 weeks after therapy	Yasui et al. (2018)

ICRS, International Cartilage Regeneration & Joint Preservation Society; MOCART, magnetic resonance observation of cartilage repair tissue scoring.

Passage 0 cells are also safer than cells passaged several times, in terms of potential development of chromosome abnormalities (Ermis et al. 1995). The possibility to use allogeneic MSCs might offer huge advantages in terms of the concentration of potent cells in suspension, although further studies are needed to demonstrate the safety of this alternative to autologous products (Vangsness Jr et al. 2014).

More recently, Shimomura et al. (2018) reported on a pilot study of implantation of a scaffold-free tissue-engineered construct that was generated from autologous synovium-derived MSCs, for the repair of focal chondral lesions in the knee (Shimomura et al. 2018). Although this study only included five patients, their 2-year follow-up revealed evidence of secure defect filling. The histology of the biopsy specimens obtained during second-look arthroscopy indicated repair of the lesion by hyaline-like constructs.

Additionally, a case report was published by Yasui et al. (2018). They showed that high-dose steroid therapies can compromise synovial MSCs. Hence, they suggested that the drug-use profiles of MSC donors and recipients must be carefully monitored to optimize the opportunities for successful repair of damaged tissues.

Due to the scarce clinical experience and poor quality of available reports, at this stage it is not possible to form any conclusions on the clinical potential of synovium-derived MSCs in regeneration or repair of chondral lesions in human joints. However, these early clinical experiences have yielded positive data with minimal adverse effects.

Just recently, an innovative approach for cartilage regeneration was proposed based on articular injection of the bioactive cell-free formulation BIOF2, which can promote expansion and chondrogenic differentiation of endogenous synovial MSCs. However, the clinical relevance of this new concept remains to be determined (Delgado-Enciso et al. 2018).

Randomised controlled trials are required to objectively compare clinical efficacy and long-term safety of various treatment protocols. With clinical research continuing to evolve and address these challenges, it is likely that MSCs will become integrated into routine clinical practice in the near future (Kon et al. 2015). Randomised control trials are required to objectively compare clinical efficacy and long-term safety of various treatment protocols. As clinical research continues to evolve and address these challenges, it is likely that MSCs will become integrated into routine clinical practice in the near future (Kon et al. 2015).

4 Synovium-Derived MSCs – What Lies Ahead?

Based on the current evidence from basic and clinical studies, it is reasonable to have expectations that MSCs from synovium and from other tissue sources will be (part of) the future therapy for degenerative joint disorders. Their advantages provide solid grounds for further clinical trials, which include ease of access, isolation, cultivation, and expansion, along with their regenerative, anti-inflammatory, and immunomodulatory properties. However, before reaching this stage, several problems need to be addressed.

First, we need to translate the knowledge from animal studies to humans. There is evidence from animal studies that specific populations of MSCs exist in adult synovium that can repair cartilage, and can even form new joints. In humans, these populations are largely undefined, and further studies to identify human markers of these populations are awaited.

Secondly, future studies will need to focus on the development of new techniques for the minimal invasive harvesting of these cells, and for their transplantation to damaged joints. The majority of studies to date require cell harvesting and transplantation that are associated with high costs, even if culture expansion and good manufacturing practice can be avoided. Currently, intra-operative cell therapies represent the easiest route of cell implantation, although more remains to be done for better selection of the MSC-like cells for these therapies (Coelho et al. 2012). Alternatively, less invasive and more cost effective ways need to be found to stimulate the endogenous MSCs toward tissue regeneration. Biophysical stimulation such as extracorporeal shock-wave treatments and pulsed electromagnetic fields can be used to enhance endogenous MSCs to maintain healthy joints and prevent osteoarthritis (Viganò et al. 2016). Another possibility is to stimulate the differentiation potential of the endogenous MSCs with pharmacological approaches, as has been shown for bone-marrow MSCs (Johnson et al. 2012; Heck et al. 2017). The prerequisite to elicit repair through activation of endogenous MSCs is, of course, to understand them fully first, both in health and disease.

Thirdly, MSC identification needs to be standardized, particularly for clinical use. Currently, hundreds of clinics and clinical trials are using the term 'human MSCs' with very few, if any, that have focused on the *in-vitro* multipotent capacities of these cells. Hence the term 'stem cells' is easily and largely misused for the direct-to-consumer marketing of unapproved stem-cell treatments for numerous medical conditions (Sipp et al. 2018). The term 'stem cells' is indeed misleading, as it implies that the patients will receive direct medical benefits because these cells will differentiate into regenerating tissue-producing cells. This has been acknowledged recently by the very father of the name MSCs, Arnold Caplan (Caplan 2017). Based on his suggestions, here the term MSCs should be replaced by 'medicinal signaling cells', to more accurately reflect that these cells home in on sites of injury and disease and secrete bioactive factors that are immunomodulatory and trophic (i.e., regenerative). This thus means that these cells release therapeutic drugs *in situ* that are medicinal. It is, indeed, the patient's own site-specific and tissue-specific resident stem cells that construct the new tissue following stimulation by the bioactive factors secreted by the exogenously supplied MSCs (Caplan 2017).

Fourthly, and certainly not the last of the issues associated with MSCs, the procedures and methods of their *in-vitro* expansion, storage, and transport need to be further optimized (and standardized) to develop safe and effective cell therapies.

To summarize, MSCs are truly the 'newcomers to the Club' (Bianco et al. 2013). More time and serious efforts are needed to produce evidence-based regenerative medicine for their use in degenerative joint disorders. Furthermore, synovium-derived MSCs represent the most recent newcomers to the 'MSC Club', and more basic and clinical studies are awaited before we can anticipate the exploitation of their full potential as a treatment for regeneration of cartilage and other joint structures.

Acknowledgments Janja Zupan was funded by UK Arthritis Research (2016–2018) and is currently part of the P3-0298 Research Programme 'Genes, hormones and personality changes in metabolic disorders', funded by the Slovenian Research Agency, and ARTE Project EU INTERREG Italia-Slovenija 2014–2020. The authors would like to thank Chris Berrie for scientific English editing of the manuscript.

References

Arthritis Research UK. (2018). *State of musculoskeletal health 2018*

Baboolal TG, Khalil-Khan A, Theodorides AA, Wall O, Jones E, McGonagle D (2018) A novel arthroscopic technique for intraoperative mobilization of synovial mesenchymal stem cells. Am J Sports Med 46:1–9. https://doi.org/10.1177/0363546518803757

Bianco P, Cao X, Frenette PS, Mao JJ, Robey PG, Simmons PJ, Wang CY (2013) The meaning, the sense and the significance: translating the science of mesenchymal stem cells into medicine. Nat Med 19:35–42. https://doi.org/10.1038/nm.3028

Čamernik K, Barlič A, Drobnič M, Marc J, Jeras M, Zupan J (2018) Mesenchymal stem cells in the musculoskeletal system: from animal models to human tissue regeneration? Stem Cell Rev Rep 14(3):346–369. https://doi.org/10.1007/s12015-018-9800-6

Campbell TM, Churchman SM, Gomez A, Mcgonagle D, Conaghan PG, Ponchel F, Jones E (2016) Mesenchymal stem cell alterations in bone narrow lesions in patients with hip osteoarthritis. Arthritis Rheumatol 68:1648–1659. https://doi.org/10.1002/art.39622

Caplan AI (2017) Mesenchymal stem cells: time to change the name! Stem Cells Transl Med 6:1445–1451. https://doi.org/10.1002/sctm.17-0051

Chan CKF, Gulati GS, Sinha R, Tompkins JV, Lopez M, Carter AC, Ransom RC, Reinisch A, Wearda T, Murphy M, Brewer RE, Koepke LS, Marecic O, Manjunath A, Seo EY, Leavitt T, Lu W-J, Nguyen A, Conley SD, Salhotra A, Ambrosi TH, Borrelli MR, Siebel T, Chan K, Schallmoser K, Seita J, Sahoo D,

Goodnough H, Bishop J, Gardner M, Majeti R, Wan DC, Goodman S, Weissman IL, Chang HY, Longaker MT (2018) Identification of the human skeletal stem cell. Cell 175:43–56.e21. https://doi.org/10.1016/j.cell.2018.07.029

Coelho MB, Cabral JMS, Karp JM (2012) Intraoperative stem cell therapy. Annu Rev Biomed Eng 14:325–349. https://doi.org/10.1146/annurev-bioeng-071811-150041

De Bari C, Roelofs AJ (2018) Stem cell-based therapeutic strategies for cartilage defects and osteoarthritis. Curr Opin Pharmacol 40:74–80. https://doi.org/10.1016/J.COPH.2018.03.009

De Bari C, Dell'Accio F, Tylzanowski P, Luyten FP (2001) Multipotent mesenchymal stem cells from adult human synovial membrane. Arthritis Rheum 44:1928–1942. https://doi.org/10.1002/1529-0131(200108)44:8<1928::AID-ART331>3.0.CO;2-P

de Sousa E, Casado P, Neto V, Duarte ME, Aguiar D (2014) Synovial fluid and synovial membrane mesenchymal stem cells: latest discoveries and therapeutic perspectives. Stem Cell Res Ther 5:112. https://doi.org/10.1186/scrt501

Delgado-Enciso I, Paz-Garcia J, Valtierra-Alvarez J, Preciado-Ramirez J, Almeida-Trinidad R, Guzman-Esquivel J, Mendoza-Hernandez MA, Garcia-Vega A, Soriano-Hernandez AD, Cortes-Bazan JL, Galvan-Salazar HR, Cabrera-Licona A, Rodriguez-Sanchez IP, Martinez-Fierro ML, Delgado-Enciso J, Paz-Michel B (2018) A phase I-II controlled randomized trial using a promising novel cell-free formulation for articular cartilage regeneration as treatment of severe osteoarthritis of the knee. Eur J Med Res 23(1):52. https://doi.org/10.1186/s40001-018-0349-2

Dominici M, Le Blanc K, Mueller I, Slaper-Cortenbach I, Marini F, Krause D, Deans R, Keating A, Prockop D, Horwitz E (2006) Minimal criteria for defining multipotent mesenchymal stromal cells. The International Society for Cellular Therapy position statement. Cytotherapy 8:315–317. https://doi.org/10.1080/14653240600855905

Enomoto T, Akagi R, Ogawa Y, Yamaguchi S, Hoshi H, Sasaki T, Sato Y, Nakagawa R, Kimura S, Ohtori S, Sasho T (2018) Timing of intra-articular injection of synovial mesenchymal stem cells affects cartilage restoration in a partial thickness cartilage defect model in rats. Cartilage:194760351878654. https://doi.org/10.1177/1947603518786542

Ermis A, Henn W, Remberger K, Hopf C, Hopf T, Zang KD (1995) Proliferation enhancement by spontaneous multiplication of chromosome 7 in rheumatic synovial cells in vitro. Hum Genet 96(6):651–654

Fellows CR, Matta C, Zakany R, Khan IM, Mobasheri A (2016) Adipose, bone marrow and synovial joint-derived mesenchymal stem cells for cartilage repair. Front Genet 7:1–20. https://doi.org/10.3389/fgene.2016.00213

Ferro T, Santhagunam A, Madeira C, Salgueiro JB, da Silva CL, Cabral JMS (2019) Successful isolation and ex-vivo expansion of human mesenchymal stem/stromal cells obtained from different synovial tissue-derived (biopsy) samples. J Cell Physiol 234 (4):3973–3984. https://doi.org/10.1002/jcp.27202

Friedenstein AJ, Chailakhjan RK, Lalykina KS (1970) The development of fibroblast colonies in monolayer cultures of Guinea-pig bone marrow and spleen cells. Cell Prolif 3:393–403. https://doi.org/10.1111/j.1365-2184.1970.tb00347.x

Fuchs E, Horsley V (2011) Ferreting out stem cells from their niches. Nat Cell Biol 13:513–518. https://doi.org/10.1038/ncb0511-513

Futami I, Ishijima M, Kaneko H, Tsuji K, Ichikawa-Tomikawa N, Sadatsuki R, Muneta T, Arikawa-Hirasawa E, Sekiya I, Kaneko K (2012) Isolation and characterization of multipotential mesenchymal cells from the mouse synovium. PLoS One 7(9):e45517. https://doi.org/10.1371/journal.pone.0045517

Garcia J, Wright K, Roberts S, Kuiper JH, Mangham C, Richardson J, Mennan C (2016) Characterisation of synovial fluid and infrapatellar fat-pad-derived mesenchymal stromal cells: the influence of tissue source and inflammatory stimulus. Sci Rep 6:24295. https://doi.org/10.1038/srep24295

Ha CW, Park YB, Kim SH, Lee HJ (2019) Intra-articular mesenchymal stem cells in osteoarthritis of the knee: a systematic review of clinical outcomes and evidence of cartilage repair. Arthroscopy 35(1):277–288.e2. https://doi.org/10.1016/j.arthro.2018.07.028

Harrell CR, Markovic BS, Fellabaum C, Arsenijevic A, Volarevic V (2019) Mesenchymal stem cell-based therapy of osteoarthritis: current knowledge and future perspectives. Biomed Pharmacother 109:2318–2326. https://doi.org/10.1016/j.biopha.2018.11.099

Heck BE, Park JJ, Makani V, Kim EC, Kim DH (2017) PPAR-δ agonist with mesenchymal stem cells induces type II collagen-producing chondrocytes in human arthritic synovial fluid. Cell Transplant 26:1405–1417. https://doi.org/10.1177/0963689717720278

Hermida-Gómez T, Fuentes-Boquete I, Gimeno-Longas MJ, Muiños-López E, Díaz-Prado S, De Toro FJ, Blanco FJ (2011) Quantification of cells expressing mesenchymal stem cell markers in healthy and osteoarthritic synovial membranes. J Rheumatol 38:339–349. https://doi.org/10.3899/jrheum.100614

Im GI, Shin YW, Lee KB (2005) Do adipose tissue-derived mesenchymal stem cells have the same osteogenic and chondrogenic potential as bone marrow-derived cells? Osteoarthr Cartil 13(10):845–853. https://doi.org/10.1016/j.joca.2005.05.005

Jia Z, Liang Y, Li X, Xu X, Xiong J, Wang D, Duan L (2018a) Magnetic-activated cell sorting strategies to isolate and purify synovial-fluid-derived mesenchymal stem cells from a rabbit model. J Vis Exp:e57466–e57466. https://doi.org/10.3791/57466

Jia Z, Liu Q, Liang Y, Li X, Xu X, Ouyang K, Xiong J, Wang D, Duan L (2018b) Repair of articular cartilage defects with intra-articular injection of autologous rabbit synovial fluid-derived mesenchymal stem cells. J Transl Med 16:123. https://doi.org/10.1186/s12967-018-1485-8

Jo CH, Ahn HJ, Kim HJ, Seong SC, Lee MC (2007) Surface characterization and chondrogenic differentiation of mesenchymal stromal cells derived from synovium. Cytotherapy 9:316–327. https://doi.org/10.1080/14653240701291620

Johnson K, Zhu S, Tremblay MS, Payette JN, Wang J, Bouchez LC, Meeusen S, Althage A, Cho CY, Wu X, Schultz PG (2012) A stem-cell-based approach to cartilage repair. Science 336:717–721. https://doi.org/10.1126/science.1215157

Katagiri K, Matsukura Y, Muneta T, Ozeki N, Mizuno M, Katano H, Sekiya I (2017) Fibrous synovium releases higher numbers of mesenchymal stem cells than adipose synovium in a suspended synovium culture model. Arthrosc J Arthrosc Relat Surg 33:800–810. https://doi.org/10.1016/J.ARTHRO.2016.09.033

Koga H, Muneta T, Nagase T, Nimura A, Ju Y-J, Mochizuki T, Sekiya I (2008) Comparison of mesenchymal tissues-derived stem cells for in-vivo chondrogenesis: suitable conditions for cell therapy of cartilage defects in rabbit. Cell Tissue Res 333:207–215. https://doi.org/10.1007/s00441-008-0633-5

Kohno Y, Mizuno M, Ozeki N, Katano H, Komori K, Fujii S, Otabe K, Horie M, Koga H, Tsuji K, Matsumoto M, Kaneko H, Takazawa Y, Muneta T, Sekiya I (2017) Yields and chondrogenic potential of primary synovial mesenchymal stem cells are comparable between rheumatoid arthritis and osteoarthritis patients. Stem Cell Res Ther 8:1–5. https://doi.org/10.1186/s13287-017-0572-8

Kon E, Roffi A, Filardo G, Tesei G, Marcacci M (2015) Scaffold-based cartilage treatments: with or without cells? A systematic review of preclinical and clinical evidence. Arthrosc J Arthrosc Relat Surg 31:767–775. https://doi.org/10.1016/J.ARTHRO.2014.11.017

Kurth TB, Dell'Accio F, Crouch V, Augello A, Sharpe PT, De Bari C (2011) Functional mesenchymal stem cell niches in adult mouse knee joint synovium in vivo. Arthritis Rheum 63:1289–1300. https://doi.org/10.1002/art.30234

Kuyinu EL, Narayanan G, Nair LS, Laurencin CT (2016) Animal models of osteoarthritis: classification, update, and measurement of outcomes. J Orthop Surg Res 11:19. https://doi.org/10.1186/s13018-016-0346-5

Lee WY, Wang B (2017) Cartilage repair by mesenchymal stem cells: clinical trial update and perspectives. J Orthop Transl 9:76–88. https://doi.org/10.1016/J.JOT.2017.03.005

Lv F-J, Tuan RS, Cheung KMC, Leung VYL (2014) Concise review: the surface markers and identity of human mesenchymal stem cells. Stem Cells 32:1408–1419. https://doi.org/10.1002/stem.1681

Mak J, Jablonski CL, Leonard CA, Dunn JF, Raharjo E, Matyas JR, Biernaskie J, Krawetz RJ (2016) Intra-articular injection of synovial mesenchymal stem cells improves cartilage repair in a mouse injury model. Sci Rep 6:23076. https://doi.org/10.1038/srep23076

Matsukura Y, Muneta T, Tsuji K, Koga H, Sekiya I (2014) Mesenchymal stem cells in synovial fluid increase after meniscus injury. Clin Orthop Relat Res 472:1357–1364. https://doi.org/10.1007/s11999-013-3418-4

Mizuno M, Katano H, Mabuchi Y, Ogata Y, Ichinose S, Fujii S, Otabe K, Komori K, Ozeki N, Koga H, Tsuji K, Akazawa C, Muneta T, Sekiya I (2018) Specific markers and properties of synovial mesenchymal stem cells in the surface, stromal, and perivascular regions. Stem Cell Res Ther 9:123. https://doi.org/10.1186/s13287-018-0870-9

Morito T, Muneta T, Hara K, Ju Y-J, Mochizuki T, Makino H, Umezawa A, Sekiya I (2008) Synovial-fluid-derived mesenchymal stem cells increase after intra-articular ligament injury in humans. Rheumatology 47:1137–1143. https://doi.org/10.1093/rheumatology/ken114

Murata Y, Uchida S, Utsunomiya H, Hatakeyama A, Nakashima H, Chang A, Sekiya I, Sakai A (2018) Synovial mesenchymal stem cells derived from the cotyloid fossa synovium have higher self-renewal and differentiation potential than those from the paralabral synovium in the hip joint. Am J Sports Med 46:2942–2953. https://doi.org/10.1177/0363546518794664

Murphy JM, Dixon K, Beck S, Fabian D, Feldman A, Barry F (2002) Reduced chondrogenic and adipogenic activity of mesenchymal stem cells from patients with advanced osteoarthritis. Arthritis Rheum 46(3):704–713. https://doi.org/10.1002/art.10118

Nakamura T, Sekiya I, Muneta T, Hatsushika D, Horie M, Tsuji K, Kawarasaki T, Watanabe A, Hishikawa S, Fujimoto Y, Tanaka H, Kobayashi E (2012) Arthroscopic, histological and MRI analyses of cartilage repair after a minimally invasive method of transplantation of allogeneic synovial mesenchymal stromal cells into cartilage defects in pigs. Cytotherapy 14:327–338. https://doi.org/10.3109/14653249.2011.638912

Neybecker P, Henrionnet C, Pape E, Mainard D, Galois L, Loeuille D, Gillet P, Pinzano A (2018) In-vitro and in-vivo potentialities for cartilage repair from human advanced knee osteoarthritis synovial fluid-derived mesenchymal stem cells. Stem Cell Res Ther 9:329. https://doi.org/10.1186/s13287-018-1071-2

Nimura A, Muneta T, Koga H, Mochizuki T, Suzuki K, Makino H, Umezawa A, Sekiya I (2008) Increased proliferation of human synovial mesenchymal stem cells with autologous human serum: comparisons with bone marrow mesenchymal stem cells and with fetal bovine serum. Arthritis Rheum 58(2):501–510. https://doi.org/10.1002/art.23219

Ogata Y, Mabuchi Y, Yoshida M, Suto EG, Suzuki N, Muneta T, Sekiya I, Akazawa C (2015) Purified human synovium mesenchymal stem cells as a good resource for cartilage regeneration. PLoS One 10(6):e0129096. https://doi.org/10.1371/journal.pone.0129096

Ozeki N, Muneta T, Koga H, Nakagawa Y, Mizuno M, Tsuji K, Mabuchi Y, Akazawa C, Kobayashi E, Matsumoto K, Futamura K, Saito T, Sekiya I (2016) Not single but periodic injections of synovial mesenchymal stem cells maintain viable cells in knees and inhibit osteoarthritis progression in rats. Osteoarthr Cartil 24:1061–1070. https://doi.org/10.1016/J.JOCA.2015.12.018

Partridge L, Deelen J, Slagboom PE (2018) Facing up to the global challenges of ageing. Nature 561:45–56. https://doi.org/10.1038/s41586-018-0457-8

Reissis D, Tang QO, Cooper NC, Carasco CF, Gamie Z, Mantalaris A, Tsiridis E (2016) Current clinical evidence for the use of mesenchymal stem cells in articular cartilage repair. Expert Opin Biol Ther 16:535–557. https://doi.org/10.1517/14712598.2016.1145651

Roelofs AJ, Zupan J, Riemen AHK, Kania K, Ansboro S, White N, Clark SM, De Bari C (2017) Joint morphogenetic cells in the adult mammalian synovium. Nat Commun 8:15040. https://doi.org/10.1038/ncomms15040

Roffi A, Nakamura N, Sanchez M, Cucchiarini M, Filardo G (2018) Injectable systems for Intra-Articular Delivery of mesenchymal stromal cells for cartilage treatment: a systematic review of preclinical and clinical evidence. Int J Mol Sci 25(19):11. https://doi.org/10.3390/ijms19113322.

Sacchetti B, Funari A, Remoli C, Giannicola G, Kogler G, Liedtke S, Cossu G, Serafini M, Sampaolesi M, Tagliafico E, Tenedini E, Saggio I, Robey PG, Riminucci M, Bianco P (2016) No identical "mesenchymal stem cells" at different times and sites: human committed progenitors of distinct origin and differentiation potential are incorporated as adventitial cells in microvessels. Stem Cell Rep 6:897–913. https://doi.org/10.1016/j.stemcr.2016.05.011

Sakaguchi Y, Sekiya I, Yagishita K, Muneta T (2005) Comparison of human stem cells derived from various mesenchymal tissues: superiority of synovium as a cell source. Arthritis Rheum 52:2521–2529. https://doi.org/10.1002/art.21212

Sekiya I, Muneta T, Horie M, Koga H (2015) Arthroscopic transplantation of synovial stem cells improves clinical outcomes in knees with cartilage defects. Clin Orthop Relat Res 473:2316–2326. https://doi.org/10.1007/s11999-015-4324-8

Shimomura K, Yasui Y, Koizumi K, Chijimatsu R, Hart DA, Yonetani Y, Ando W, Nishii T, Kanamoto T,

Horibe S, Yoshikawa H, Nakamura N, Sakaue M, Sugita N, Moriguchi Y (2018) First-in-human pilot study of implantation of a scaffold-free tissue-engineered construct generated from autologous synovial mesenchymal stem cells for repair of knee chondral lesions. Am J Sports Med 46:2384–2393. https://doi.org/10.1177/0363546518781825

Sipp D, Robey PG, Turner L (2018) Clear up this stem-cell mess. Nature 561:455–457

Smith MD (2011) The normal synovium. Open Rheumatol J 5:100–106. https://doi.org/10.2174/1874312901105010100

Vangsness CT Jr, Farr J 2nd, Boyd J, Dellaero DT, Mills CR, LeRoux-Williams M (2014) Adult human mesenchymal stem cells delivered via intra-articular injection to the knee following partial medial meniscectomy: a randomized, double-blind, controlled study. J Bone Joint Surg Am 96(2):90–98. https://doi.org/10.2106/JBJS.M.00058

Viganò M, Sansone V, d'Agostino MC, Romeo P, Perucca Orfei C, de Girolamo L (2016) Mesenchymal stem cells as therapeutic target of biophysical stimulation for the treatment of musculoskeletal disorders. J Orthop Surg Res 11(1):163. https://doi.org/10.1186/s13018-016-0496-5

Waldner M, Zhang W, James IB, Allbright K, Havis E, Bliley JM, Almadori A, Schweizer R, Plock JA, Washington KM, Gorantla VS, Solari MG, Marra KG, Rubin JP (2018) Characteristics and Immunomodulating functions of adipose-derived and bone marrow-derived mesenchymal stem cells across defined human leukocyte antigen barriers. Front Immunol 24(9):1642. https://doi.org/10.3389/fimmu.2018.01642

Worthley DL, Churchill M, Compton JT, Tailor Y, Rao M, Si Y, Levin D, Schwartz MG, Uygur A, Hayakawa Y, Gross S, Renz BW, Setlik W, Martinez AN, Chen X, Nizami S, Lee HG, Kang HP, Caldwell JM, Asfaha S, Westphalen CB, Graham T, Jin G, Nagar K, Wang H, Kheirbek MA, Kolhe A, Carpenter J, Glaire M, Nair A, Renders S, Manieri N, Muthupalani S, Fox JG, Reichert M, Giraud AS, Schwabe RF, Pradere JP, Walton K, Prakash A, Gumucio D, Rustgi AK, Stappenbeck TS, Friedman RA, Gershon MD, Sims P, Grikscheit T, Lee FY, Karsenty G, Mukherjee S, Wang TC (2015) Gremlin 1 identifies a skeletal stem cell with bone, cartilage, and reticular stromal potential. Cell 160:269–284. https://doi.org/10.1016/j.cell.2014.11.042

Xing D, Kwong J, Yang Z, Hou Y, Zhang W, Ma B, Lin J (2018) Intra-articular injection of mesenchymal stem cells in treating knee osteoarthritis: a systematic review of animal studies. Osteoarthr Cartil 26:445–461. https://doi.org/10.1016/j.joca.2018.01.010

Yan M, Liu X, Dang Q, Huang H, Yang F, Li Y (2017) Intra-articular injection of human synovial membrane-derived mesenchymal stem cells in murine collagen-induced arthritis: assessment of immunomodulatory capacity *in vivo*. Stem Cells Int 2017:9198328. https://doi.org/10.1155/2017/9198328

Yao Y, Li ZY, Zhang H, Zheng YH, Mai LX, Liu WJ, Zhang ZG, Sun YP (2018) Synovial fluid-derived synovial fragments represent an improved source of synovial mesenchymal stem cells in the temporomandibular joint. Int J Mol Med 41:173–183. https://doi.org/10.3892/ijmm.2017.3210

Yasui Y, Hart DA, Sugita N, Chijimatsu R, Koizumi K, Ando W, Moriguchi Y, Shimomura K, Myoui A, Yoshikawa H, Nakamura N (2018) Time-dependent recovery of human synovial membrane mesenchymal

stem cell function after high-dose steroid therapy: case report and laboratory study. Am J Sports Med 46:695–701.   https://doi.org/10.1177/0363546517741307

Zayed M, Newby S, Misk N, Donnell R, Dhar M (2018) Xenogenic implantation of equine synovial fluid-derived mesenchymal stem cells leads to articular cartilage regeneration. Stem Cells Int 2018:1–9. https://doi.org/10.1155/2018/1073705

Zhou BO, Yue R, Murphy MM, Peyer JG, Morrison SJ (2014) Leptin-receptor-expressing mesenchymal stromal cells represent the main source of bone formed by adult bone marrow. Cell Stem Cell 15:154–168. https://doi.org/10.1016/j.stem.2014.06.008

Adv Exp Med Biol – Cell Biology and Translational Medicine (2019) 6: 107–126
https://doi.org/10.1007/5584_2019_380
© Springer Nature Switzerland AG 2019
Published online: 8 May 2019

Skin Stem Cells, Their Niche and Tissue Engineering Approach for Skin Regeneration

Nur Kübra Çankirili, Ozlem Altundag, and Betül Çelebi-Saltik

Abstract

Skin is the main organ that covers the human body and acts as a protective barrier between the human body and the environment. Skin tissue as a stem cell source can be used for transplantation in therapeutic application in terms of its properties such as abundant, easy to access, high plasticity and high ability to regenerate. The immunological profile of these cells makes it a suitable resource for autologous and allogeneic applications. The lack of major histo-compatibility complex 1 is also advantageous in its use. Epidermal stem cells are the main stem cells in the skin and are suitable cells in tissue engineering studies for their important role in wound repair. In the last 30 years, many studies have been conducted to develop substitutions that mimic human skin. Stem cell-based skin substitutions have been developed to be used in clinical applications, to support the healing of acute and chronic wounds and as test systems for dermatological and pharmacological applications. In this chapter, tissue specific properties of epidermal stem cells, composition of their niche, regenerative approaches and repair of tissue degeneration have been examined.

Keywords

Epidermal stem cells · Niche · Skin · Stem cells · Tissue engineering

Abbreviations

3D	Three dimensional
ATRA	All-Trans Retinoic Acid
BM	Basement Membrane
BM MSC	Bone Marrow Mesenchymal Stem Cell
BMP	Bone Morphogenic Protein
CD	Cluster of Differentiation
DNA	Deoxyribonucleic acid
DP	Dermal Papilla
ECM	Extracellular matrix
EB	Epidermolysis Bullosa
EGF	Epidermal Growth Factor
EPU	Epidermal Proliferative Unit
FDA	Food and Drug Administration
FGF	Fibroblast Growth Factor
GAG	Glycosaminoglycan
hASCs	Human Adipose Tissue Derived Stem/Stromal Cells
HF	Hair Follicle Bulge
IFE	Interfollicular Epidermis
IRS	Inner Root Sheath
Krt15+	Keratin15

N. K. Çankirili, O. Altundag, and B. Çelebi-Saltik (✉)
Department of Stem Cell Sciences, Hacettepe University Graduate School of Health Sciences, Ankara, Turkey

Center for Stem Cell Research and Development, Hacettepe University, Ankara, Turkey
e-mail: betul.celebi@hacettepe.edu.tr

MRNA	Messenger RNA
miRNAs	MicroRNAs
MMP	Matrix Metalloproteinase
Muse	Multilineage Differentiating Stress Enduring
ORS	Outer Root Sheath
Ptch	Patch
RER	Rough Endoplasmic Reticulum
RNA	Ribonucleic acid
Shh	Sonic Hedgehog
SSEA	Stage-Spesific Embryonic Antigen
TGF-β	Transforming Growth Factor-beta
UCPC	Umbilical cord pericyte cell

1   Introduction

Human skin is composed of two layers; epidermis and dermis and dermo-epidermal junction which is usually known as subcutaneous tissues or hypodermis (Proksch et al. 2008). The outher part of the skin, epidermis, is an important part of the skin that prevents water and body fluid loss, protects the body from bacteria, viruses and parasitic infections and shows resistance to mechanical and chemical injuries. The major cell group inside of the epidermis is keratinocytes. They form a sequence of layers during movement from the basal layer to the skin surface. These layers are arranged from top to bottom; stratum corneum, stratum granulosum, stratum spinosum and basal layer. Other prominent cells in the epidermis layer are melanocytes, merkel cells, and immune cells such as langerhans cells. The basal epidermal layer contains undifferentiated proliferative progenitors expressing keratin K14 and K5. In addition to the regeneration of the basal layer, these progenitor cells also form outer layers of terminally differentiated dead stratum corneum cells and non-proliferative, transcriptionally active spinous and granular layers expressing K1, K10 (Figs. 1 and 2). These progenitors also form the dead layer barrier between the granular layer and the terminally differentiated cells. Second part of the skin that is named dermis composed of fibroblasts, which

are involved in the construction of collagen, elastin and other structural molecules. The dermis consists of physically and functionally two layers. These are; *stratum papillare* and *stratum reticulare*. Dermis provides architecture and support against mechanical injuries. Instead of providing static support, the dynamic interaction between cells and extracellular matrix (ECM) affects cell behavior and fate.

In the dermis, collagen type I and type III are the common fibrous forming collagens within the ECM. They form a network with other proteins in the ECM such as laminins, nidogens, fibronectin, proteoglycans etc.. Hyaluronan, is one type of proteoglycan proteins found in the basal lamina of the skin, especially in the dermis in large amounts. It is also found more in basal regions containing the proliferating cells, but gradually decreasing in the upper layers of the skin, towards the stratum corneum (Brizzi et al. 2012; Chermnykh et al. 2018). Hypodermis is under the dermis. There is no clear boundary between these two layers, extending to the underlying muscle layer. Basic function of the hypodermis is carrying and connect. It also acts as an energy store and mechanical buffer and protects the body from temperature fluctuations.

2   Embryology of the Skin

After the fertilization, the zygote is divided into successive cleavages, forming blastomeres and blastocyst structure. During the gastrulation event which starts in the third week of early development, epiblast cells invade through the primitive streak and organize the formation of three germ layers, ectoderm, mesoderm, and endoderm. The ectoderm then form structures of nervous system (central nervous system from neural plate and pheripheral nervous system from the neural crest) and the skin epidermis is developed from the surface ectoderm. During the embryogenesis, a multistep process consist of epidermal specification, epidermal commitment, stratification, terminal differentiation and generation of epidermal appendages is required for skin development (Fig. 3) (Hu et al. 2018).

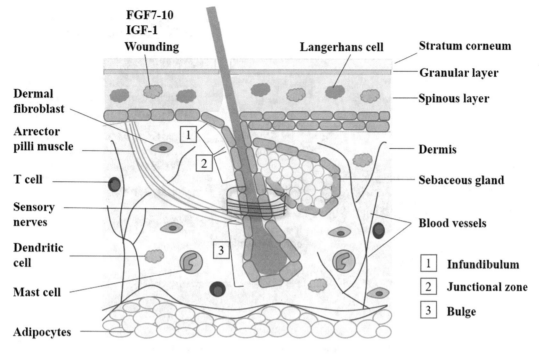

Fig. 1 Composition of the skin. (It is adapted from Hsu et al. (2014))

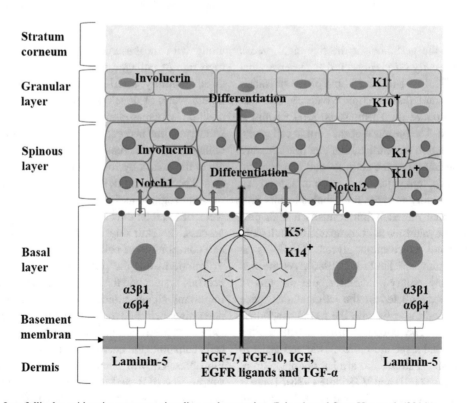

Fig. 2 Interfollicular epidermis: structure, signaling and progenies. (It is adapted from Hsu et al. (2014))

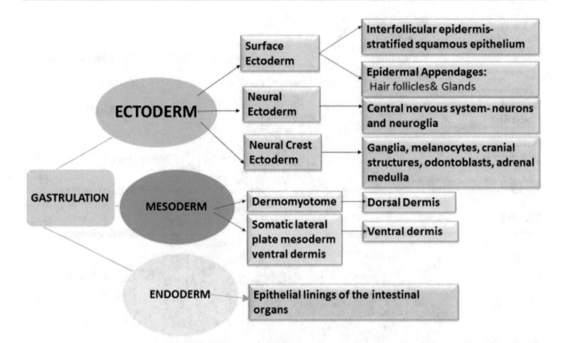

Fig. 3 Embryonic development of ectodermal lineages. After fertilization and the successive cleavages, at the third week of the embryonic development, the epiblast cells of the embryo give rise to three different germ layers (Ectoderm, Mesoderm, Endoderm) by gastrulation. From mesoderm, dorsal and ventral dermis of the embryo is formed. Ectoderm lineage has potential to generate both nervous system and the stratified epithelium of the skin&epidermal appendages, from the surface ectoderm

During the gestation, at the five-eight weeks period, the early skin is observed as mesenchymal dermis. The first trimester is the phase of transition to fibrous dermal connective structure and during second& third trimesters, fetal dermis is specialized. The presence of the certain cell types, the distribution of extracellular matrix, structure and organization of the cellullar and biochemical components, enzymes, and structural proteins, especially collagen types contributes to this process. Dermal mesenchyma contains a high proportion of hyaluronic acid compared to adult skin, and sulfated glycosaminoglycan (GAG) is in scarce (Smith and Holbrook 1986). As the age of the early embryo increases, the amount of collagen accumulated in the extracellular space increases and the cells with the round shape lose their shape. The collagen fibers in the dermal mesenchymal stage are more elastic and resist to the mechanical stress, and are largely made up of collagen types I, III and V (Riddle 1986). There is no presence of a clear distinction of dermis, and

along with the basement membrane, there is a high rate of collagen type IV around the small-sized capillary arterioles and venules. At the nineth week of development, the fibrillary composition is increased in the matrix and the dermal-subdermal line becomes evident. At this stage, the collagen fibers increase their diameters and they are in the matrix by forming the bundle organization. The advanced development of papillary and reticular regions in the dermis is seen at the 14th week of the fetal development. As a result of this process, the skin thickens with the organization and contribution of collagen fibers.

Ectoderm has the potential to form both the nervous system and the skin epithelium. There are important signal molecules and pathways that determine cell fate in early embryonic period. During gastrulation, neuroectoderm-derived cells constitute the nervous system with increased Fibroblast Growth Factor (FGF) expression and suppression of Bone Morphogenic Protein (BMP) expression (Gaspard and Vanderhaeghen 2010).

In addition, BMP and Wnt expressions are necessary for the formation of skin epithelium (Wilson and Hemmati-Brivanlou 1995). As a result of this process, multipotent cells are epithelialized as a single layer. At fifth-sixth weeks of embryonic development, the embryonic dermis, called cellular dermis, consists of mesenchymal cells and there is no clear boundary with the dermis. These multicellular cells are characterized by higher nucleus/cytoplasmic volume ratio and have cell-to-cell contacts, tight connections, glycogen stores and a less developed Rough Endoplasmic Reticulum (RER)-golgi complex (Riddle 1986). The keratinocyte cells found in the skin produce keratin, and the type of keratin varies depending on the embryonic period. Initially generated multipotent-single layer epithelial cells express largely K8 and K18, while upon the stratification, the major keratin constituent is replaced with K5 and K14 (Byrne et al. 1994). Epithelial cells originating from the surface ectoderm generates epidermis and with further differentiation, it gives rise to the formation of future epidermal appendages.

The stratification of the hair follicle occurs in three stages, namely, the placode formation, organogenesis of the hair follicle and cytodifferentiation. During the first stage, the keratinocytes show clustering, extend and expand and form hair placodes, the precursor of the hair follicle, at E14.5 (Embryonic day 14.5) (Driskell et al. 2013). After the formation of specialized placode constructs, proliferation is observed as a result of cross-talk signaling with fibroblast cells (Michno et al. 2003). In the second stage, rapidly proliferating cells show downward growth and form the dermal papilla structure. At E15.5, hair germ is formed. Following E16.5–17.6, terminal differantiation of the inner root sheath and formation of hair shaft is observed. Formation of outer root sheath occurs at E18.5 (Muller-Rover et al. 2001). The dermis has multiple embryonic origins and is a result of 3 different anatomical organizations, anterior-posterior, proximal-distal and dermal-nondermal, fibroblasts are directed to different lineages. Each fibroblast type produces, secretes and contributes to the protein composition of the dermis (Driskell et al. 2013).

Sebacous glands are located at the end of the hair follicle and prevent the water loss of the skin with the sebum it produces. Apocrine sweat glands are found near the hair follicles and are only seen in these regions throughout the body. Eccrine sweat glands are another epidermal derivative, spread across various parts of the body and are surruounded by adipose tissue. In human, both sebaceous glands and sweat glands develops at 13 and 14 weeks of gestation, from the ectoderm.

3 Skin Stem Cells and Their Niche

The mammalian skin consists of two structural layers: the outermost structural component is epidermis and the collagen-rich dermis (rich connective tissue) separated from it by a Basement Membrane (BM). Stem cells are in close contact with the BM, which has specialized sheat-like structure plus with ECM for the epidermal cells. As stem cells deteached from the BM and migrate towards the overlying zones, they become progenitor cells and undergo terminal differantiation (*cornification*) (Chermnykh et al. 2018; Tadeu and Horsley 2014). Epidermal cells shed and replaced with new ones thanks to the multipotent and unipotent differentiation potential of skin stem cells (Tadeu and Horsley 2014). The microenvironmental components of the BM is provided by the secretuar activity of basal keratinocytes and dermal fibroblasts within the skin (Blanpain et al. 2004). Beneath the BM, firstly uppermost papillary dermis comprises delicate matrix fibers. Secondly reticular dermis includes large fibers and thirdly dermal adipocytes are found to be within the intradermal adipocytes (Arwert et al. 2012; Zouboulis et al. 2008). Among these layers, except from acellular the matrix fibers, there are additional cell types that support the epidermal stem cells by providing a microenvironment: inflammatory cells, neurons, muscle cells and the blood vessels of the skin (Brizzi et al. 2012; Kretzschmar and Watt 2014). Keratinocyte is the epidermal cell lineage which mainly constitutes the skin. Other cell types such as Merkel cells, melanocytes, Langerhan cells found in mammalian epidermis (Kretzschmar and Watt 2014). Merkel cells reside in specialized compartments of the IFE, called touch domes, and function in the touch sensation. Langerhans cells are the dendritic cells of the epidermis and contribute to the adaptive immune responses. Melanocyte cells are responsible for the formation of melanin granules, and these granules provide protection to the keratinoctyes against Deoxyribonucleic Acid (DNA) damage from sun light (Sorg et al. 2017). Merkel cells reside in the touch domes in the Interfollicular Epidermis (IFE) and function in the touch sensation of skin. They are in close contact with somatosensory nerve fibers in the specialized region of skin, which is the borderline between dermal-epidermal regions (Brizzi et al. 2012; Arwert et al. 2012). Melanocytes are responsible for the protection of skin against exposue of sun light ultraviolet, radiation by expression of melanin granules/production of melanin pigment which are transported to keratinocytes and function against DNA damage. Langerhans cells, epidermal dendritic cells: part of adaptive immune response (Chermnykh et al. 2018). Arrector pili muscle function as protector against the heat loss and it interconnects Hair Follicle Bulge (HF) and IFE (Sorg et al. 2017; Wong et al. 2012).

There seems to exist three different niche regions defined for epidermal stem cells to date: **interfollicular epidermis, hair follicle bulge and the sebaceous gland** (Chermnykh et al. 2018; Levy et al. 2007) (Table 1). In the skin, each niche regions counted previously contain stem cells at the basal layer, in association with the basement membrane (Sadowski et al. 2017). The sebaceous gland and hair follicle forming basal cells within the **"pilosebaceous unit"** is formed during epidermal development (Levy et al. 2007).

3.1   Interfollicular Epidermis (IFE)

The mammalian skin outermost region: IFE is tightly connected to its underlying basement membrane via integrin-receptor and ligand adhesion. Its top layer is called stratum corneum and constantly shed and proliferate. The cells which are anchored and dinamically interacting the BM are basal keratinoctyes. These cells are continually shed and proliferate in order to regenerate the skin. Keratinocytes are formed from the surface ectoderm during embriyonic development and cells further differentiate to form later epidermal constituents of the skin via asymmetric cell divisions. Each asymmetric division gives rise to a progenitor suprabasal and basal daughter cell (Blanpain et al. 2004; Gillespie and Owens 2018). The interfollicular epidermis, which is located between the hair follicles, is important in the replacement of terminally differentiated cells that continuously spilled from the skin surface during adult life and in the regeneration of the hair follicle (Abbas and Mahalingam 2009). The hair region of the hair follicle represents the most well-characterized epidermal stem cell population today (Abbas and Mahalingam 2009). Lineage-tracing experiments have done to clarify the proposal of the existence of Epidermal Proliferative Unit (EPU) in the IFE. EPU suggests that there are central slow cycling cells serve as pre-progenitor cells for later rapidly cycling progenitor cells to differantiate as units. These units are resembled as a single stem cell and its surruounder committed progenitor cells (Transit amplifying cells) (Purba et al. 2014; Terskikh et al. 2012). Thus, there is a heterogeneity and hierarchy in the basal cells. Merkel cells within the IFE has a seperate specialized compartment called "touch domes". Merkel cells function to serve touch sensation and response to mechanical stimuli to an organism. The afferent somatosensory nerve fibers at dermal-epidermal border provide the mechanical stimulus to Merkel cells (Doucet et al. 2013).

3.2   Pilosebaceous Unit

Several lineage tracing experiments illustrates that several type of cells are capable of production of the lineages of hair follicle and sebaceous gland as well as the epidermis after wounding (Levy et al. 2007). Sebaceous gland produces specialized lipid sebum for skin via lysis of

Table 1 Skin stem cells inside of the niche and their markers

Stem Cells	Niche	Markers
Interfollicular epidermal stem cells	Epidermal basal layer	β1high/melanoma chondroitin sulfate
		Proteoglycan+(MCSP+), P63 α6high/CD71dim
Hair follicle stem cells	Bulge region	CD34, Lgr5, Sox9, CD200 Lhx2, NFATC1, Bromodeoxyuridine dye retention
		NFIB, K15, PHLDA1, Lhx2, K19.
	Isthmus	Lrig1, MST24, Lgr6, Gli1
	Hair germ at base of hair follicle	Gli1, Lgr5, K15
Melanocyte stem cells	Hair follicle bulge region, hair germ	Pax3, Dct, Sox
Sebaceous gland stem cells	Sebacesus glands, infundibulum	Blimp1Pax3, Dct, Sox
Neuronal progenitor cells	Bulge region	Nestin

It is adapted from Ojeh et al. (2015)

sebocytes and the lipid produced is excreted to the skin surface via hair canal. Hair follicle generates hair in a three-step cycle firstly begins with hair follicle growth, secondly destruction of its lower portion remaining with bulge, and the last step is the resting phase of hair follicle for later regenerative cycles; anagen, catagen and telogen, respectively (Tadeu and Horsley 2014).

3.3 Hair Follicle

3.3.1 Hair Follicle Embryogenesis and Hair Cycle in Adults

The hair follicle maintains its development with a series of epithelial-mesenchymal interplays. First, the dermis sends a signal to the upper epidermis to form a supplement, and in response to the cpidcrmis, conducting Dermal Papilla (DP) signaling to the underlying dermal cells. The DP contains a densely packed small mesenchymal cell cluster. The first morphological symptom of hair folliculc development is the formation of a hair plate where the basal epithelium is prolonged, and the dermal density occurs.

The developing follicle extends downwards to surround the DP and the underlying cells (matrix) begin to multiply. These cells, which proliferate during follicle maturation, begin to differentiate into the Inner Root Sheath (IRS) of the hair tissue to be formed in later stages. The outer cell layer became the Outer Root Sheath (ORS) which is wrapped by the basement membrane, which expresses ORS, K5 and K14. As the follicle expands, a new inner core is formed in the cells and the keratin genes of the hair begin to appear. The IRS, which is degenerated closer the skin surface of the hair root, allows the hair strand to overflow itself. On the 16th day of the cycle, proliferation is terminated in the matrix and a rapid deterioration occurs with apoptosis (catagen stage) in a large region beneath the hair follicle. DP is stimulated with the withdrawal of the epithelial hairs encircled by the basement membrane and enters the resting phase. This resting phase is called telogen. In the first hair cycle, the telogen lasts for one day, but in later cycles, this phase is gradually extended. The molecular mechanisms involved in the pathogenesis of hair follicles are not clearly understood, but genetic studies in mice have shown that Wnt/β-catenin, BMP, FGF, Sonic Hedgehog (Shh), Epidermal Growth Factor (EGF), NFkB and Notch indicates the importance of the signal path. Hair follicle occurs in the process of epithelial-mesenchymal interactions that begin with the formation of hair plaques in dermal mesenchymal cell regions (Hardy 1992).

This hair placode allows hair to grow and differentiate. An adult hair follicle consists of a stable top part and an under portion which is in continuous construction. The bulge region of the hair follicle is currently the best characterized site of epidermal stem cell populations (Abbas and Mahalingam 2009). Laminins, together with collagen type IV are essential components of the BM and Laminin-511 (laminin-10) and Laminin-332 (laminin 5) are mostly found in the adult skin and are important for the regeneration of hair folliculle and its development. Hair follicles are particularly well-defined niche regions with their known molecular and developmental mechanisms (Wong et al. 2012). The regeneration of the hair follicle comprises three phases: (1) growth phase, (2) regression phase (3) rest phase. Each cycle results in the production of the hair shaft that is grown over the skin surface (Rompolas and Greco 2014).

3.3.2 Bulge

Bulge is the region of hair follicle where the stem cells reside. In vivo genetic lineage tracing experiments have clarified that bulge stem cell progeny is Keratin15 (Krt15+). Bulge stem cells are known as their quiescence and long-lived nature compared with the other microenvironments within the skin (Ito et al. 2005). For mouse hair follicle stem cells, bulge markers are Krt19, Lgr5, Cluster of Differentiation (CD) 34 together with the transcription factors Gli1, Hopx, Lhx2, Sox9, Tcf3 and Nfatc1 (Rompolas and Greco 2014). Hair folliculle stem cells demonstrate regenerative response after wounding by re-construction of epithelium (temporary function) and capacity of generating both hair folliculle and sebaceous gland lineages (primary function). The multipotency of hair follicle bulge is related with its stem cell heterogeneity and it is demonstrated recently via lineage tracing experiments. The resident stem cells have slow-cycling property compared with other epithelial cells of the skin. Bulge stem cells are sorted by alpha 6-integrin and CD34+ identity (Brizzi et al. 2012; Choi et al. 2015).

3.3.3 Isthmus, Infundibulum

The compartment between the hair follicle bulge and the base of the sebaceous gland is called isthmus and cells inthere are Krt15- and CD34- yet they express Gli1, MTS24 and Lgr6 in high amounts. Ishtmus stem cells have capabilty to generate hair follicle lineages in homeostasis

and have regenerative power in the case of injury (Gaspard and Vanderhaeghen 2010). Within isthmus, there are Lrig1+ cells which do not contribute to hair follicle formation in normal conditions but promote maintenance of infundibulum. After wounding, cells of the infundibulum (upper follicle region) migrate towards the upper layers for regeneration, and in order to replace them, the stem cells in the bulge migrate to infundibulum (Rompolas and Greco 2014; Ito et al. 2005; Clevers et al. 2014). In contrast with the hair follicle bulge stem cells, hair germ cells express P-cadherin instead of CD34 and Nfatc1 (Kretzschmar and Watt 2014; Nowak et al. 2008).

3.4 Sweat Gland

Sweat glands are another form of the epidermal appandages of the skin which function in thermo-regulation (Ji et al. 2017). The secretory structure of the sweat gland consists of outer basal layer of myoepithelial cells expressing K5, K14 and smooth muscle actin. Inner suprabasal layer of luminal cells is positive for K8, K18 and K19 (Gillespie and Owens 2018). Unlike the mammalian gland, sweat gland has a little ability for regeneration (Tadeu and Horsley 2014). In the wounding, the progenitor cells of sweat gland participate in the replenishment of damaged tissue and regeneration of skin epithelium however, sweat gland itself is quiescent during this period (Wong et al. 2012).

3.5 Sebaceous Gland

The underlying mechanism of the maintanance of stem cells in sebaceous gland is an issue to be further investigation.The peripheral unipotent stem cells or the hair follicle bulge stem cells function in the renewal of the gland (Chermnykh et al. 2018). Sebocytes are marked by the presence of B lymphocyte-induced maturation protein (Blimp1) and this protein is expressed by the *Prdm1* gene. It is demonstrated by the lineage tracing experiments that Blimp1+ cells are generating the sebum-producing sebocyte lineage (Sorg et al. 2017).

The basic stem cells in the skin can be listed as follows; keratinocytes, melanocytes, follicular stem cells, sebaceous stem stem cells, mesenchymal stem cell-like stem cells, nerve progenitor cells, and hematopoietic stem cells. Epidermal stem cells, which are among these different subtypes of skin stem cells, are the most effective cells in tissue repair and skin regeneration. Research shows that the epidermis has a small number of stem cells, is uncommonly divided and short-lived. The majority of epidermal stem cells are found in the basal layer of the epidermis and the rest is located at the base of the hair follicle and sebaceous glands (Watt et al. 2006). Epidermal stem cells circulate between two different cell phases throughout their life cycle. During the slow cell phase, the epidermal stem cells are inactive, but in the transition to the transformed cell phase, these cells begin to split rapidly and increase the density of the skin cells to regenerate the skin tissue. In the final stage, it undergoes many cell divisions before it becomes terminal (Chu et al. 2018). Most of them are found in the basal layer of the epidermis and may differentiate into transient amplifying cells and terminal differentiated epidermal cells. Important cell markers are $\beta1$ integrin, $\alpha6$ integrin, K15, p63 and nestin. For the proliferative basal cells of the skin, high levels of $\alpha6$-integrin expression, Delta1 expression (Notch ligand), CD200 expression together with the low levels of CD71 expression are indicative. The skin stem cell is linked to these laminin proteins via specific integrin units; mostly via $\alpha3\beta1$, $\alpha6\beta1$ and $\alpha6\beta4$ integrins (Chermnykh et al. 2018). Among the previous receptors, $\alpha6\beta4$-integrin is the most common one for the connection between the basal membrane and the basal keratinocytes as a part of hemidesmosomes. Integrins expressed by the basal keratinocytes are categorised differently: firstly depending on their expression levels, secondly depending on whether they are expressed constitutively or transiently, and lastly depending on pathological and homeostatic conditions (Sorg et al. 2017). Also, for the quiescent basal keratinocyte cells, non-actively cycling stem cells, Lrig1 expression level is high and it is related with the promotion of epidermal stem cell maintenance. Studies demonstrated the putative contribution of with type IV collagen to basal stem cell proliferation capacity in the skin and promoting the maintanence (Hsu et al. 2014; Chermnykh et al. 2018; Kumar et al. 2017). Follicular stem cells are in the follicular bulge region. The outer root sheath, inner root sheath, and cells in the hair hole can be derived from the hair follicle epithelium. The known surface markers are K15, K19, Sox9, CD34, Lgr5, Lhx2, NFIB NFATC1, CD200, PHLDA1. CD34 expression, a specific marker for hematopoietic progenitor cells, has also been demonstrated in the protrusion region of murine hair follicles. CD34-positive cells are CK15-negative, which may represent transgenic amplifier cells or progeny of the protrusion stem cell (Abbas and Mahalingam 2009). These cells have a high proliferative potential. In the dermis, each region has fibroblast cells which are functionally specialized, especially the ones in the DP. DP cells induce the epidermal growth towards to the formation of hair follicle and once it is formed, DP stays inside the hair follicular structure throughout hair cycle (Chermnykh et al. 2018; Gattazzo et al. 2014). These cells in the dermis can be divided into mesodermal-like cells and some nerve cells and express important cell surface markers such as CD70, CD90 and CD105. The sebaceous glands, which play a role in the formation of differentiated sebocytes as terminal, begin with the formation of progenitor cells towards the end of embryogenesis, and maturation occurs in the postnatal period (Fuchs 2007). Blimp1, located in the follicle protrusion region and hair germ, represents the progenitor cell population found in the sebaceous glands of mice (Abbas and Mahalingam 2009). Melanocyte stem cell markers are Pax3, Dct, Sox and MITF, also known as melanocyte host transcription regulator. It was reported that Pax3 was involved in the initiation of melanogenic cascade and showed undifferentiated stem cell status (Abbas and Mahalingam 2009). In recent years, neural crest-derived stem cell populations have been found in the murine hair follicle bulge. These stem cells can differentiate in vitro from melanocytes, smooth muscle cells, keratinocytes, glial cells, neurons, and adipocytes. Nestin, which is the surface marker of neural stem cells, is an intermediate filament protein (Abbas and Mahalingam 2009).

4 Regulation of Stem Cell Behavior in the Skin and Signalling Pathways

Since stem cells exhibit their functional properties within the specialized microenvironments, characterizing the components of stem cell niche is important in order to understand the cell fate dynamics. There are several molecular mechanisms which mediate cell fate decisions and play a role in the skin homeostasis and morphogenesis (Tadeu and Horsley 2014; Arwert et al. 2012).

4.1 Shh Signalling

Shh is an important signaling pathway involved in cell fate and proliferation during animal development. In the absence of Shh, a mutation in the activated Patch (Ptch) resulted in basal cell carcinomas, a type of skin cancer that is commonly seen in humans, and these results suggest that Ptch has served as a tumor suppressor gene (Rubin et al. 2005). Derivation of basal cell carcinomas from the hair follicle indicates that the Shh signal is expressed in hair plaques in embryonic skin and is a signal required for follicle formation (St-Jacques et al. 1998; Oro and Higgins 2003). The Shh signal is additionally important for follicle regeneration throughout the adult hair cycle. Shh is expressed in the matrix and developing germ, where the anagen is polarized to one side during its advance (Blanpain and Fuchs 2006). Anti-Shh antibodies to postnatal follicles obstruct anagen progression (Wang et al. 2000) and hair regeneration (Silva-Vargas et al. 2005). In contrast, Shh or small molecule

Shh agonists accelerate the make progress from the telogene to the anagen (Paladini et al. 2005).

4.2 Wnt Signalling

The Wnt/β-catenin signaling pathway controls cellular events along embryonic and postnatal developmental processes.Wnt deregulation often give a leads to cancer-induced proliferation and differentiation instability (Reya et al. 2001). Wnt and β-catenin in the skin induce hair follicle morphogenesis, stem cell maintenance or activation, hair shaft differentiation (Alonso and Fuchs 2003). It is thought that the Wnt signalling might be the first dermal stimulus to instruct the epidermal stem cells to produce hair. Wnt signals that temporarily regulate follicle stem cell lines, promote β-catenin stabilization that encourage stem cell activation, proliferation and stimulation of follicle regeneration; supports the morphogenesis of de novo hair follicle; enhances the specification of matrix cells to terminally differentiate across the hair cell line (Blanpain and Fuchs 2006).

4.3 Bone Morphogenetic Protein Signalling

BMP binds to a transmembrane receptor complex via Bmpr1a and Bmpr1b receptors to activate signal transduction. Bmpr1a is expressed in most of the developing skin epithelium but is particularly emphasized in the hair follicle. In adult hair follicules, Bmp's also function in epithelial-mesenchymal interplays (Kratochwil et al. 1996; Lyons et al. 1989, 1990). Bmp signaling begins with neuroepithelia, where the epidermis indicates incomplete ectodermal cells (Nikaido et al. 1999). Bmp signaling was found to play an important role in the differentiation of matrix cells into IRS and hair shaft lines, and inhibition was shown to support stem cell activation (Andl et al. 2004; Kobielak et al. 2003; Ming Kwan et al. 2004). At the end of the usual hair cycle, the growth stops and the hair follicle passes through the disruptive phase (katagen). However, Bmpr1a-null ORS continues to expand, matrix cells accumulate and cause the creation of follicular tumors (Andl et al. 2004; Ming Kwan et al. 2004).

4.4   Notch Signaling

Notch signal is expressed in embryonic and adult epidermis and is involved in various cell fate processes. At the beginning of epidermal stratification, Notch1 is expressed from the basal and suprabasal cells of the epidermis and sebaceous glands (Okuyama et al. 2004; Rangarajan et al. 2001). In the final phases of epidermal stratification, Notch1 activity is reduced in the basal layer (Okuyama et al. 2004). The loss of the Notch1 function leads to problems in the IFE differentiation phase (Rangarajan et al. 2001). In hair follicule, Notch1–3 is expressed in matrix cells with proliferative capacity and in differentiating HF cells (Kopan and Weintraub 1993). When Notch 1 and Notch2 are conditionally eliminated, hair follicules disappear quantitatively and epidermal cysts emerge that emphasize the role of the Notch signal in follicle maturation and differentiation (Pan et al. 2004). In the lack of both Notch1 and Notch2, the sebaceous glands cannot form (Pan et al. 2004). The loss of Notch1 in the epidermis does not disrupt the early follicle morphogenesis, but decreases the number of hair follicule in time (Vauclair et al. 2005). Understanding the downstream genes regulated by Notch signals in the epidermis and their role in cellular functions needs further investigation.

5   Skin Tissue Regeneration

Injuries that damage the body and the internal epithelium endanger the integrity of the body. Skin wound healing is a process that results in repair of tissue integrity and tasks. A rapid reepalization is achieved by preventing or limiting threats from the environment. Reepithelization of the skin is carried out by the migration of epidermal cells and keratinocytes to the wound site. In order to ensure regeneration of the epidermis, the dermoepidermal composition, which allows the epidermis to differentiate into the dermis and keratinocytes into a protective layer, needs to be reconstructed (Stanley et al. 1981; Rigal et al. 1991; Demarchez et al. 1987; Odland and Ross 1968). In the epidermis, mitotically active keratinocytes are present in the basal layer of the dermis, where they come into direct communication with the basal lamina and proliferate slowly. Keratinocytes in the supra basal layer migrate to the surface and the terminal begins to differentiate (Tomic-Canic et al. 1998; Bartkova et al. 2003). Keratinocytes are an significant part of the wound healing response with autocrine and paracrine communications. Several hours after injury, keratinocytes begin to migrate without proliferation (Bartkova et al. 2003). If the wound is superficial, the epithelial cells will appear as normal skin within 3 days. However, if the wound is damaged deep and dermal extensions, keratinocytes and fibroblasts on the edge of the wound will migrate to the wound. During the migration processes of re-epithelialization, keratinocytes can be grouped in three different groups (1) high proliferative cells with wound edge; (2) cells expressing integrins; (3) cells that are responsible for the regeneration of the basement membrane and epidermis (Saarialho-Kere et al. 1995). In order for these cells to migrate, some chemical signaling molecules, such as nitric oxide, must be formed with the inhibition of cell contact (Lee et al. 2001).

However, fibroblasts migrate from the edges of the damaged tissue to the wound area and begin to produce collagen, thus initiating the proliferative phase. The granulation tissue begins to form with the proliferative phase. Fibronectin and HA, the major components of the granulation tissue, facilitate cell migration by providing a well-hydrated transient matrix for subsequent healing events (Krawczyk 1971). Fibroblasts adhere to fibronectin and migrate through the wound bed. An increase in chondrotin sulphate and reduction in HA production decrease the migration and proliferation of fibroblast cells. The collagen provides a scaffold to promote and facilitate the proliferation and differentiation of the cells related in inflammation angiogenesis and connective tissue making (Eichler and Carlson 2006). Fibroblasts play significant roles in the production of basal membranes by secreting cytokines such as collagen types IV and VII, laminin 5, nidogen and transforming growth factor (TGF-β).

Any complications during wound healing may cause wounds to become chronic wounds that prolong the healing process. Factors such as venous/arterial insufficiency, diabetes, kidney disease, trauma, advanced age and local pressure effects may cause delay in wound healing. In addition to these factors, systemic factors such as tissue hypoxia, ischemia, infection, dysregulation of the inflammatory process, irregular nutrition or immune status may also prevent healing. It is known that the spread of diabetes, obesity and vascular diseases cause chronic wounds. In major skin injuries that occur as a result of extensive burns, infection or trauma, the skin often cannot repair it itself and requires medical attention. Autologous skin graft is the most appropriate and aesthetically effective technique. However, this approach has several barriers, such as being effective only in a specific part of the skin and causing additionally injuries in the donor. Due to these obstacles, scientists have searched for alternate ways of treating serious skin damage and found that the in vitro colonies of adult epidermal stem cells in the epidermis and hair follicles were capable of forming a functional skin barrier (Green et al. 1979; Rheinwald and Green 1975; Sun et al. 2013).

During the inflammatory phase, the wound is closed by fibrin, which acts as a transient matrix, circulating immune cells occupy the new matrix, cleans the dead tissue and prevents infection. Proliferative fibroblasts come together and support angiogenesis by secreting collagen to form granulation tissue. Myofibroblasts derived from fibroblasts in that region express a-smooth muscle actin and initiate contraction in the wound area. Re-epithelialization begins with local stem cell populations migrating from the edge of the wound. In the last step, fibroblasts and keratinocytes provide fresh ECM components

and Matrix Metalloproteinases (MMPs) to reshape the matrix. After 3–4 months of injury, IFE cells migrate to the wound area and provide a healthy skin appearance at a rate of 80% (Ito et al. 2005; Mascre et al. 2012; Page et al. 2013). It was observed that K15-positive hair follicle stem cells temporarily contributed to wound epithelialization immediately after injury but disappeared after a few weeks. In a study by Jimenez et al. (2012), 10 patients showed that chronic leg ulcers reduce the area of ulcerative scarring of autologous scalp follicular grafts transplanted into the wound site. It is predicted that hair follicle grafts may be a hopeful treatment alternative for difficult chronic wounds due to epithelization, neovascularization and increased dermal reorganization in the wound area (Jimenez et al. 2012).

Among the important goals of stem cell applications are collecting stem cells from the patient, replicating by making modifications on it, and treating many genetic and acquired disease by giving the patient again. In short, epidermal stem cells can be used as a potential therapeutic source in regenerative medicine applications (Fig. 4). Epithelial stem cells that have been replaced in vitro to repair genetic disorders can be used as a new tool in gene therapy.

Epidermal stem cells with clonogenicity and long-term lastingness properties, can provide permanent treatment for skin diseases like Epidermolysis Bullosa (EB) (De Rosa et al. 2014) and vitiligo (Falabella et al. 1992). EB is a skin disease that occurs with a genetic mutation in any of the genes involved in the binding of basal epidermal cells and basement membrane, leading to a destructive effect on the skin and often death (Jackson et al. 2017). Clones of primary epidermal stem cells from biopsy taken from a patient with EB caused by a mutation in the Laminin-5 gene were corrected with a retrovirus and applied to the legs of nine patients (Mavilio et al. 2006). One year after the application, it was seen that the levels of laminin-5 synthesis were similar to the normal epidermis and at the end of 6.5 years there was no disease in the transplanted areas and a healthy epidermis was formed. In a study conducted by Amoh et al. in 2005, stem cells in the root sheath of hairs

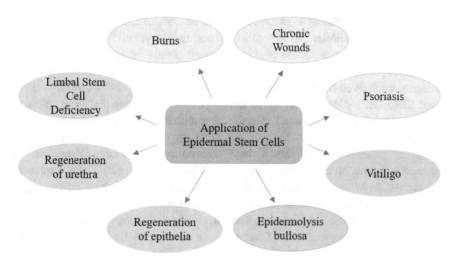

Fig. 4 Application of epidermal stem cells. (It is adapted from Jackson et al. (2017))

during anagen were cultured in the form of ex vivo organotypic culture and obtained a multi-layer epidermal-like structure. As a result of the treatment applied to 23 patients, positive responses were obtained in 18 patients. Total wound closure with a single transplantation was around 33% (4 cases) (Amoh et al. 2005, 2008). A successful in vitro epidermis was obtained by using autologous epidermal stem cells in undamaged skin biopsy to treat large skin burns (Lapouge and Blanpain 2008). The skin biopsy specimen is separated by trypsin and the epidermal stem cells are isolated. Epidermal stem cells are developed on fibroblasts which make the ECM and growth factors suitable for keratinocyte proliferation (Green et al. 1979). Keratinocytes are cultured until they form a layered epithelium covering the wound. On the other hand, this technique has two main disadvantages: (1) the long time required for in vitro expansion of keratinocytes and (2) the high price of the process (Clark et al. 2007).

Micro Ribonucleic Acids (miRNAs) are small non-coding RNAs that regulate post-transcriptional gene expression by suppressing or reversing Messenger RNA (mRNA) translation (Ambros 2004). One miRNA can target hundreds of genes, and a gene can be rearranged with various miRNAs. The miRNAs, which are the primary regulators of gene expression, play an important role in numerous biological processes such as homeostasis, differentiation and survival. Abnormal expressions of miRNAs can cause diseases. In the epidermis of mice, several miRNAs expressed as distinct from other skin lines have been identified. The miR-200 family such as a, b and c, miR-141, miR-429 and miR-19/miR-20 family (miR-19b, miR-20, miR-17-5p and miR-93) expressed in the epidermal line while the miR-199 family is present only in the hair follicles (Andl et al. 2006; Yi et al. 2006).

Hildebrand et al. (2011) compared expression profiles of calcium-derived keratinocytes with TA keratinocytes and terminally differentiated keratinocytes. At the end of the comparison, 8 up-regulated miRNAs (miR-23b, miR-95, miR-210, miR-224, miR-26a, miR-200a, miR-27b and miR-328) and one down-regulated miRNA (miR-376a) were found to be involved in the epidermal differentiation duration (Koster et al. 2004). The expression of miR-203 during acute wound healing showed that it contributed to the wound epithelization with the upward expression of p63, RAN (member of the G-protein super family), and RAPH1 (lamellipoid) at the anterior edge of the epithelial migrating extension (Viticchie et al. 2012).

miRNAs modulate cellular action of target cells by secretion by various cell types. miRNAs enter into the circulation through exosomes, where they modulate the cellular movement of the target cells (Lo Cicero et al. 2015; Shabbir et al. 2015; Valadi et al. 2007). Studies have shown that exosomes derived from Bone Marrow Mesenchymal Stem Cells (BM MSCs) show an increase in miRNAs in proliferation and migration of normal and diabetic scar fibroblasts (Shabbir et al. 2015). It has been shown that exosomes secreted by keratinocytes also have an effect on the regulation of melanin synthesis and provide important data for better understanding of the role of miRNAs mediated by exosomes in the protection of epidermal stem cells (Lo Cicero et al. 2015).

In 2010, Kuroda et al. defined Multilineage Differentiating Stress Enduring (Muse) cells (Kuroda et al. 2010) expressing pluripotency markers and being able to differentiate into three germ leaves. Muse cells are sporadically multiplied in the fascia of most organs and are not associated with a structural niche and are also found in the bone marrow cavity (Wakao et al. 2011).Muse cells are the pluripotent subpopulation of mesenchymal stem cells and express the following markers with mesenchymal markers; Stage-Spesific Embryonic Antigen (SSEA-3), CD105, CD29, CD90, Oct 3/4, Nanog, Sox2 and Rex1. When Muse cells identify damage signals and are delivered to the target tissue, they can differentiate into compatible cells and cross the mesodermal boundary between mesodermal and ectodermal cells. Thanks to these critical features, muse cells can be used to repair cells in cells, tissues and organs. Thus, Muse cells can undertake tasks such as repairing cells in various tissues and organs. Muse cells may be preferred over other cells since they are integrated in the damaged tissue or organ and have little risk of developing teratomas. So far, Muse cells have been isolated from human skin fibroblasts, adipose stem cells, and bone marrow stem cells (Yamauchi et al. 2017). In the process of skin renewal, clinicians are looking for new ways for bioengineered skin by using cells that

can be obtained more easily. Muse cells are an exciting type of cell because it has the potential to bypass certain limitations of embryonic stem cells and induced pluripotent stem cells for skin regeneration (Yamauchi et al. 2017).

It has been shown that muse cells can differentiate into melanocytes as well as fibroblasts and keratinocytes and produce reconstructed skins derived from muse cells (Yamauchi et al. 2017). Itoh and his team (2011) have managed to produce reconstructed skins containing stimulated pluripotent stem cell-derived keratinocytes and fibroblasts. For keratinocyte differentiation, we have found that BMP4 and All-Trans Retinoic Acid (ATRA) are required molecules for stem cells, including the muse, induced pluripotent stem cells and embryonic stem cells (Itoh et al. 2011). In the study, it has been declared that somatic muse cells may be a promising source for regenerative medicine in the skin. In another study, SSEA-3 + Muse cells were isolated from human Adipose Tissue Derived Stem/Stromal Cells (hASCs) (Kinoshita et al. 2015)., In the study, adipose-derived Muse cells were characterized, and their therapeutic potential was evaluated in the treatment of diabetic skin ulcers. Skin ulcers under ischemic conditions indicate long-term wound healing and use mesenchymal stem cells or endothelial progenitor cells to treat (Kinoshita et al. 2015). To investigate the efficacy of Muse cells in comparison to other cells, two separate cell populations were compared between treatments in wound healing processes in diabetic rats. Muse-derived fat-derived cells supported wound healing in DM-SCID mice, while Muse cells in ASC populations showed a more effective outcome on tissue repair. Histological analyzes showed that Muse-rich ASCs were integrated into the repaired dermis. Although the proliferative capacity of Muse cells is not high, it is thought to be a powerful vehicle for clinical use because they are nontumorogenic (Kuroda et al. 2010; Heneidi et al. 2013; Ogura et al. 2014; Wakao et al. 2014).

5.1 Advantages of Intelligent Matrices

Advances in skin reconstruction, which were developed in tissue engineering applications, were recorded along with the development of reconstructed living equivalents. The provision of tissue homoeostasis in restructured substitutions is very significant for self-renewing tissues such as the skin. Tissue-engineered substitutions are useful vehicles for learning about cell interplays and regulation of tissue homoeostasis. In addition to the use of such substitutions in the treatment of skin defects, it is important to understand the contribution of stromal-epithelial interplays to differentiation (Boa et al. 2013; Carrier et al. 2009).

Dermal matrices provide the appropriate environment and conditions to promote cell proliferation and regeneration. They are available in a variety of forms such as natural, synthetic and hybrid matrices and are arranged by different technics such as (a) electrospinning, (b) phase separation, (c) freeze drying and (d) self-assembly. Dermal matrices as simple equivalents of ECM may be cellular or acellular, biodegradable or non-degradable polymers. In natural matrices, disease transmission and immunogenic risks are higher than synthetic matrices. In addition, synthetic matrices can be produced in larger quantities, standardized and thus have a minimum risk of disease by reducing their variability. Despite all these advantages, these skin scaffolds have limitations such as the inability to perform all the functions of the skin or the inability to produce skin supplements after in vivo application. Moreover, cell-cultivated matrices did not produce the desired result in low cell proliferation, survival rates. Demarchez et al. 1987 indicated that allogenic cells in the Food and Drug Administration (FDA)-approved Apligraf, a therapeutic product for chronic wounds, were not long-lasting in vivo and acted as a temporary biological bandaging that provides growth factors to deep wounds Therefore, studies on cell survival in scaffolds formed by tissue engineering to get better results in wound treatments

are ongoing. Electrospinning and three dimensional (3D) bioprinting are new methods for cell survival and scaffold standardization. Studies have reported that electrospun scaffolds support fibroblast viability, enhance cellular organization, and reduce wound shrinkage in a mouse full-thickness wound model compared to freeze-dried scaffolds (Powell et al. 2008). Lee et al. (2014) successfully modeled human skin based on a layer of 3D bioprinting using collagen type 1, fibroblasts and keratinocytes. These cells are at a promising point in scaffold-based therapeutic research ensure new therapeutic approaches.

Commercially available epidermal substitutions are available in various forms, such as combined or pre-fluid, autologous, allogenic keratinocytes or cells used in combination with aerosol spray methods to facilitate access to the wound. EpiDex, which is used in chronic wound treatments, consists of ORS keratinocytes from hair follicles and shows positive results in the treatment of chronic leg ulcers (Ortega-Zilic et al. 2010). Examples of allogenic keratinocyte layers include Cryoskin (Beele et al. 2005). To preserve a proliferative phenotype in vitro, cells were grown up to pre-flow in delivery systems such as Laserskin® and Myskin, which facilitate transplantation (Moustafa et al. 2007; Andreassi et al. 1998). It has also been reported that fibrin glue is used in combination with aerosol method for the transmission of cell suspension to keratinocytes in BioSeed-S and CellSpray wounds. These epidermal substitutes have successful clinical applications in the treatment of diabetic ulcers, venous and burn wounds.

The curative effects of BM MSCs planted on collagen scaffolds applied to various wound styles in patients have been shown to positively affect wound healing (Yoshikawa et al. 2008). Badiava and Falang achieved successful results by applying autologous BM MSCs embedded in collagen matrix to chronic leg ulcers (Badiavas and Falanga 2003). AlloDerm and ASCs embedded in a fibrin-chitosan scaffold have been shown to increase wound healing by releasing angiogenic factors that trigger the development of the vascular system (Altman et al. 2008). In a study with newborn mice, the epidermal and dermal stem cell suspension was developed at the back of the mice with Integra Artificial Skin before implementation to full thickness injuries in vitro to form a gel-like matrix (Lee et al. 2011). The data obtained from the studies show that the prospective use of stem cells in new wound therapies based on cell/scaffold is promising.

Pericytes have major tasks in vascular development, maturation, stabilization and remodeling processes. Other important features of pericytes are wound healing, physiological repair and renewal of organs etc. (Armulik et al. 2005; Rajkumar et al. 2006; Corselli et al. 2010; Davidoff et al. 2009). The biggest obstruction to in the clinical applications of tissue engineering structures is the formation of necrosis or ischemic conditions due to lack of oxygen and nutrition. This is a very important limitation for major and vital organs such as the kidney, heart and liver. Considering that the diffusion limit of oxygen is 150–200 µm, appropriate vascularity support should be provided for survival in large structures (Avolio et al. 2017). The growth of the vessels in vivo, the colonization of the structure by host cells providing the construction is a very slow process. Thus, the cellular combination of fibroblasts, smooth muscle cells and pericytes in the vascular niche, which physiologically constitute blood vessels, is a preferred solution to support vascularization. Because of the failure to achieve optimal improvement in the treatment of skin diseases, direct injection or scaffold cell based applications are used in the treatment of cutaneous wounds by dermal tissue engineering. The treatment feature of pericyte cells obtained from human umbilical cord (UCPC) in skin wounds is very interesting today (Zebardast et al. 2010). UCPCs showed a higher proliferative rate compared to human BM MSC used in clinical therapies. Placing UCPCs on the wound site as a polymeric fibrin patch promoted the cure of full-thickness mouse skin defects and thus showed that UCPCs could be a potential source of cells for skin tissue engineering and repair (Zebardast et al. 2010).

Wound healing is a dynamic cyclic event including the migration of different cell types such as keratinocytes, macrophages, fibroblasts, extracellular matrix, leukocytes, endothelial cells and pericytes (Bodnar et al. 2016). Coagulation occurs after skin injury to provide homeostasis. The platelets and fibrinogens play a critical role in the development of the inflammatory response. The platelets stimulate the pericyt activity by releasing platelet derived growth factor and Transforming Growth Factor-beta (TGF-β) from the endothelial cells that support the separation of pericytes and migration into the parenchyma (Bodnar et al. 2016). Pericytes provide the secretion of many chemokines, growth factors and matrix proteins involved in the functioning of parenchymal cells in the skin (Dulmovits and Herman 2012). Paquet-Fifield et al. (2009) investigated pericytes associated with the proliferative basal layer of the epidermis and showed that pericytes were mesenchymal stem cells. They found that pericytes form a strong cell population in the skin and are important micro-environment regulators in skin regeneration. In a recent study by the same group, the use of pericytes in murine wounds did not significantly affect the re-epithelization of the dermal wound. The team stated that this may be related to the pro-inflammatory effects of pericytes in paracrine signaling and that the experiments should be performed more repetitive. Pericytes are an important group of cells with properties such as the umbilical cord, adipose tissue and placenta, which can be obtained from many different tissues, improve re-epithelialization and rebuild dermis (Zebardast et al. 2010). In one study, pericytes were applied to full-thickness wounds on a fibrin gel, and full thickness wounds formed a higher tensile strength, denser collagen fibers and a dermis with enhanced angiogenic effect (Zebardast et al. 2010). In another study using an organotypic skin culture, it was observed that pericytes increased epidermal regeneration through the secretion of laminin α5 and thus a potential cell in the therapy of non-healing wounds (Paquet-Fifield et al. 2009).

In a study by Tomoko and his team (Yamazaki et al. 2017), tissue localized myeloid progenitors have been shown to contribute to the development of pericytes in the embryonic skin vessels.

Myeloid cells isolated by flow cytometry and their progenitors in embryonic skin have been shown to induce pericidal differentiation in culture with TGF-β. The role of TGF-β signaling in pericyte development will not be fully known but will provide important information to better understand the mechanisms controlling neovascularization. Evidence has been found in BMP-2 organotypic cultures, a protein secreted by pericytes, that provides cell polarity and planar sections on epidermal cells (Paquet-Fifield et al. 2009). At the end of his study of Zhuang et al. (2018), It was found that the extrinsic dermal signals provided by the pericytes have the ability to regulate the orientation of the keratinocyte layer in the proliferative basal cell section of the epidermis. Asymmetric divisions in the epidermis cause signals to support cell proliferation against differentiation. Therefore, the balance of symmetrical and asymmetrical sections in the basal layer is of great importance for the preservation of the homeostatic balance in the interfollicular epidermis (Zhuang et al. 2018).

6   Conclusions and Perspectives

Skin tissue as a stem cell source can be used for transplantation in terms of properties such as abundant, easy to access, high plasticity and high ability to regenerate. The immunological profile of these cells makes it a suitable resource for autologous and allogeneic applications. There is evidence that they generate strong immunosuppressants such as α-melanocyte-stimulating hormone, interleukin 10 and transforming growth factor beta 1. The lack of major histocompatibility complex 1 is also advantageous in its use. Epidermal stem cells are the main stem cells in the skin and are suitable cells in tissue engineering studies for their important role in wound repair. However, epidermal stem cells do not have specific markers, and there is no effective method for its purification. Therefore, its use in tissue engineering studies is limited. Other stem cells can be used as an alternative. Recently, organ on a chip technology has been started to be used both as skin substitute and drug testing. It

is estimated that the studies in this area will gain even more success in the future.

References

Abbas O, Mahalingam M (2009) Epidermal stem cells: practical perspectives and potential uses. Br J Dermatol 161(2):228–236

Alonso L, Fuchs E (2003) Stem cells in the skin: waste not, Wnt not. Genes Dev 17(10):1189–1200

Altman AM et al (2008) Dermal matrix as a carrier for in vivo delivery of human adipose-derived stem cells. Biomaterials 29(10):1431–1442

Ambros V (2004) The functions of animal microRNAs. Nature 431(7006):350–355

Amoh Y et al (2005) Implanted hair follicle stem cells form Schwann cells that support repair of severed peripheral nerves. Proc Natl Acad Sci U S A 102 (49):17734–17738

Amoh Y et al (2008) Multipotent hair follicle stem cells promote repair of spinal cord injury and recovery of walking function. Cell Cycle 7(12):1865–1869

Andl T et al (2004) Epithelial Bmpr1a regulates differentiation and proliferation in postnatal hair follicles and is essential for tooth development. Development 131 (10):2257–2268

Andl T et al (2006) The miRNA-processing enzyme dicer is essential for the morphogenesis and maintenance of hair follicles. Curr Biol 16(10):1041–1049

Andreassi L et al (1998) A new model of epidermal culture for the surgical treatment of vitiligo. Int J Dermatol 37 (8):595–598

Armulik A, Abramsson A, Betsholtz C (2005) Endothelial/pericyte interactions. Circ Res 97(6):512–523

Arwert EN, Hoste E, Watt FM (2012) Epithelial stem cells, wound healing and cancer. Nat Rev Cancer 12 (3):170–180

Avolio E et al (2017) Perivascular cells and tissue engineering: current applications and untapped potential. Pharmacol Ther 171:83–92

Badiavas EV, Falanga V (2003) Treatment of chronic wounds with bone marrow-derived cells. Arch Dermatol 139(4):510–516

Bartkova J et al (2003) Cell-cycle regulatory proteins in human wound healing. Arch Oral Biol 48(2):125–132

Beele H et al (2005) A prospective multicenter study of the efficacy and tolerability of cryopreserved allogenic human keratinocytes to treat venous leg ulcers. Int J Low Extrem Wounds 4(4):225–233

Blanpain C, Fuchs E (2006) Epidermal stem cells of the skin. Annu Rev Cell Dev Biol 22:339–373

Blanpain C et al (2004) Self-renewal, multipotency, and the existence of two cell populations within an epithelial stem cell niche. Cell 118(5):635–648

Boa O et al (2013) Prospective study on the treatment of lower-extremity chronic venous and mixed ulcers using tissue-engineered skin substitute made by the self-assembly approach. Adv Skin Wound Care 26 (9):400–409

Bodnar RJ et al (2016) Pericytes: a newly recognized player in wound healing. Wound Repair Regen 24 (2):204–214

Brizzi MF, Tarone G, Defilippi P (2012) Extracellular matrix, integrins, and growth factors as tailors of the stem cell niche. Curr Opin Cell Biol 24(5):645–651

Byrne C, Tainsky M, Fuchs E (1994) Programming gene expression in developing epidermis. Development 120 (9):2369–2383

Carrier P et al (2009) Impact of cell source on human cornea reconstructed by tissue engineering. Invest Ophthalmol Vis Sci 50(6):2645–2652

Chermnykh E, Kalabusheva E, Vorotelyak E (2018) Extracellular matrix as a regulator of epidermal stem cell fate. Int J Mol Sci 19(4)

Choi HR et al (2015) Niche interactions in epidermal stem cells. World J Stem Cells 7(2):495–501

Chu GY et al (2018) Stem cell therapy on skin: mechanisms, recent advances and drug reviewing issues. J Food Drug Anal 26(1):14–20

Clark RA, Ghosh K, Tonnesen MG (2007) Tissue engineering for cutaneous wounds. J Invest Dermatol 127 (5):1018–1029

Clevers H, Loh KM, Nusse R (2014) Stem cell signaling. An integral program for tissue renewal and regeneration: Wnt signaling and stem cell control. Science 346 (6205):1248012

Corselli M et al (2010) Perivascular ancestors of adult multipotent stem cells. Arterioscler Thromb Vasc Biol 30(6):1104–1109

Davidoff MS et al (2009) The neuroendocrine Leydig cells and their stem cell progenitors, the pericytes. Adv Anat Embryol Cell Biol 205:1–107

De Rosa L et al (2014) Long-term stability and safety of transgenic cultured epidermal stem cells in gene therapy of junctional epidermolysis bullosa. Stem Cell Rep 2(1):1–8

Demarchez M et al (1987) Wound healing of human skin transplanted onto the nude mouse. II An immunohistological and ultrastructural study of the epidermal basement membrane zone reconstruction and connective tissue reorganization. Dev Biol 121(1):119–129

Doucet YS et al (2013) The touch dome defines an epidermal niche specialized for mechanosensory signaling. Cell Rep 3(6):1759–1765

Driskell RR et al (2013) Distinct fibroblast lineages determine dermal architecture in skin development and repair. Nature 504(7479):277–281

Dulmovits BM, Herman IM (2012) Microvascular remodeling and wound healing: a role for pericytes. Int J Biochem Cell Biol 44(11):1800–1812

Eichler MJ, Carlson MA (2006) Modeling dermal granulation tissue with the linear fibroblast-populated collagen matrix: a comparison with the round matrix model. J Dermatol Sci 41(2):97–108

Falabella R, Escobar C, Borrero I (1992) Treatment of refractory and stable vitiligo by transplantation of in vitro cultured epidermal autografts bearing melanocytes. J Am Acad Dermatol 26(2. Pt 1):230–236

Fuchs E (2007) Scratching the surface of skin development. Nature 445(7130):834–842

Gaspard N, Vanderhaeghen P (2010) Mechanisms of neural specification from embryonic stem cells. Curr Opin Neurobiol 20(1):37–43

Gattazzo F, Urciuolo A, Bonaldo P (2014) Extracellular matrix: a dynamic microenvironment for stem cell niche. Biochim Biophys Acta 1840(8):2506–2519

Gillespie SR, Owens DM (2018) Isolation and characterization of cutaneous epithelial stem cells. Methods Mol Biol

Green H, Kehinde O, Thomas J (1979) Growth of cultured human epidermal cells into multiple epithelia suitable for grafting. Proc Natl Acad Sci U S A 76 (11):5665–5668

Hardy MH (1992) The secret life of the hair follicle. Trends Genet 8(2):55–61

Heneidi S et al (2013) Awakened by cellular stress: isolation and characterization of a novel population of pluripotent stem cells derived from human adipose tissue. PLoS One 8(6):e64752

Hildebrand J et al (2011) A comprehensive analysis of microRNA expression during human keratinocyte differentiation in vitro and in vivo. J Invest Dermatol 131 (1):20–29

Hsu YC, Li L, Fuchs E (2014) Emerging interactions between skin stem cells and their niches. Nat Med 20 (8):847–856

Hu MS et al (2018) Embryonic skin development and repair. Organogenesis 14(1):46–63

Ito M et al (2005) Stem cells in the hair follicle bulge contribute to wound repair but not to homeostasis of the epidermis. Nat Med 11(12):1351–1354

Itoh M et al (2011) Generation of keratinocytes from normal and recessive dystrophic epidermolysis bullosa-induced pluripotent stem cells. Proc Natl Acad Sci U S A 108(21):8797–8802

Jackson CJ, Tonseth KA, Utheim TP (2017) Cultured epidermal stem cells in regenerative medicine. Stem Cell Res Ther 8(1):155

Ji J et al (2017) Aging in hair follicle stem cells and niche microenvironment. J Dermatol 44(10):1097–1104

Jimenez F et al (2012) A pilot clinical study of hair grafting in chronic leg ulcers. Wound Repair Regen 20(6):806–814

Kinoshita K et al (2015) Therapeutic potential of adipose-derived SSEA-3-positive muse cells for treating diabetic skin ulcers. Stem Cells Transl Med 4(2):146–155

Kobielak K et al (2003) Defining BMP functions in the hair follicle by conditional ablation of BMP receptor IA. J Cell Biol 163(3):609–623

Kopan R, Weintraub H (1993) Mouse notch: expression in hair follicles correlates with cell fate determination. J Cell Biol 121(3):631–641

Koster MI et al (2004) p63 is the molecular switch for initiation of an epithelial stratification program. Genes Dev 18(2):126–131

Kratochwil K et al (1996) Lef1 expression is activated by BMP-4 and regulates inductive tissue interactions in tooth and hair development. Genes Dev 10 (11):1382–1394

Krawczyk WS (1971) A pattern of epidermal cell migration during wound healing. J Cell Biol 49(2):247–263

Kretzschmar K, Watt FM (2014) Markers of epidermal stem cell subpopulations in adult mammalian skin. Cold Spring Harb Perspect Med 4(10)

Kumar A, Placone JK, Engler AJ (2017) Understanding the extracellular forces that determine cell fate and maintenance. Development 144(23):4261–4270

Kuroda Y et al (2010) Unique multipotent cells in adult human mesenchymal cell populations. Proc Natl Acad Sci U S A 107(19):8639–8643

Lapouge G, Blanpain C (2008) Medical applications of epidermal stem cells. In: StemBook. Harvard Stem Cell Institute, Cambridge, MA

Lee CH, Singla A, Lee Y (2001) Biomedical applications of collagen. Int J Pharm 221(1–2):1–22

Lee LF et al (2011) A simplified procedure to reconstitute hair-producing skin. Tissue Eng Part C Methods 17 (4):391–400

Lee V et al (2014) Design and fabrication of human skin by three-dimensional bioprinting. Tissue Eng Part C Methods 20(6):473–484

Levy V et al (2007) Epidermal stem cells arise from the hair follicle after wounding. FASEB J 21 (7):1358–1366

Lo Cicero A et al (2015) Exosomes released by keratinocytes modulate melanocyte pigmentation. Nat Commun 6:7506

Lyons KM, Pelton RW, Hogan BL (1989) Patterns of expression of murine Vgr-1 and BMP-2a RNA suggest that transforming growth factor-beta-like genes coordinately regulate aspects of embryonic development. Genes Dev 3(11):1657–1668

Lyons KM, Pelton RW, Hogan BL (1990) Organogenesis and pattern formation in the mouse: RNA distribution patterns suggest a role for bone morphogenetic protein-2A (BMP-2A). Development 109(4):833–844

Mascre G et al (2012) Distinct contribution of stem and progenitor cells to epidermal maintenance. Nature 489 (7415):257–262

Mavilio F et al (2006) Correction of junctional epidermolysis bullosa by transplantation of genetically modified epidermal stem cells. Nat Med 12 (12):1397–1402

Michno K et al (2003) Shh expression is required for embryonic hair follicle but not mammary gland development. Dev Biol 264(1):153–165

Ming Kwan K et al (2004) Essential roles of BMPR-IA signaling in differentiation and growth of hair follicles and in skin tumorigenesis. Genesis 39(1):10–25

Moustafa M et al (2007) Randomized, controlled, single-blind study on use of autologous keratinocytes on a transfer dressing to treat nonhealing diabetic ulcers. Regen Med 2(6):887–902

Muller-Rover S et al (2001) A comprehensive guide for the accurate classification of murine hair follicles in distinct hair cycle stages. J Invest Dermatol 117 (1):3–15

Nikaido M et al (1999) In vivo analysis using variants of zebrafish BMPR-IA: range of action and involvement of BMP in ectoderm patterning. Development 126 (1):181–190

Nowak JA et al (2008) Hair follicle stem cells are specified and function in early skin morphogenesis. Cell Stem Cell 3(1):33–43

Odland G, Ross R (1968) Human wound repair. I. Epidermal regeneration. J Cell Biol 39(1):135–151

Ogura F et al (2014) Human adipose tissue possesses a unique population of pluripotent stem cells with nontumorigenic and low telomerase activities: potential implications in regenerative medicine. Stem Cells Dev 23(7):717–728

Ojeh N et al (2015) Stem cells in skin regeneration, wound healing, and their clinical applications. Int J Mol Sci 16 (10):25476–25501

Okuyama R et al (2004) High commitment of embryonic keratinocytes to terminal differentiation through a Notch1-caspase 3 regulatory mechanism. Dev Cell 6 (4):551–562

Oro AE, Higgins K (2003) Hair cycle regulation of Hedgehog signal reception. Dev Biol 255(2):238–248

Ortega-Zilic N et al (2010) EpiDex(R) Swiss field trial 2004-2008. Dermatology 221(4):365–372

Page ME et al (2013) The epidermis comprises autonomous compartments maintained by distinct stem cell populations. Cell Stem Cell 13(4):471–482

Paladini RD et al (2005) Modulation of hair growth with small molecule agonists of the hedgehog signaling pathway. J Invest Dermatol 125(4):638–646

Pan Y et al (2004) Gamma-secretase functions through notch signaling to maintain skin appendages but is not required for their patterning or initial morphogenesis. Dev Cell 7(5):731–743

Paquet-Fifield S et al (2009) A role for pericytes as microenvironmental regulators of human skin tissue regeneration. J Clin Invest 119(9):2795–2806

Powell HM, Supp DM, Boyce ST (2008) Influence of electrospun collagen on wound contraction of engineered skin substitutes. Biomaterials 29 (7):834–843

Proksch E, Brandner JM, Jensen JM (2008) The skin: an indispensable barrier. Exp Dermatol 17 (12):1063–1072

Purba TS et al (2014) Human epithelial hair follicle stem cells and their progeny: current state of knowledge, the widening gap in translational research and future challenges. BioEssays 36(5):513–525

Rajkumar VS et al (2006) Platelet-derived growth factor-beta receptor activation is essential for fibroblast and pericyte recruitment during cutaneous wound healing. Am J Pathol 169(6):2254–2265

Rangarajan A et al (2001) Notch signaling is a direct determinant of keratinocyte growth arrest and entry into differentiation. EMBO J 20(13):3427–3436

Reya T et al (2001) Stem cells, cancer, and cancer stem cells. Nature 414(6859):105–111

Rheinwald JG, Green H (1975) Serial cultivation of strains of human epidermal keratinocytes: the formation of keratinizing colonies from single cells. Cell 6 (3):331–343

Riddle CV (1986) Focal tight junctions between mesenchymal cells of fetal dermis. Anat Rec 214(2):113–117

Rigal C et al (1991) Healing of full-thickness cutaneous wounds in the pig. I. Immunohistochemical study of epidermo-dermal junction regeneration. J Invest Dermatol 96(5):777–785

Rompolas P, Greco V (2014) Stem cell dynamics in the hair follicle niche. Semin Cell Dev Biol 25-26:34–42

Rubin AI, Chen EH, Ratner D (2005) Basal-cell carcinoma. N Engl J Med 353(21):2262–2269

Saarialho-Kere UK et al (1995) Interstitial collagenase is expressed by keratinocytes that are actively involved in reepithelialization in blistering skin disease. J Invest Dermatol 104(6):982–988

Sadowski T et al (2017) Large-scale human skin lipidomics by quantitative, high-throughput shotgun mass spectrometry. Sci Rep 7:43761

Shabbir A et al (2015) Mesenchymal stem cell exosomes induce proliferation and migration of normal and chronic wound fibroblasts, and enhance angiogenesis in vitro. Stem Cells Dev 24(14):1635–1647

Silva-Vargas V et al (2005) Beta-catenin and Hedgehog signal strength can specify number and location of hair follicles in adult epidermis without recruitment of bulge stem cells. Dev Cell 9(1):121–131

Smith LT, Holbrook KA (1986) Embryogenesis of the dermis in human skin. Pediatr Dermatol 3(4):271–280

Sorg H et al (2017) Skin wound healing: an update on the current knowledge and concepts. Eur Surg Res 58 (1–2):81–94

Stanley JR et al (1981) Detection of basement membrane zone antigens during epidermal wound healing in pigs. J Invest Dermatol 77(2):240–243

St-Jacques B et al (1998) Sonic hedgehog signaling is essential for hair development. Curr Biol 8 (19):1058–1068

Sun X et al (2013) Epidermal stem cells: an update on their potential in regenerative medicine. Expert Opin Biol Ther 13(6):901–910

Tadeu AM, Horsley V (2014) Epithelial stem cells in adult skin. Curr Top Dev Biol 107:109–131

Terskikh VV, Vasiliev AV, Vorotelyak EA (2012) Label retaining cells and cutaneous stem cells. Stem Cell Rev 8(2):414–425

Tomic-Canic M et al (1998) Epidermal signal transduction and transcription factor activation in activated keratinocytes. J Dermatol Sci 17(3):167–181

Valadi H et al (2007) Exosome-mediated transfer of mRNAs and microRNAs is a novel mechanism of genetic exchange between cells. Nat Cell Biol 9 (6):654–659

Vauclair S et al (2005) Notch1 is essential for postnatal hair follicle development and homeostasis. Dev Biol 284(1):184–193

Viticchie G et al (2012) MicroRNA-203 contributes to skin re-epithelialization. Cell Death Dis 3:e435

Wakao S et al (2011) Multilineage-differentiating stress-enduring (Muse) cells are a primary source of induced pluripotent stem cells in human fibroblasts. Proc Natl Acad Sci U S A 108(24):9875–9880

Wakao S et al (2014) Muse cells, newly found non-tumorigenic pluripotent stem cells, reside in human mesenchymal tissues. Pathol Int 64(1):1–9

Wang LC et al (2000) Regular articles: conditional disruption of hedgehog signaling pathway defines its critical role in hair development and regeneration. J Invest Dermatol 114(5):901–908

Watt FM, Lo Celso C, Silva-Vargas V (2006) Epidermal stem cells: an update. Curr Opin Genet Dev 16 (5):518–524

Wilson PA, Hemmati-Brivanlou A (1995) Induction of epidermis and inhibition of neural fate by Bmp-4. Nature 376(6538):331–333

Wong VW et al (2012) Stem cell niches for skin regeneration. Int J Biomater 2012:926059

Yamauchi T et al (2017) The potential of muse cells for regenerative medicine of skin: procedures to reconstitute skin with muse cell-derived keratinocytes, fibroblasts, and melanocytes. J Invest Dermatol 137 (12):2639–2642

Yamazaki T et al (2017) Tissue myeloid progenitors differentiate into pericytes through TGF-beta signaling in developing skin vasculature. Cell Rep 18 (12):2991–3004

Yi R et al (2006) Morphogenesis in skin is governed by discrete sets of differentially expressed microRNAs. Nat Genet 38(3):356–362

Yoshikawa T et al (2008) Wound therapy by marrow mesenchymal cell transplantation. Plast Reconstr Surg 121(3):860–877

Zebardast N, Lickorish D, Davies JE (2010) Human umbilical cord perivascular cells (HUCPVC): a mesenchymal cell source for dermal wound healing. Organogenesis 6(4):197–203

Zhuang L et al (2018) Pericytes promote skin regeneration by inducing epidermal cell polarity and planar cell divisions. Life Sci Alliance 1(4):e201700009

Zouboulis CC et al (2008) Human skin stem cells and the ageing process. Exp Gerontol 43(11):986–997

Adv Exp Med Biol – Cell Biology and Translational Medicine (2019) 6: 127–153
https://doi.org/10.1007/5584_2019_398
© Springer Nature Switzerland AG 2019
Published online: 25 July 2019

Neurological Regulation of the Bone Marrow Niche

Fatima Aerts-Kaya, Baris Ulum, Aynura Mammadova,
Sevil Köse, Gözde Aydin, Petek Korkusuz, and
Duygu Uçkan-Çetinkaya

Abstract

The bone marrow (BM) hematopoietic niche is the microenvironment where in the adult hematopoietic stem and progenitor cells (HSPCs) are maintained and regulated. This regulation is tightly controlled through direct cell-cell interactions with mesenchymal stromal stem (MSCs) and reticular cells, adipocytes, osteoblasts and endothelial cells, through binding to extracellular matrix molecules and through signaling by cytokines and hematopoietic growth factors. These interactions provide a healthy environment and secure the maintenance of the HSPC pool, their proliferation, differentiation and migration. Recent studies have shown that innervation of the BM and interactions with the peripheral sympathetic neural system are important for maintenance of the hematopoietic niche, through direct interactions with HSCPs or via interactions with other cells of the HSPC microenvironment. Signaling through adrenergic receptors (ARs), opioid receptors (ORs), endocannabinoid receptors (CRs) on HSPCs and MSCs has been shown to play an important role in HSPC homeostasis and mobilization. In addition, a wide range of neuropeptides and neurotransmitters, such as Neuropeptide Y (NPY), Substance P (SP) and Tachykinins, as well as neurotrophins and neuropoietic growth factors have been shown to be involved in regulation of the hematopoietic niche. Here, a comprehensive overview is given of their role and interactions with important cells in the hematopoietic niche, including HSPCs and MSCs, and their effect on HSPC maintenance, regulation and mobilization.

F. Aerts-Kaya (✉), A. Mammadova, G. Aydin, and
D. Uçkan-Çetinkaya
Graduate School of Health Sciences, Department of Stem Cell Sciences, Hacettepe University, Ankara, Turkey

Center for Stem Cell Research and Development,
Hacettepe University, Ankara, Turkey
e-mail: fatimaaerts@yahoo.com; fatima.aerts@hacettepe.edu.tr

B. Ulum
Center for Stem Cell Research and Development,
Hacettepe University, Ankara, Turkey

Faculty of Arts and Sciences, Department of Biological Sciences, Middle East Technical University, Ankara, Turkey

S. Köse
Faculty of Health Sciences, Department of Medical Biology, Atilim University, Ankara, Turkey

P. Korkusuz
Graduate School of Health Sciences, Department of Stem Cell Sciences, Hacettepe University, Ankara, Turkey

Faculty of Medicine, Department of Histology and Embryology, Hacettepe University, Ankara, Turkey

Keywords

Bone Marrow Niche · Endocannabinoids · Hematopoiesis · Neuropeptides · Opioids · Tachykinins

Abbreviations

2-AG	2-ArachidonoylGlycerol
ACh	acetylcholine
AEA	Anandamide
AGM	aorta-gonad-mesonephros
ARs	adrenergic receptors
BDNF	Brain-Derived Growth factor
BM	bone marrow
CBD	Cannabidiol
CD271	Low affinity nerve growth factor receptor
CFU-F	Colony Forming Unit-Fibroblast
CFU-GEMM	Colony Forming Unit-Granulocyte/Erythrocyte/Monocyte/Megakaryocyte
ChAT	choline acetyltransferase
CKs	cytokines
CNS	central nervous system
CNTF	ciliary neurotrophic factor
CRs	endocannabinoid receptors
CT-1	cardiotrophin-1
CXCL12/SDF1	Stromal Derived Factor-1
D	Dopamine
DCs	Dendritic cells
DRs	dopamine receptors
E	Epinephrine/Adrenaline
ECB	Endocannabinoid
ECM	extracellular matrix
ECs	endothelial cells
EK	Endokinin
FAAH	fatty acid amide hydrolase
FGF	Fibroblasts growth factor
FTOC	fetal thymus organ cultures
G-CSF	Granulocyte-Colony Stimulating Factor
GDNF	glial cell-line derived neurotrophic factor
GDNF	Glial-derived Neurotrophic Factor
GFLs	GDNF family of ligands
GM-CSF	Granulocyte/Macrophage-Colony Stimulating Factor
GPCRs	G-protein coupled receptors
HGFs	hematopoietic growth factors
HK-1	Hemokinin-1
HSCs	hematopoietic stem cells
HSPCs	hematopoietic stem and progenitor cells
IL	Interleukin
LIF	Leukemia inhibiting factor
LSK cells	Lin$^-$Sca$^+$c-kit$^+$ cells
MIP1α	Macrophage Inflammatory Protein-1alpha
MMP	Metalloproteinase
MSCs	mesenchymal stromal/stem cells
NE	Norepinephrine/Noradrenaline
NGF	Nerve growth factor
NK cell	Natural Killer cell
NKA	Neurokinin A
NKB	Neurokinin B
NK-Rs	Neurokinin receptors
NPY	Neuropeptide Y
NTs	Neurotrophins
OBs	osteoblasts
ORs	opioid receptors
OSM	Oncostatin M
PAA	periarterial adventitial cells
PDGF	Platelet-derived growth factor
RET	rearranged during transfection receptor
RTK	receptor tyrosine kinase
SCF	Stem Cell Factor
SNS	sympathetic nervous system
SP	Substance P
TGFβ	Transforming Growth Factor
TH	tyrosine hydrolase
THC	Tetrahydrocannabinol
TK	Tachykinins
TLRs	Toll-like receptors
TNFR	Tumor necrosis factor receptor
TNFα	Tumor Necrosis Factor alpha
TPO	Thrombopoietin
Trk	tropomyosin receptor kinase
UCB	Umbilical Cord Blood
VEGF	Vascular endothelial growth factor
WT	Wild type

1   Introduction

In the adult, Hematopoietic Stem Cells (HSCs) and progenitor cells are located in a specialized microenvironment called the bone marrow (BM) niche. This microenvironment contains all the requirements for the support and maintenance of healthy hematopoiesis. The functions of these HSCs, as well as their fate are determined by tightly regulated interactions with this BM niche, which directs HSC survival, quiescence, self-renewal, proliferation, differentiation and migration. The BM niche contains many different types of supporting cells, such as Nestin+ Mesenchymal Stem Cells (MSCs), osteoblasts (OBs), macrophages, adipocytes, fibroblasts, osteoclasts, CXCL12-abundant reticular (CAR) cells, endothelial cells (ECs) and perivascular cells, but also sympathetic nerves and non-myelinated GFAP+ Schwann cells (Morrison and Scadden 2014). In addition to cell-cell interactions, regulation and maintenance of HSCs is heavily affected by signals transferred from the extracellular matrix (ECM), such as fibronectin, collagen, laminin and proteoglycans; by gradients of hormones, hematopoietic growth factors (HGFs), cytokines (CKs) and chemokines; by pH, O_2 tension and calcium gradients; as well as by interactions with adhesion molecules for anchoring and retention in and release of HSCs from the niche. Thus, the BM niche provides a complex shelter to protect HSCs from harmful conditions and environmental damage.

The sympathetic nervous system (SNS) has also been implicated in the physiological regulation of the BM niche and direction of hematopoietic stress responses. Anatomically, both myelinated and unmyelinated nerve fibers enter the marrow cavity alongside nutrient blood vessels (Kuntz and Richins 1945) and provide an extensive network of innervation of the BM (Artico et al. 2002). The nerve fibers can be visualized using staining for tyrosine hydrolase (TH), which is involved in catecholamine synthesis. TH+ noradrenergic nerve fibers have been found to be closely associated with blood vessels, but can also be found in the periosteum, mineralized bone and BM (Jung et al. 2017). The nerve fibers secrete different classes of neurotransmitters, including the catecholamines Norepinephrine (NE) and Dopamine (D), and a range of neurotransmitters, neuropeptides and neurotrophic factors (Tabarowski et al. 1996). Both afferent and efferent sympathetic nerves can be found adjacent to the vasculature of the BM (Calvo and Forteza-Vila 1969). Furthermore, sympathetic nerves synapse with perivascular cells to regulate BM homeostasis and are associated with HSC quiescence and regulation of self-renewal (Park et al. 2015a; Yamazaki et al. 2011). These perivascular cells include a specialized group of periarterial adventitial (PAA) cells, that are attached to each other through gap-junctions and are found in close proximity of nerve terminals (Yamazaki and Allen 1990). This network was appropriately termed the "neuro-reticular complex". The role of the parasympathetic innervation of the BM is less well established. However, parasympathetic choline acetyltransferase (ChAT) + nerve fibers which synthesize acetylcholine (ACh) have been detected around hematopoietic islets in the BM of rats (Artico et al. 2002). The significance of BM innervation gains even more importance after realizing that neuronal control of hematopoiesis already commences during early embryogenesis, and that only after the beginning of BM innervation the onset of hematopoiesis is initiated (Fitch et al. 2012).

Nerve endings releasing Substance P (SP) and Neurokinin A (NKA) have been shown to control secretion of hematopoietic growth factors and cytokines from stromal and parenchymal cells, BM stromal cells and monocytes (Rameshwar and Gascon 1995, 1996), whereas Granulocyte-Colony Stimulating Factor (G-CSF) and Granulocyte/Macrophage-Colony Stimulating Factor (GM-CSF) were shown to upregulate neuronal receptor expression on human HSCs, further increasing sensitivity of these cells for neuronal cues (Kalinkovich et al. 2009). BM nerve damage and physical or chemical denervation of the BM was shown to severely impact the survival of HSCs and BM regeneration and treatment with neuroprotective agents, such as Neuropeptide Y(NPY), or neuroregeneration supporting growth factors, such as Glial-derived Neurotrophic Factor (GDNF) restored BM dysfunction and promoted hematopoietic recovery (Park et al. 2015a; Lucas et al. 2013). Non-myelinating Schwann cells were shown to maintain HSC quiescence through production of TGFβ and induction of Smad signaling. Autonomic denervation reduced the number of these glial cells, resulting in rapid loss of HSCs from the BM (Yamazaki et al. 2011; Bruckner 2011).

The SNS also plays an important role in HSC mobilization (Katayama et al. 2006). Studies using the UDP-galactose ceramide galactosyl transferase-deficient ($Cgt^{-/-}$) mouse model, elegantly showed that defects in nerve conduction and absence of adrenergic neurotransmission are directly responsible for the lack of HSC egress from BM following G-CSF stimulation (Katayama et al. 2006). Furthermore, important evidence that supports the role of the SNS in HSC mobilization came from a study that showed that HSC release from the BM is regulated by circadian oscillations (Mendez-Ferrer et al. 2008). In mice, numbers of circulating HSCs were shown to fluctuate in antiphase with BM Cxcl12 concentrations. These circadian changes in levels of Cxcl12 and circulating HSCs was shown to be synchronized through Norepinephrine (NE) secretion. NE secreted from BM-SNS fibers interacted with β-adrenergic receptors expressed by osteoblasts, MSCs and HSCs, and caused a decrease in the expression of the transcription factor Sp1 and downregulation of Cxcl12 (Mendez-Ferrer et al. 2008; Mendez-Ferrer et al. 2010). G-CSF and GM-CSF treatment has been shown to increase the expression of DR3 and DR5 dopamine receptors and b2-adrenergic receptors on immature CD34+ cells in mice and humans (Spiegel et al. 2007; Kose et al. 2018). Furthermore, treatment with neurotransmitters increased the motility, proliferation and colony formation of human HSCs, as well as expression and activity of the metalloproteinases MT1-MMP and MMP-2, respectively (Spiegel et al. 2007).

Thus, the regulation of HSC homeostasis, self-renewal, differentiation and mobilization appears to be a cooperative interaction between the nervous system and the bone marrow, through direct communication with HSPCs or indirect effects through other cells of the bone marrow niche, as well as the involvement of hormones, cytokines, neurotrophic factors, neurotransmitters, and neuropeptides, which will be discussed in detail below. In line with the above data, Lapidot and colleagues proposed the presence of a blood-brain-blood triad (Lapidot and Kollet 2010).

2 Adrenergic Regulation of the Hematopoietic Bone Marrow Niche

Interactions between neural signals and hematopoiesis occur early during embryogenesis where sympathetic signaling contributes to the emergence of primitive HSCs from the aorta-gonad-mesonephros (AGM) region (Fitch et al. 2012). Microarray analysis showed that expression of the transcription factor *Gata3* was upregulated in the AGM and deletion of *Gata3* caused a significant reduction in the numbers of AGM resident HSPCs at E10.5 and E11.5 days. Interestingly, local *Gata3* expression was found not to co-localize with AGM-HSPCs, but with SNS fibers and *Gata3* was shown to regulate HSC numbers and development via the production of catecholamines. Similarly, deficiency of the sympathoadrenal marker Tyrosine Hydroxylase (Th) which is normally expressed in the SNS and is responsible for the production of catecholamines, was shown to result in a similar phenotype with a dramatic reduction in the numbers of AGM-HSPCs (Fitch et al. 2012).

Dopamine (D) is an important neurotransmitter in the central nervous system, where it mainly regulates movement, cognition, motivation-reward responses. However, it has also been shown to be released from sympathetic nerve endings in the peripheral nervous system. Epinephrine (E, adrenaline) is mainly derived from the adrenal medulla, whereas norepinephrine (NE, noradrenaline) is secreted from sympathetic nerves. All three neurotransmitters are closely related molecules, that are derived from the non-essential amino acid Tyrosine and act through activation of 7-transmembrane G-protein coupled receptors

(GPCRs). Dopaminergic receptors (DR) consist of five different subtypes, i.e. the D1-like (D1 and D5) receptors, which activate adenylate cyclase and the D2-like (D2, D3 and D4) receptors, which inhibit adenylate cyclase (Cosentino et al. 2015). DRs are expressed on most hematopoietic cells, including T and B cells, dendritic cells, macrophages, neutrophils and NK cells (Cosentino et al. 2015) and have been shown to play an important role in the regulation of immune responses. The adrenergic receptors (AR) are classified into alpha-adrenergic receptors (α-AR) and beta-adrenergic receptors (β-AR). The effects of the SNS on HSCs have been suggested to be mediated by signaling via both α-AR and β-AR. Surface expression of α1-AR, α2-AR, β1-AR, β2-AR and β3-AR was detected on human adult CD34+ HSCs; on murine embryonic HSCs β2-AR was found to be predominantly expressed; expression of α1-AR, α2-AR and β2-AR was shown on different subsets of adult murine myeloid progenitors, multipotent progenitors and HSCs, and β3-AR expression was found on BM-MSCs (Kose et al. 2018; Spiegel et al. 2007; Mendez-Ferrer et al. 2008; Fitch et al. 2012; Muthu et al. 2007).

Maestroni and colleagues were among the first to show the involvement of the adrenal system in the modulation of hematopoiesis (Maestroni et al. 1992). They showed that chemical sympathectomy using B-hydroxydopamine (6-OHDA) and the α-adrenergic antagonist prazosin significantly increased the number of PB granulocytes, monocytes and platelet counts after BMT, whereas addition of the β-adrenergic blocker propranolol completely abolished the effects on blood platelets, but not on white blood cell count (Maestroni et al. 1992). In contrast, α-adrenergic agonists seem to exert an inhibitory effect on myelopoiesis (Maestroni and Conti 1994) and it has been suggested that inhibition of the α1-AR might increase interactions of NE with beta-adrenergic receptors. Not only do catecholamines play an important role in the regulation of hematopoiesis through α1-AR, they are also secreted in high levels by BM sympathetic postganglionic fibers and BM resident cells (Marino et al. 1997).

Also (circadian) release of HSCs from the BM has been shown to be regulated through involvement of the adrenergic system (Katayama et al. 2006). Adrenergic signals are directly delivered to the hematopoietic niche through secretion from sympathetic nerve endings and transmitted to β3-AR expressing BM-MSCs. This results in a decrease of the transcription factor Sp1 and rapid downregulation of SDF-1 (Mendez-Ferrer et al. 2008). Furthermore, these circadian oscillations were shown to occur through changes in parasympathetic and sympathetic activity. At night, when parasympathetic cholinergic signals are dominant, the sympathetic noradrenergic tone is suppressed and egress of HSPCs from the BM decreases. In contrast, during daytime, increased sympathetic noradrenergic activity causes HSPC mobilization through activation of β3-AR (Garcia-Garcia et al. 2019). β2-AR, but not β3-AR stimulation was shown to induce sequential gene expression of clock genes *per1*, *bmal1* and *clock,* showing that although both β2-AR and β3-AR are both required for G-CSF induced mobilization of HSPCs, their roles are functionally distinct. In addition, a double deficiency of both β2-AR and β3-AR negatively affects HSPCs mobilization, showing that although β2- and β3-ARs have distinct roles in stromal cells, during HSPC mobilization both are needed (Mendez-Ferrer et al. 2010).

Further proof that the adrenergic system is heavily involved in modulation of HSPCs trafficking was shown using the UDP-galactose ceramide galactosyltransferase-deficient ($Cgt^{-/-}$) mouse model. These mice were shown to exhibit impaired nerve conduction and lack of HSPC mobilization in response to G-CSF stimulation. In addition, the adrenergic tone, OB function, and bone SDF-1 were all shown to be dysfunctional in this mouse model (Katayama et al. 2006). Inhibition of adrenergic signaling showed that NE regulates G-CSF-induced OB suppression, bone SDF-1 downregulation, and HSPC mobilization, and administration of a β2-AR agonist enhanced mobilization in both control and NE-deficient mice (Katayama et al. 2006).

3   Opioids

Opiates were first isolated as natural extracts from the poppy plant (*Papaver Somniverum*) and have been used medically for centuries for mediation of nociception and sedation. The most well-known opiate Morphine binds to Opioid receptors (Opr) in the central and peripheral nervous system. Opioid receptors belong to the superfamily of the GPCRs (Satoh and Minami 1995). In vertebrates, the opioid receptors are encoded by the μ (Oprm1), κ (Oprk1) and δ (Oprd1) genes and are respectively named MOR, KOR and DOR (Waldhoer et al. 2004). Endogenous opioid signaling plays an important role in the regulation of analgesia, euphoria, homeostasis, anorexia/obesity, immune responses and the cardiovascular system. The opioid peptides are further classified in three genetically distinct families, i.e. the endorphins, enkephalins and dynorphins (Kieffer and Evans 2009). These endogenous opioids function as neurotransmitters or neuromodulators. Similar to other small peptide molecules, endogenous opioids are synthesized through proteolytic cleavage of large protein precursors. These precursors include pro-opiomelanocortin (POMC), the precursor of β-endorphin; pre-pro-enkephalin (PENK), the precursor of leucine (Leu)- and methionine (Met)-enkephalins; and pre-pro-dynorphin (PDYN), the precursor of the dynorphins (Benarroch 2012). Each opioid peptide is derived from a pre-pro and a pro-form, which is then further cleaved and processed through post-translational modifications. As a result of these modifications, receptor binding, affinity and potency of the responses towards opioids is context and tissue-dependent. All opioids share a common five amino acid motif at their N-terminus consisting of Tyr-Gly-Gly-Phe-Met/ Leu. POMC is first cleaved into adrenocorticotropin-releasing hormone (ACTH) and β-Lipotropin (βLPH). The latter is further cleaved into α-melanocyte-stimulating hormone (αMSH) and β-endorphin. Beta-endorphins have been shown to both serve as locally active neuromodulators and as paracrine hormones after release into the peripheral circulation. Cleavage of pro-enkephalin results in the release of four

met-enkephalins, one leu-enkephalin, one octapeptide and one heptapeptide. As a consequence of the rapid metabolism by peptidases enkephalinase-A and enkephalinase-B, the enkephalins are known to have a short half-life. Cleavage of pro-dynorphin results in the formation of dynorphin A, dynorphin B and neoendorphin (Benarroch 2012).

Although opioid peptides such as β-endorphin and the dynorphin have been shown to play a role in the modulation of the immune system and function as cytokines (Bidlack 2000), not much is known about their role in hematopoiesis. Opioid receptor expression has been detected on many hematopoietic cells, including DOR expression by T-cells, B-cells, macrophages and BM dendritic cells (DCs) and BM-MSCs (Benard et al. 2008; Higuchi et al. 2012); KOR expression by lymphocytes, PB-CD34+ cells (Steidl et al. 2004), BM neutrophils (Kulkarni-Narla et al. 2001), BM stromal cells and BM macrophages (Maestroni 1998, 1999); and MOR expression by lymphocyte subsets, monocytes/macrophages, granulocytes (Sharp et al. 1998) and PB-CD34+ (Steidl et al. 2004) and UCB-CD34+ cells (Rozenfeld-Granot et al. 2002). Agonists of in particularly KOR and to a lesser extent MOR and DOR were shown to inhibit chemotaxis of BM neutrophils towards macrophage inflammatory protein-2 (MIP-2) through downregulation of β2-integrin CD11b/CD18 in a dose-dependent, and naloxone-reversible fashion (Kulkarni-Narla et al. 2001). However, enkephalins themselves appear to have chemotactic properties on DOR expressing mature BM-DCs, although the migratory effect on mature DCs decreased when the concentration of other chemokines, such as CCL19 and CCL21 increased (Benard et al. 2008). When the effects of opioid receptor agonists on colony forming capacity of BM were tested, agonists of KOR, such as Dynorphin A, increased numbers of colony forming unit-granulocyte/macrophage (CFU-GM) in synergy with GM-CSF (Maestroni et al. 1999). In contrast, DOR agonists, such as met-enkephalin, decreased numbers of CFU-GM in a circadian pattern when HSPCs were cultured in presence of adherent BM stromal cells, whereas leu-enkephalin predominantly suppressed lymphoid cell proliferation (Krizanac-Bengez et al. 1996). Mice subjected to stress induced by immobilization, were shown to develop BM hyperplasia with increased production of neutrophils, monocytes and erythrocytes, which could be completely inhibited by injection of the DOR agonists Leu-enkephalin and its synthetic analog Dalargin (Gol'dberg et al. 1987). MOR-deficient mice displayed increased proliferation of granulocyte-macrophage, erythroid, and hematopoietic progenitor cells in both BM and spleen (Tian et al. 1997). The neutral endopeptidase CD10 (NEP, CALLA, enkephalinase) is expressed by immature lymphoid, myeloid and BM stromal cells and through cleavage can activate or inactivate opioid peptides and a range of neuropeptides, such as Substance P and tachykinins (Boranic et al. 1997).

4 Endocannabinoids

The endocannabinoid (ECB) system consists of two G protein-coupled classical Cannabinoid Receptors (CB1 and CB2), the endogenous Cannabis ligands Anandamide (AEA) and 2-ArachidonoylGlycerol (2-AG), who have an unsaturated fatty acid structure, and the enzymes fatty acid amide hydrolase (FAAH) and monoacylglycerol lipase, respectively (Diaz-Laviada and Ruiz-Llorente 2005). ECBs and their receptors are distributed and act on many systems, predominantly, but not exclusively, the central nervous system (CNS) (Diaz-Laviada and Ruiz-Llorente 2005). Members of the ECB system are involved in many physiological processes, such as hunger control, pain perception, motor function, bone metabolism and regulation of the immune response and may be involved in the development of a range of pathological conditions, including certain inflammatory diseases and the development or progression of cancer (Maccarrone et al. 2015). Furthermore, ECBs play important roles in the regulation of proliferation, differentiation, migration and apoptosis of germ cells and many somatic cells.

4.1 Endocannabinoids and Hematopoietic Stem Cells

The expression of ECBs in hematopoietic tissues, including BM, spleen and thymus, has been reported in a limited number of studies. CB receptor expression has been detected in macrophages, erythroid, B and T lymphoid, mast cell lines (Valk et al. 1997), murine BM HSCs and MSCs (Scutt and Williamson 2007). CB receptor mediated cell migration has been demonstrated in rodent myeloid leukemia cells (Jiang et al. 2011b), mouse mononuclear cells (MNCs) (Patinkin et al. 2008) and endothelial cells (Mo et al. 2004). The interactions between the ECB system and BM cells were investigated by assessing the presence of CB receptors on rodent BM cells and hematopoietic growth factor (HGF, IL-3, GM-CSF, G-CSF, EPO) dependent cell lines. Expression of the CB1 receptor was found on a T lymphoid cell line, whereas expression of the CB2 receptor could be detected more broadly on a variety of myeloid, macrophage, erythroid, B-lymphoid, T-lymphoid and mast cell lines (Valk et al. 1997). The stimulatory effect of AEA on proliferation of these cell lines was further enhanced by addition of IL-3, GM-CSF, G-CSF and EPO. Therefore, it was proposed that AEA could be used to synergistically to enhance proliferation of hematopoietic cells in combination with known hematopoietic growth factors and cytokines (Yamaguchi and Levy 2016). Furthermore, stimulation of CB2 overexpressing rodent myeloid leukemia cells was shown to induce migration of the cells in the direction of AEA, 2-AG and other cannabinoids, thus revealing a role for ECBs in migration and chemotaxis of hematopoietic cells (Jiang et al. 2011a). It has also been suggested that the primary function of CB2 receptor expression on B lymphocytes might be regulation of migration of these cells (Jorda et al. 2002). Administration of AEA and 2-AG to mice was shown to increase the numbers of BM CFU-GEMM colonies in culture two-fold. Furthermore, an increase in circulating mouse MNCs was shown within 4 h of administration of 2-AG and AEA. Studies using CB1 and CB2 receptor antagonists showed that the effect of 2-AG is mainly mediated by interaction with CB1 (Patinkin et al. 2008). The importance of CB2 for hematopoiesis was emphasized in a study where hematopoietic recovery was assessed in CB2 knockout ($Cnr2^{-/-}$) mice and WT mice subjected to sublethal irradiation. In comparison to the WT mice, the $Cnr2^{-/-}$ mice displayed considerably decreased hematopoietic recovery with low colony formation and a prolonged period of low PB counts. When the WT mice were administered CB2 agonists after sublethal irradiation, HSC survival increased and resulted in accelerated BM recovery. Additionally, when WT mice were administered CB2 agonists 12 days prior to sublethal irradiation, HSC apoptosis decreased, c-kit positive and total BM cell numbers increased, resulting in an overall increased survival of the mice (Jiang et al. 2011b). Interestingly, during and after transplantation of HSCs, peripheral blood levels of IL-6 and 2-AG increased. This increase in 2-AG was linked to a general increase in stress and inflammation in the receiver (Knight et al. 2015).

4.2   Endocannabinoids and Mesenchymal Stem Cells

In addition to their effect on HSC, ECBs have also been shown to be involved in the regulation of migration, differentiation, survival and activity of bone cells. In addition, the ECB system has been shown to play a critical role in the regulation of bone mass maintenance (Gowran et al. 2013; Idris et al. 2005; Idris et al. 2008; Idris et al. 2009). *Ex vivo* studies have shown that bone cells express both CB1 and CB2 receptors, as well as the Transient Receptor Potential Vanilloid type 1 (TRPV1) receptors and the G-coupled protein receptor 55 (GPR55), which have been shown to oppositely modulate the effects of ECBs *in vitro* human osteoblast activity (Whyte et al. 2009). In addition, they were also shown to possess the machinery required for synthesis and metabolism of ECBs (Smith et al. 2015). Although CB1 and CB2 receptors could not be detected on the general

population of BM stromal cells; both receptors have been shown to be present in a small fraction of the BM stromal cell side population, which is typically enriched for stem cells (Scutt and Williamson 2007; Kose et al. 2018).

In mice, CB1 receptor inactivation was shown to initially result in increased bone mass and protection from ovariectomy-induced bone loss due to increased osteoclast activity (Idris et al. 2005). However, prolonged CB1 receptor deficiency eventually results in osteoporosis due to preferential differentiation of the BM stromal cells towards adipogenic lineage (Idris et al. 2009). CB2 receptor activation plays a role in bone turnover and CB2 agonists were shown to enhance bone mass by increasing osteoblasts numbers and activity, by inhibition of osteoclast proliferation and by increasing CFU-F numbers (Ofek et al. 2006; Scutt and Williamson 2007; Idris et al. 2008). To further assess the effect of ECBs on BM migration and colony forming unit-fibroblast (CFU-F) formation, 2-AG was applied to rat BM cultures. Addition of 2-AG to these cultures resulted in a dose-dependent increase in colony size and colony formation. In addition, it is thought that activation of the CB2 receptor plays an important role in osteogenic differentiation of BM-MSCs (Yamaguchi and Levy 2016). Tetrahydrocannabinol (THC) is the psychoactive component of cannabis and functions as a partial agonist of the CB1 receptor and an antagonist of the CB2 receptor. Addition of THC *in vitro* to cultures of rat BM-MSCs resulted in a negative effect on MSC survival and osteogenic capacity of the cells (Gowran et al. 2013). BM-MSCs have been shown to secrete the CB ligands AEA and 2-AG (Kose et al. 2018). In response to the locally produced AEA and 2-AG, activation of in particular CB2 signaling has been shown to result in an altered cytokine release profile in immune cells and immunosuppressive effects on activated T lymphocytes (Rossi et al. 2013; Xie et al. 2016). The non-psychoactive phytocannabinoid Cannabidiol (CBD) displays a suppressive effect on inflammatory processes through activation of both CR-dependent and independent mechanisms. In an *in vitro* model of inflammation where extended exposure to LPS was used activate

Toll-like receptors (TLRs) on adipose tissue derived MSCs, application of CBD was shown to decrease oxidative stress and restore adipogenic and chondrogenic differentiation potential of these cells (Ruhl et al. 2018). Similarly, the CB2 agonist AM1241 was shown to protect adipose tissue derived MSCs from H_2O_2 induced oxidative stress and promote their regenerative potential when transplanted *in vivo* into the ischemic myocardium of mice (Han et al. 2017).

Using Boyden chamber assays, both CBD and THC were found to increase the migration of adipose tissue-derived MSCs in a time- and concentration-dependent manner. This effect of ECBs on MSC migration was shown to be modulated by activation of the p42/44 MAPK pathway. Furthermore, CBD-induced migration inhibited by the CB2 antagonist AM-630 and GRP55 agonist O-1602 and PD98059, an inhibitor of the p42/44 MAPK pathway (Luder et al. 2017), whereas THC-induced migration was almost completely blocked by addition of CB1 receptor antagonist AM-251 and AM-630 (Schmuhl et al. 2014). These data were further confirmed by other studies that showed that inhibition of FAAH, which catalyzes the degradation of AEA and 2-AG, increased stem cell migration via PPARα (Wollank et al. 2015). Similarly, CB1 receptor agonist ACEA promoted migration of murine BM-MSCs *in vitro* and *in vivo*, whereas CB2 agonist JWH133 had no effect. Accordingly, pharmacological or genetic ablation of CB1 reduced ACEA-induced migration, whereas inhibition of CB2 activation did not noticeably affect migration (Wang et al. 2017).

4.3 The Role of Endocannabinoids in the Brain-Bone-Blood Triad

Recent studies have established that HPSCs/HSCs dynamically change their features and location, shifting from quiescent and stationary cells anchored in the BM to cycling and motile cells entering the circulation. These changes are controlled by brain-bone-blood triad via stress signals controls (Lapidot and Kollet 2010). However, how and why HSCs enter and exit the BM is not fully understood. ECBs may act as mobilizers of HSC from the BM under stress conditions similar to stimulation of beta-adrenergic receptors (β-AR). Our group previously demonstrated that BM-MSCs secrete both AEA and 2-AG, and AEA and 2-AG was detected in both PB and BM plasma samples of healthy donors (Kose et al. 2018). BM mononuclear cells (MNCs) and CD34+ HSCs were both shown to express CB1, CB2 and β-AR subtypes. When compared to BM-MSCs, CD34+ HSCs showed a higher CB1 and CB2 receptor expression. AEA and 2-AG induced HSC migration was blocked by ECB receptor antagonists in *in vitro* migration assays. In addition, components of the ECB system and β-AR subtypes were shown to interact with HSCs and MSCs of G-CSF treated and untreated healthy donors in vitro, revealing that ECBs might be potential candidates to enhance or facilitate G-CSF-mediated HSC migration under stress conditions (Kose et al. 2018). The effect of the ECB system in the migration/mobilization of HSC has been summarized in Fig. 1.

Thus, ECB stimulation increases proliferation, migration and especially osteogenic differentiation of human and rodent MSCs, albeit with different potencies, depending on the agonist used, and promotes their anti-inflammatory properties.

5 Neuropeptides

Neuropeptides have been defined as small peptides produced and secreted by neurons and acting on neural substrates. The neuropeptides consist of a large group of diverse signaling molecules that can act directly as neurotransmitters, as modulators of neurotransmission, as autocrine or paracrine regulators of a close microenvironment or as hormones, affecting other systems at a long distance (Burbach 2011; Burbach 2010).

5.1 Neuropeptide Y (NPY)

Neuropeptide Y (NPY) is a neurotransmitter composed of 36 amino acids that is secreted from the

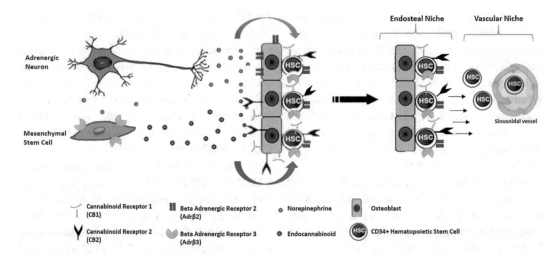

Fig. 1 The effect of the ECB system in the migration/mobilization of HSC. (i) ECBs, AEA and 2-AG are secreted from BM-MSCs; (ii) BM-MNCs and CD34 + HSCs express the cannabinoid receptors CB1, CB2 and beta adrenergic receptors β2-AR, β3-AR; (iii) CD34 + HSCs migrate towards SDF-1, Norepinephrine, AEA and 2-AG is mediated by CB receptors; (iv) CD34 + HSCs migration toward MSCs is inhibited by cannabinoid receptors CB1, CB2 and β-AR antagonists. Courtesy of S. Köse

central nervous system and sympathetic nerve ends (Park et al. 2015b). NPY regulates a range of physiological processes such as appetite, energy storage, emotional regulation and pain through activation of 5 different G-protein coupled receptors (Y1R-Y5R), which are highly expressed in mammals (Brothers and Wahlestedt 2010). While the expression levels of the NPY receptors in the CNS have been extensively studied, their distribution and function in peripheral tissues remains largely unexplored (Table 1). NPY has been shown to be involved in the regulation of immune cell homeostasis (Wheway et al. 2005), bone homeostasis (Sousa et al. 2009; Baldock et al. 2009) and vascular remodeling (Kuo et al. 2007) through NPY receptors expressed in BM cells, osteoblasts, macrophages and endothelial cells (Lee and Herzog 2009; Singh et al. 2017; Park et al. 2016). Furthermore, the fact that NPY deficient mice have decreased numbers of HSCs and a significantly impaired bone marrow function (Lin et al. 2004; Park et al. 2015a), as well as the fact that treatment with NPY through activation of the Y1-receptor results in hematopoietic regeneration, suggest that NPY plays a protective role in the bone marrow microenvironment. Even more, HSC transplantation into NPY deficient mice was shown to result in delayed and impaired engraftment, whereas treatment with NPY restored bone marrow function. However, the differentiation capacity and the maturation of HSCs from NPY deficient mice were not affected (Park et al. 2015a).

5.1.1 NPY and Mesenchymal Stem Cells

NPY was shown to exert multiple regulatory effects on a range of stem cells, as reviewed by Peng et al. (2017). Outside of the CNS, NPY plays an important role on bone homeostasis through direct stimulation of proliferation (via the NPY-Y5 receptor), migration (CXCR4) and differentiation (via the NPY-Y1 and Y2 receptors) of BM-MSCs (Peng et al. 2017).

NPY significantly increased rat BM-MSC proliferation, prevented apoptosis and promoted osteogenic differentiation in a dose-dependent through upregulation of β-catenin, p-GSK-3β and c-myc and activation of canonical Wnt signaling (Liu et al. 2016; Wu et al. 2017). Furthermore, NPY treatment of rat BM-MSCs also caused an upregulation in the expression of VEGF and SDF-1α and genes involved in the regulation of proliferation, including aurora B

Table 1 Tissue expression of NPY receptors and their function

NPY receptor	Alternative name	Chromo-some	Ligand	Tissue expression	Function
Y1		4q32.2	PYY > NPY	**Spleen, placenta, bone marrow, lymphnodes,** adipose tissue, adrenal glands, kidney, brain, heart	Modulation of anxiety, depression and nociception, regulation of the circadian rhythm and cardiovascular sympathetic tone, bone homeostasis, proliferation of neuronal stem cells and smooth muscle cells, angiogenesis, nutrient absorption.
Y2		4q32.1	NPY = PYY	Brain, testis, gastrointestinal system, adipose tissue	Modulation of anxiety, depression, nociception and addiction, regulation of the circadian rhythm and cardio-vascular sympathetic tone, bone homeostasis, proliferation of neuronal stem cells and smooth muscle cells, angiogenesis.
Y3	CXCR4, CD184	2q22.1	CXCL12, NPY > > PYY	**Bone marrow, lymphnodes, spleen, placenta**	**Role of NPY on CXCR4 signaling is currently unknown**
Y4	PYY-R1, PPR-1	10q11.22	PP > PYY > NPY	Gastrointestinal system, skin, prostate, lungs, pancreas	Nutrient absorption, regulation of acid secretion
Y5		4q32.2	NPY > PYY	**Spleen, placenta, lymphnodes,** adipose tissue, testis, kidneys, brain, adrenal glands	Cardiac hypertrophy, regulation of food intake and obesity, modulation of addiction, proliferation of mesenchymal stem cells

kinase, bFGF, cyclin A2 and EIF-4E (Wang et al. 2010; Liu et al. 2016). The proliferative effects of NPY on BM-MSCs were shown to be directly mediated through interactions with the NPY-Y5 receptor: BM-MSCs from old rats showed decreased expression of Y5 and a lower proliferative response to NPY than BM-MSCs from young rats, who had high Y5 expression and an increased proliferative response (Igura et al. 2011).

In addition to proliferation, NPY also plays a role in the regulation of BM-MSC differentiation. For example, NPY-deficient murine BM-MSCs showed an increase in osteogenic differentiation as apparent from prominent alkaline phosphatase staining and increased *bsp* (Bone Sialoprotein) and *ocn* (Osteocalcin) gene expression (Wee et al. 2019). Y2 receptor knockout (Y2$^{-/-}$) mice were shown to control bone volume via modulation of osteoblastic activity (Baldock et al. 2002). *In vivo* Y2$^{-/-}$ mice displayed an increase in the numbers of BM osteoprogenitor cells and *in vitro* these BM-MSCs showed increased mineralization and adipocyte formation (Lundberg et al. 2007). Similarly, antagonists of the NPY Y1 receptor promoted anabolic activity of osteoblasts in mice resulting in increased bone mass (Sousa et al. 2012) and BM-MSCs isolated from Y1 knockout (Y1$^{-/-}$) mice showed increased mRNA levels of the osteogenic transcription factors, *runx2* (Runx2) and *osx* (Osterix), and adipogenic transcription factor *pparg* (PPAR-γ) (Lee et al. 2010). Thus, Y2 and Y1 receptor signaling appears to have similar effects on the regulation of bone mass and osteoprogenitor cell numbers. Interestingly, Y2 receptor deletion resulted in a simultaneous decrease in the expression of Y1 transcripts, possibly through a lack of feedback inhibition, thus further increasing bone density and aggravating the skeletal phenotype in these mice (Lundberg et al. 2007). Chronic exposure to NPY resulted in downregulation of both Y2 and Y1 receptor expression (Teixeira et al. 2009).

Furthermore, NPY was shown to promote migration of BM-MSCs through upregulation of CXCR4 expression and induce endothelial differentiation and tube formation of BM-MSCs (Wang et al. 2010).

5.1.2 NPY and Hematopoietic Stem Cells

NPY was shown to play an important role in the regulation and maintenance of HSPCs (Ulum 2019). Using the Y1 knockout ($Y1^{-/-}$) mice Park and colleagues showed that NPY deficiency causes a severe impairment of HSC survival and BM regeneration (Park et al. 2015a). In the absence of NPY/NPY-Y1 signaling, the researchers found increased SNS nerve fiber destruction and a reduction in the numbers of ECs. Treatment with NPY or other Y1 agonists was shown to protect the BM nerves from Cisplatin-induced toxicity. In addition, TGF-β secreted by NPY-mediated Y1 receptor stimulation in macrophages was shown to play a key role in neuroprotection by NPY and HSC survival in the BM (Park et al. 2015b). Further evidence of the importance of NPY for the maintenance of HSPCs came from another study using $NPY^{-/-}$ and $NPY-Y1^{-/-}$ mouse models. $NPY^{-/-}$ mice did not exhibit HSPC mobilization in response to common mobilizing agents such as AMD3100 or G-CSF (Park et al. 2016). This lack of mobilization was found to be associated with an increased level of SDF-1α in the BM. Since NPY deficient HSCPs were shown to exhibit a normal migratory activity *in vitro* (Park et al. 2015a), the lack of mobilization *in vivo* was thought to be related to the effects of NPY on the BM niche. Indeed, $NPY^{-/-}$ mice were shown to have decreased levels of MMP-9, which is required for the degradation of SDF-1α in the BM by MMP-9 (Park et al. 2016) and NPY treatment of WT mice resulted in increased MMP-9 levels and rapid mobilization of HSPCs. In contrast, HSPCs from WT mice did not display significant migration in response to NPY *in vitro*, thus

confirming that NPY has no chemotactic effect itself and requires a functional BM niche to exert its effects on HSPCs. Similar to the $NPY^{-/-}$ mice, WT mice treated with NPY-Y1 receptor antagonists showed inhibition of NPY induced mobilization, whereas treatment with NPY or NPY-Y1 receptor agonist resulted in increased HSPC mobilization. The effects of NPY on HSPC mobilization was shown to be mediated by Y1 expressing OBs, and NPY treatment of WT mice led to an increase in MMP-9 activation by OBs and a subsequent reduction in the levels of SDF-1α followed by rapid HSPC mobilization (Park et al. 2016). Others pathways through which NPY appears to affect mobilization is through regulation of transendothelial migration of HSPCs (Singh et al. 2017). NPY was shown to reduce VE-Cadherin and CD31 expression along EC junctions, thus causing increased vascular permeability and HSPC egress. Regulation of vascular integrity was shown to be mediated by interactions with NPY-Y2 and Y5 receptors, and Y2 and Y5 receptor antagonists restored vascular integrity and prevented HSPC egress from the BM (Singh et al. 2017). The CD26 (DPPIV/dipeptidyl-peptidase IV) antigen plays an important role in mobilization of HSCPs. CD26 is a membrane-bound peptidase that is expressed on the surface of a subpopulation of stem and mononuclear cells and degrades several chemokines, including SDF-1. In response to G-CSF the HSPCs are stimulated to expand and differentiate into neutrophils. The latter produce a variety of enzymes including Neutrophil Elastase, Pepsin G and MMP-9, which are activated inside the BM and cleave the SDF-1 protein, inhibiting its ability to interact with HSPCs. When the interaction between SDF-1 and CXCR4 is broken, the HSPCs are no longer held in place inside the BM and are released into the PB. Recently it was shown that CD26 on ECs increases in response to G-CSF and cleaves the NPY peptide, which through signaling via NPY-Y2 and Y5 receptors increased vascular permeability of the BM niche (Singh et al. 2017). In addition, mice lacking either CD26 or NPY were shown to exhibit impaired HSPC mobilization, that could be restored by treatment with truncated NPY (Singh et al. 2017).

The alternative name for the NPY-Y3 receptor is CXCR4. CXCR4 is a chemokine receptor crucial for the anchoring of HSCs in the BM and necessary for homing after HSPC mobilization and transplantation. The actual ligand of CXCR4 is SDF-1 and there is no homology at the nucleotide and amino acid level between SDF1 and NPY (NCBI BLAST and NCBI pBLAST). Due to SDF-1/CXCR4 interactions, HSCs are retained in the BM niche and HSC mobilization occurs when G-CSF or SDF-1 antagonists, such as AMD3100 are administered. Although NPY increases CXCR4 in BM-MSCs, NPY does not appear to affect CXCR4 expression by HSPCs (Park et al. 2016). The effect of NPY on CXCR4-expressing hematopoietic cells is not yet known, but studies indicate that NPY may not effectively bind to or activate the chemokine receptor (Herzog et al. 1993). However, NPY/NPY-Y3 interactions have been shown to increase vascular permeability in the pulmonary circulation and in rat aorta endothelial cells (Nan et al. 2004; Hirabayashi et al. 1994). It is highly likely that a similar effect of NPY may be found on the BM sinus endothelium. The NPY-Y4 receptor has not been shown to play an important role in the hematopoietic and immune system until now and is expressed mostly in the gastrointestinal tract. NPY-Y5 is expressed in hematopoietic tissues such as spleen, lymph nodes and placenta, and its role in hematopoiesis has not yet been elucidated.

Interestingly, it has been recently shown that all NPY receptors are present in varying degrees on HSPCs, although the highest expression was observed for NPY Y1 (Ulum 2019). Thus far the effects of NPY on maintenance, self-renewal, proliferation and migration have been attributed to interactions with cells from the hematopoietic niche. However, new data indicate that NPY may directly interact with NPY receptors on HSCs as well and affect their potential for self-renewal through modulation of quiescence (Ulum 2019). Current knowledge of NPY receptor-mediated effects on hematopoiesis and their role in the regulation of the hematopoietic niche are summarized in Fig. 2. The effects of NPY on hematopoiesis under physiological conditions and after G-CSF mobilization are depicted in Fig. 3.

5.2 Tachykinins

Tachykinins belong to a peptide family that is characterized by a common C-terminal sequence, consisting of Phe-X-Gly-Leu-Met-NH$_2$. This family of peptides is predominantly active in the brain and gut and owns its name to their ability to rapidly induce contraction of gut and bladder smooth muscle. Similar to the opioids, the tachykinin genes encode large precursor proteins called pre-pro-tachykinins, that are further spliced into sets of peptides and precursor proteins and posttranslationally processed by several proteases. The Neurokinin family includes substance P (SP), the two structurally related peptides neurokinin A (NKA) and neurokinin B (NKB), and Hemokinin-1 (HK-1). These peptides are encoded by three different pre-pro-tachykinin genes, i.e. the *TAC1* (PPT-A) gene, which encodes the sequences of SP, NKA, and neuropeptide K; the *TAC3* (PPT-B) gene, which encodes NKB and the *TAC4* (PPT-C) gene, which encodes HK-1, Endokinin A and B (EKA and EKB) and the Tachykinin-related peptides Endokinin C and D (EKC and EKD) (Nowicki et al. 2007; Klassert et al. 2010). The effects of the neurokinins are mediated through the family 1 (rhodopsin-like) GPCR neurokinin receptors NK-1R, NK-2R and NK-3R (Gerard et al. 1993; Klassert et al. 2010). Whereas SP and HK-1 bind preferentially to NK-1R, NK-A binds with highest affinity to NK-2R (Pennefather et al. 2004). NK-Rs are broadly expressed in both neural and non-neural tissues, including hematopoietic cells, immune cells and BM stroma (Beaujouan et al. 2004). However, the distribution of the NK receptors is quite different, with the NK-1R being ubiquitously expressed in vascular endothelium, BM, immune cells, muscle and neurons, NK-2R expression predominantly being found in peripheral tissues, while NK-3R is preferentially expressed in the CNS (Pennefather et al. 2004). Similarly, expression of the neurokinins has also been detected in many non-neural tissues, such as the cardiovascular gastrointestinal and immune system, although neurons are the most prominent source of SP,

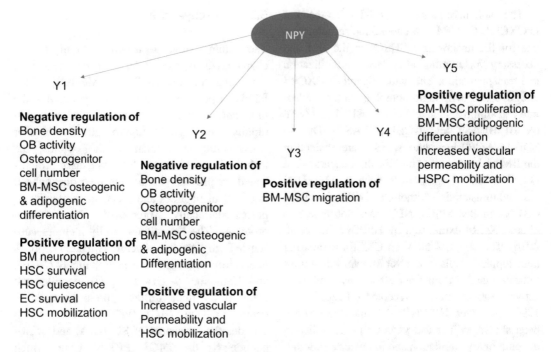

Fig. 2 Role of the NPY receptors in the regulation of the hematopoietic niche. NPY-Y1 and Y2 show overlapping actions and are highly expressed in the BM on MSCs, OBs, ECs and HSCs. NPY-3 is highly expressed on BM-MSCs and HSCs, but appears to be playing a minor role in NPY signaling. The function of NPY-Y4 in the BM has not been assessed and NPY-Y5 appear to play an important role in regulation of HSC egress

NKA and NKB (Nowicki et al. 2007). Whereas NKA, SP and HK-1 are known to be involved in the regulation of hematopoiesis, NKB plays a predominant role in nociception.

5.2.1 Substance P

Substance P is not only widely expressed throughout the central and peripheral nervous system, expression and local production of SP was also be detected in bone marrow innervating fibers, T and B cells, (vascular) endothelial cells, eosinophils, fibroblasts and monocytes/macrophages (Rameshwar et al. 1993a). SP was shown to induce the production of Interleukin-1 (IL-1), IL-6, and Tumor Necrosis Factor-alpha (TNFα) by monocytes and mast cells, IL-3 and GM-CSF production from BM-MNCs, IL-2 expression by activated T-cells, Interferon gamma (IFNγ) by PB-MNCs, and IL-1 and SCF synthesis by BM stromal cells (Nowicki et al. 2007). In addition, BM stromal cells were shown to participate in the local synthesis of IL-1 and SCF (Rameshwar and Gascon 1995).

In vitro, SP alone, in the absence of HGFs, was able to support hematopoiesis and formation of myeloid and erythroid colonies in semi-solid cultures. Using SP antagonists, the effect of SP on hematopoietic cultures was completely abrogated, whereas addition of antibodies against IL-3 and GM-CSF resulted in a reduction of the effect of SP, indicating that the effects of SP on hematopoiesis are mediated by IL-3 and GM-CSF (Rameshwar et al. 1993a). This and other studies show that SP affects both murine and human HSPCs during the gradual stages of differentiation (Liu et al. 2007; Greco et al. 2004). In response to treatment with SP, expression of IL-2 was induced in murine T-cell lines, PB lymphocytes and spleen lymphocytes and purified CD4+ T cells (Rameshwar et al. 1993b). Degradation of SP by neuroendopeptidase (NEP/CD10) expressed by hematopoietic cells into the smaller SP(1–4) has

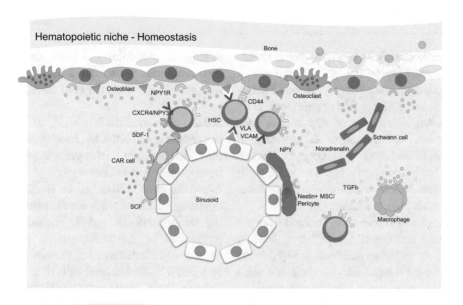

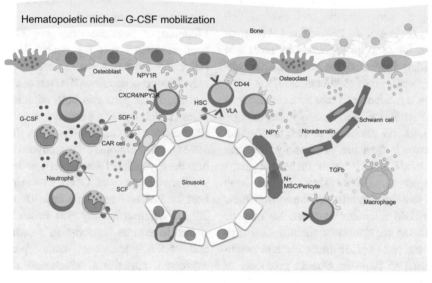

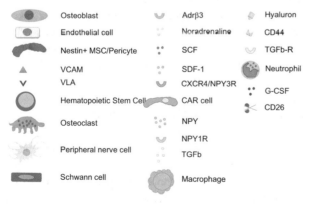

Fig. 3 The role of NPY in the hematopoietic bone marrow niche. Upper panel: NPY protects the hematopoietic environment through interactions with Y1, Y2 receptors on BM-MSCs, OBs, macrophages and ECs. NPY is secreted from sympathetic nerve endings, macrophages, OBs and ECs. Lower panel: HSPCs are retained in the BM by firm interactions between CXCR4 and SDF-1. G-CSF induced mobilization causes upregulation of CD26 and MMP-9, cleavage of SDF-1/CXCR4 interactions, cleavage and binding of truncated NPY to NPY-Y2 and Y5, increased vascular permeability and mobilization of HSPCs

been shown to modify its function, inhibit proliferation of mature myeloid progenitors and induce the production of Transforming Growth Factor-beta (TGFβ) and TNFα in BM stroma (Joshi et al. 2001). Further stimulation of BM stroma with SCF, increases NEP transcripts and degradation of SP. Since SP and SP(1–4) compete for binding to NK-1R but have opposite functions, SCF stimulation of BM stroma results in a negative feedback on hematopoiesis mediated by SP.

In vivo, SP was shown to be released from the sensory nerve endings of C-type nerve fibers in response to stress factors and tissue injury (Holzer 1988). These peptidergic nerve endings are often found in close proximity to immune cells in mucosal membranes (Stead et al. 1987) and in BM, affecting both hematopoietic and stromal cell populations (Felten et al. 1992). The SP receptor NK-1R was found to be expressed by most BM cells, including hematopoietic cells, stromal cells and T-lymphocytes (Nowicki et al. 2007).

SP plays a fundamental role as a regulator of synaptic transmission in sympathetic peptidergic nerve endings. As discussed above, the sympathetic nervous system has been shown to steer mobilization of HSPCs from their BM niche (Katayama et al. 2006). Thus, it has been suggested that by modulating the sympathetic tone of the BM microenvironment, SP may be involved in the regulation of mobilization efficacy in response to G-CSF treatment and potentially be used to improve clinical protocols of mobilization and stem cell harvests. These intriguing thoughts are further supported by the fact that SDF-1/CXCR4 interactions are essential for HSPC retaining, quiescence and mobilization (Petit et al. 2002) and that SDF-1 may induce expression of the *TAC1* gene and production of SP and NKA *in vitro* (Klassert et al. 2010). As mentioned above, SP was also shown to stimulate production of GM-CSF by BM-MNCs, which also plays a role in mobilization of HSPCs (Rameshwar et al. 1993a). In addition, SP was shown to upregulate gene and cell surface expression of CXCR4 on CD34+ cord blood cells (Shahrokhi et al. 2010).

In addition to its effect on hematopoiesis, SP has also been shown to have important angiogenic properties and stimulate endothelial cell proliferation through interaction with its receptor NK-1R (Pelletier et al. 2002)

5.2.2 Neurokinin A

In contrast to SP, NKA has been shown to inhibit the development of granulocyte/monocyte precursors, but stimulates erythrocyte progenitors (Nowicki et al. 2007; Liu et al. 2007). These opposing effects of NKA are the result of signaling through NK-2R and its stimulation of the production of the hematopoietic suppressors, Macrophage Inflammatory Protein-1a (MIP1α) and TGFβ by BM stromal cells (Rameshwar and Gascon 1996). Under physiologic conditions, SP and NKA, promote the production of distinct cytokines, exerting contrasting effects on the proliferation of hematopoietic progenitors. It has been suggested that the contrasting effects of NKA and SP may serve to protect HSPCs from exhaustive proliferation (Klassert et al. 2010) and that the *TAC1* gene can both stimulate and suppress HSPCs through a negative feedback system (Rameshwar et al. 1997; Rameshwar et al. 1993a). Furthermore, a protective role for NKA has been predicted based on the levels of the predominant types of TAC1 transcripts found in healthy BM and leukemic cells (Nowicki et al. 2003): Whereas healthy BM stromal cells were shown to express transcript b, encoding both SP and NKA, leukemic cells predominantly expressed transcript a, which only produces SP (Nowicki et al. 2007). In leukemic cells, where predominantly SP production takes place, the autoregulatory feedback regulated by NKA is absent, and as a consequence cells may not be able to inhibit their proliferative responses (Nowicki and Miskowiak 2003).

5.2.3 Hemokinin-1

HK-1 is encoded by the *TAC4* (*PPT-C*) gene and its expression was exclusively outside the neural system and predominantly detected in hematopoietic cells, in contrast to PPT-A and B genes (Zhang et al. 2000). HL-1 was shown to have a particularly prominent role in the regulation of lymphopoiesis. Based on the structural similarity of SP and HK-1, it was suggested that HK-1 might exert its hematopoietic effects through binding of the NK-1R. Indeed, it was shown that HK-1 could bind to NK-1R with equal affinity to that of SP (Morteau et al. 2001). Thus, as expected, HK-1 has been shown to display a range of biological activities that closely resemble those of SP. HK-1 was shown to be a critical regulator of B cell development in mice, and induced proliferation of IL-7 expanded B cell precursors, whereas SP was not effective. In addition, HK-1, but not SP, promoted the survival and expansion, and suppressed apoptosis of fresh and cultured B cell precursors (Zhang et al. 2000; Morteau et al. 2001). Similarly, HK-1 was shown to support and regulate T cell development and use of antagonists in fetal thymus organ cultures (FTOC) resulted in an early blockage of differentiation of T cell precursors at the double negative (DN, CD44-,CD25+) or double positive (DP, CD4 +/CD8+) stage (Zhang and Paige 2003). Thus, HK-1 may provide a proliferative stimulus for T and B cell precursors, and decreased TAC4 mRNA expression has been suggested to support progression into more mature stages. Expression of HK-1 was observed in murine monocyte and macrophage cells lines and freshly isolated human PB granulocyte, eosinophil, monocyte and lymphocyte subsets. Stimulation of these cells with either proinflammatory cytokines or PHA decreased HK-1 and NK-1 receptor mRNA (Klassert et al. 2008; Berger et al. 2013). Interestingly in the same cell subsets also NKB expression was detected.

5.2.4 Effect of NK Receptor Signaling on Hematopoiesis

Using competitive BM repopulation assays it was shown that BM from NK-1R deficient mice (Tacr1$^{-/-}$) showed decreased lymphoid lineage specific engraftment potential, whereas myeloid and erythroid lineages were not affected. However, BM cells from Tachykinin-knockout mice lacking either SP, HK-1 or both (Tac1$^{-/-}$, Tac4$^{-/-}$ and Tac1$^{-/-}$/Tac4$^{-/-}$ mice) were able to repopulate lethally irradiated recipients with normal efficiency, presumably due to Tachykinin peptide release by the host or competitor cells, indicating that Tachykinin

signaling plays an important role in engraftment and normal hematopoiesis and functions through a paracrine or endocrine mode of action (Berger et al. 2013). However, as discussed above, aberrant regulation of TAC1 expression with predominant expression of SP and loss of NKA, may contribute to the phenotype of leukemic transformation (Nowicki et al. 2007).

The effects of NK-1R and NK-2R agonists on hematopoiesis are distinctly different (Klassert et al. 2010; Rameshwar et al. 1997), with stimulation of NK-1R mediating production of hematopoiesis-supporting cytokines, and NK-2R stimulation resulting in the synthesis of inhibitory cytokines (Greco et al. 2004; Rameshwar and Gascon 1996). The balance between the expression and activation of NK-1R and NK-2R in the hematopoietic microenvironment is most likely to direct the effects and outcome for the HSPCs. The opposite nature of the effects of NK-1R and NK-2R stimulation makes it unlikely that both receptors are present on the same cells at the same time. Indeed, co-expression of these receptors could not be detected on BM stromal cells (Klassert et al. 2010). Evenmore, BM stromal cells expressing high levels of NK-1R showed downregulation of NK-2R expression and *vice versa* (Kang et al. 2004; Rameshwar et al. 1997; Bandari et al. 2003). These opposing expressions of the NK-1R and NK-2R are thus in line with the stimulatory effects of NK-1R agonists SP and HK-1 and the suppressive effects of NK-2R ligand NKA (Bandari et al. 2003; Kang et al. 2004).

The role of tachykinins on hematopoiesis during stress or injury (model 1), during SCF treatment (model 2), during mobilization (model 3) and during physiological situations, where the majority of HSCs are quiescent (model 4), are summarized in Fig. 4.

6 Neurotrophins and Neuropoietic Cytokines

The neurotrophins (NTs) are a family of growth and survival factors of which the members are related to Nerve Growth Factor (NGF). NTs were initially thought to mainly regulate the

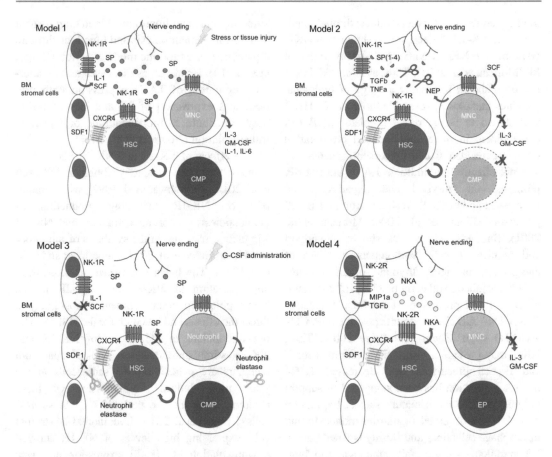

Fig. 4 **Effects of Tachykinins on hematopoiesis.** *Model 1:* In response to stress or tissue injury, SP is secreted from BM nerve endings. SP stimulates NK-1Rs on HSCs and BM stromal cells, leading in increased self-renewal of HSCs and secretion of IL-1 and SCF by BM stromal cells. Stimulation of NK-1R on BM-MNCs results in local production of IL-1, IL-3, IL-6 and GM-CSF and differentiation of HSCs into common myeloid progenitors (CMP). *Model 2:* SCF production increases expression of neuroendopeptidase (NEP), which cuts SP into the smaller peptide SP(1–4). SP(1–4) induces TGFβ and TNFα production by BM stromal cells. Since SP and SP(1–4) compete for the same receptor, but have opposite actions, differentiation of HSCs into CMPs is inhibited. *Model 3:*

SDF-1 binding to CXCR4 retains HSCs in the BM and induces SP production by HSCs. In response to G-CSF, HSCs produce neutrophils. G-CSF also increases CXCR4 expression by HSCs. SDF-1 degradation by neutrophil elastase results in loss of SP production and release of the HSCs from the BM. *Model 4:* NKA binds to NK-2R on BM stromal cells and HSCs. NKA stimulates erythrocyte progenitors at the expense of the granulocyte/monocyte precursors and induces the production of hematopoiesis suppressing cytokines MIP1α and TGFβ by BM stromal cells. HSCs enter quiescence and are protected from exhaustive proliferation. Leukemic cells do not express NKA and lose their ability to suppress their proliferative responses

development and the maintenance of neuronal cells (McAllister 2001; Ip 1998), but some of its members have now been shown to play a significant role in hematopoiesis as well. Similar to opioids and neurokinins, the NTs are generated by enzymatic processing of their proneurotrophin precursors. Their effects are mediated by a

subgroup of the receptor tyrosine kinase (RTK) family of receptors, known as the tropomyosin receptor kinases (Trk). The Trk receptors function similarly to other subgroups of the RTK family including PDGF, VEGF and FGF. Members of the NT family include NGF, Brain-Derived Growth factor (BDNF), Neurotrophin 3 (NT3) and Neurotrophin 4/5 (NT4/5) (Ip 1998). Although the Trk receptors TrkA, TrkB and TrkC display overlapping specificities for these NTs, NGF preferentially binds to TrkA (high affinity NGF receptor), BDNF and NT4/5 to TrkB, and NT3 to TrkC (Bothwell 1995; Ip et al. 1993).

CD271 (p75NTR, low affinity NGF receptor) is a member of the Tumor Necrosis Factor Receptor (TNFR) superfamily of transmembrane proteins (Rogers et al. 2008). Members of the TNFR family consist of a ligand-binding extracellular domain and an intracellular death domain (Lu et al. 2005; Nykjaer et al. 2005). CD271 is an atypical member of the TNFR family because distinct from the other members, 1) CD271 has a different intracellular structure and downstream signaling partners; 2) it binds dimeric, rather than trimeric ligands; and 3) CD271 can bind both pro-neurotrophins and mature NTs (Rogers et al. 2008). CD271 is involved in both the regulation of growth and differentiation, and death of cells in the nervous system and can associate with any of the Trk proteins TrkA, TrkB and TrkC. These opposing effects of CD271 are mediated through interactions with different receptor partners. When CD271 forms a complex with Trk it supports cell survival and growth. However, in the absence of Trk expression and signaling it induces cell death or apoptosis (Lu et al. 2005; Nykjaer et al. 2005; Hempstead et al. 1991). Thus interaction between CD271 and NGF plays an important role in the survival and protection of sympathetic and sensory neurons (Levi-Montalcini and Angeletti 1963).

The CD271 antigen is expressed in both neuronal and non-neuronal tissues. CD271 is expressed in the bone marrow (Cattoretti et al. 1993; Caneva et al. 1995), in BM-MSCs (Kuci et al. 2010; Quirici et al. 2002) and in the trabecular bone cavity (Jones et al. 2010; Cox et al. 2012). CD271 is also expressed by other BM resident cells, including endothelial cells, perivascular fibroblasts and B cells (Chesa et al. 1988), but its exact role in the bone marrow has not yet been revealed. Nevertheless, the role of

CD271 must be substantial, given that expression of CD271+ stromal reticular cells already appears in the fetal BM in a perivascular pattern and lining the sinus endothelial cells even before the beginning of hematopoietic activity (Cattoretti et al. 1993). Furthermore, the presence and distribution of CD271+ cells in the BM stroma also supports evidence of a co-stimulatory effect of NGF on early hematopoiesis (Caneva et al. 1995).

Prospective isolation of multipotent stromal BM progenitor cells using anti-CD271 has been shown to be feasible and result in the enrichment of a population of MSCs with a high clonogenic and proliferative potential, and similar if not better immune modulatory properties and hematopoiesis supporting activity (Kuci et al. 2010; Aydin 2018). However, since murine CD271 knockout models (Ngfr$^{tm1.1Vk}$ mice) were not associated with a major deficit of hematopoiesis, the role of NGF signaling in hematopoietic differentiation has been thought to be mediated through activation of the Trk receptors, rather than through CD271 (Bracci-Laudiero et al. 2003). Indeed, although high expression of CD271, TrkA, TrkB and TrkC has been observed on BM stromal cells (Rezaee et al. 2010), CD34+ HSCs do not express the pan-neurotrophin receptor CD271. Even more, human umbilical cord blood HSCs were shown to highly express both the Trk receptor and NGF, and their expression was strongly downregulated during differentiation of cells (Bracci-Laudiero et al. 2003). NGF is also synthesized by neurons, Schwann cells, fibroblasts, smooth muscle cells and B-cells in the bone marrow niche and has been shown to support human hematopoietic colony growth and differentiation in vitro (Matsuda et al. 1988) and to have a synergistic effect on the development of basophilic cells when used in combination with GM-CSF (Tsuda et al. 1991). Further evidence of the effect of NGF is provided by a study that showed that both normal primitive human BM CD34^{+}/CD38^{-} HSCs and a HGF-dependent leukemic cell line UT-7 could be maintained in the presence of stem cell factor (SCF) and NGF only, and that NGF promoted survival and differentiation of CD34$^+$/CD38$^-$ mature erythroid progenitors (Auffray et al. 1996). Additionally, NGF was also shown to promote human granulopoiesis in synergy with GM-CSF, IL-1, M-CSF, IL-3 and SCF (Simone et al. 1999).

Another neurotrophic factor receptor implicated in hematopoiesis is the RET (rearranged during transfection) receptor (Fonseca-Pereira et al. 2014). The RET receptor is activated after interaction with the glial cell-line derived neurotrophic factor (GDNF) family of ligands (GFLs). Fonseca-Pereira and colleagues showed that RET and its co-receptors Gfra1, Gfra2 and Gfra3 were highly expressed on murine fetal liver Lin$^-$Sca$^+$c-kit$^+$ (LSK) cells, but less on multipotent myeloid progenitor cells. However, whereas RET expression was found to be substantial in fetal liver LSK cells, BM HSCs were shown to express only low levels of RET (Fonseca-Pereira et al. 2014). Interestingly, in both fetal liver and the bone marrow niche, high expression of GFLs was detected, indicating an active interaction between GFL expressing niche cells and RET expressing HSCs. Although RET$^{-/-}$ LSK cells were shown to normally differentiate, they completely lost their potential for long-term engraftment (Fonseca-Pereira et al. 2014).

In addition to the neurotrophins, also the neuropoietic cytokines play a dual role with effects on the neural system and neural development and direct effects on the hematopoietic system and hematopoietic stem cells. The neuropoietic cytokines share their gp130 chain with a number of known hematopoietic cytokines belonging to the class I Hematopoietic growth factor/cytokine receptor family. The gp130 chain is used for transduction of the signals of IL-6, IL-11, leukemia inhibitory factor (LIF), oncostatin M (OSM), ciliary neurotrophic factor (CNTF), and cardiotrophin-1 (CT-1) (Taga 1996). The functions of these cytokines are all pleiotropic, displaying overlapping biological activities for hematopoietic and neuronal regulation (Ip 1998). IL-6, CT-1, CNTF and LIF have all been shown to possess the ability to stimulate hematopoiesis and neuropoiesis and their effects are mainly directed by the expression of their ligand-specific chain.

7 Conclusions

The data summarized here are merely a reflection of the wide impact of the neural system on the bone marrow niche and do not give a complete overview of all the factors and players involved. However, it can be easily understood that the involvement of the brain and in particular the sympathetic nervous system in the regulation of hematopoiesis does not involve a single neurotransmitter or neurotrophic factor, but the interaction between the innervation of the bone marrow, a multitude of factors, including catecholamines, opioids, endocannabinoids, neurotransmitters, neuropeptides and neurotrophic factors on one side and the bone marrow niche stromal cells, its vasculature and the hematopoietic stem and progenitor cells on the other side. It also becomes clear that not only maintenance of the stem cell pool through retention and self-renewal of HSPCs, but also their physiological circadian and forced mobilization in response to G-CSF and GM-CSF are affected and regulated by neuronal involvement. Chemical or physical denervation of the bone marrow have shown a direct effect on HSC function, HSPC numbers, hematopoietic regeneration and BM impairment. In conclusion, a better understanding of the innervation of the bone marrow and the role of the nervous system on the hematopoietic niche will pave the way for the development of new procedures for *ex vivo* expansion of HSPCs, *in vivo* mobilization of HSPCs and development of new drugs that can be used in the treatment of BM failure syndromes.

Acknowledgements This manuscript was supported by grants from the Scientific and Technological Research Council of Turkey, project no 318S073 and the Hacettepe University, Scientific Research Project Coordination Unit TYL-2018-17435 and THD-2018-17209.

Conflicts of Interest The authors declare no conflict of interest.

References

Artico M, Bosco S, Cavallotti C, Agostinelli E, Giuliani-Piccari G, Sciorio S, Cocco L, Vitale M (2002) Noradrenergic and cholinergic innervation of the bone marrow. Int J Mol Med 10(1):77–80

Auffray I, Chevalier S, Froger J, Izac B, Vainchenker W, Gascan H, Coulombel L (1996) Nerve growth factor is involved in the supportive effect by bone marrow-derived stromal cells of the factor-dependent human cell line UT-7. Blood 88(5):1608–1618

Aydin G (2018) Detailed characterization of MSCs from different sources, their ability to support lymphohematopoiesis and the importance of CD271 antigen for their isolation. Hacettepe University, Ankara

Baldock PA, Sainsbury A, Couzens M, Enriquez RF, Thomas GP, Gardiner EM, Herzog H (2002) Hypothalamic Y2 receptors regulate bone formation. J Clin Invest 109(7):915–921. https://doi.org/10.1172/JCI14588

Baldock PA, Lee NJ, Driessler F, Lin S, Allison S, Stehrer B, Lin EJ, Zhang L, Enriquez RF, Wong IP, McDonald MM, During M, Pierroz DD, Slack K, Shi YC, Yulyaningsih E, Aljanova A, Little DG, Ferrari SL, Sainsbury A, Eisman JA, Herzog H (2009) Neuropeptide Y knockout mice reveal a central role of NPY in the coordination of bone mass to body weight. PLoS One 4(12):e8415. https://doi.org/10.1371/journal.pone.0008415

Bandari PS, Qian J, Oh HS, Potian JA, Yehia G, Harrison JS, Rameshwar P (2003) Crosstalk between neurokinin receptors is relevant to hematopoietic regulation: cloning and characterization of neurokinin-2 promoter. J Neuroimmunol 138(1–2):65–75

Beaujouan JC, Torrens Y, Saffroy M, Kemel ML, Glowinski J (2004) A 25 year adventure in the field of tachykinins. Peptides 25(3):339–357. https://doi.org/10.1016/j.peptides.2004.02.011

Benard A, Boue J, Chapey E, Jaume M, Gomes B, Dietrich G (2008) Delta opioid receptors mediate chemotaxis in bone marrow-derived dendritic cells. J Neuroimmunol 197(1):21–28. https://doi.org/10.1016/j.jncuroim.2008.03.020

Benarroch EE (2012) Endogenous opioid systems: current concepts and clinical correlations. Neurology 79(8):807–814. https://doi.org/10.1212/WNL.0b013e3182662098

Berger A, Frelin C, Shah DK, Benveniste P, Herrington R, Gerard NP, Zuniga-Pflucker JC, Iscove NN, Paige CJ (2013) Neurokinin-1 receptor signalling impacts bone marrow repopulation efficiency. PLoS One 8(3):e58787. https://doi.org/10.1371/journal.pone.0058787

Bidlack JM (2000) Detection and function of opioid receptors on cells from the immune system. Clin Diagn Lab Immunol 7(5):719–723. https://doi.org/10.1128/cdli.7.5.719-723.2000

Boranic M, Krizanac-Bengez L, Gabrilovac J, Marotti T, Breljak D (1997) Enkephalins in hematopoiesis. Biomed Pharmacother 51(1):29–37

Bothwell M (1995) Functional interactions of neurotrophins and neurotrophin receptors. Annu Rev Neurosci 18:223–253. https://doi.org/10.1146/annurev.ne.18.030195.001255

Bracci-Laudiero L, Celestino D, Starace G, Antonelli A, Lambiase A, Procoli A, Rumi C, Lai M, Picardi A, Ballatore G, Bonini S, Aloe L (2003) CD34-positive cells in human umbilical cord blood express nerve growth factor and its specific receptor TrkA. J Neuroimmunol 136(1–2):130–139

Brothers SP, Wahlestedt C (2010) Therapeutic potential of neuropeptide Y (NPY) receptor ligands. EMBO Mol Med 2(11):429–439. https://doi.org/10.1002/emmm.201000100

Bruckner K (2011) Blood cells need glia, too: a new role for the nervous system in the bone marrow niche. Cell Stem Cell 9(6):493–495. https://doi.org/10.1016/j.stem.2011.11.016

Burbach JP (2010) Neuropeptides from concept to online database www.neuropeptides.nl. Eur J Pharmacol 626(1):27–48. https://doi.org/10.1016/j.ejphar.2009.10.015

Burbach JPH (2011) Chapter 1: What are neuropeptides? In: Merighi A (ed). https://doi.org/10.1007/978-1-61779-310-3

Calvo W, Forteza-Vila J (1969) On the development of bone marrow innervation in new-born rats as studied with silver impregnation and electron microscopy. Am J Anat 126(3):355–371. https://doi.org/10.1002/aja.1001260308

Caneva L, Soligo D, Cattoretti G, De Harven E, Deliliers GL (1995) Immuno-electron microscopy characterization of human bone marrow stromal cells with anti-NGFR antibodies. Blood Cells Mol Dis 21(2):73–85

Cattoretti G, Schiro R, Orazi A, Soligo D, Colombo MP (1993) Bone marrow stroma in humans: anti-nerve growth factor receptor antibodies selectively stain reticular cells in vivo and in vitro. Blood 81(7):1726–1738

Chesa PG, Rettig WJ, Thomson TM, Old LJ, Melamed MR (1988) Immunohistochemical analysis of nerve growth factor receptor expression in normal and malignant human tissues. J Histochem Cytochem 36(4):383–389. https://doi.org/10.1177/36.4.2831267

Cosentino M, Marino F, Maestroni GJ (2015) Sympathoadrenergic modulation of hematopoiesis: a review of available evidence and of therapeutic perspectives. Front Cell Neurosci 9:302. https://doi.org/10.3389/fncel.2015.00302

Cox G, Boxall SA, Giannoudis PV, Buckley CT, Roshdy T, Churchman SM, McGonagle D, Jones E (2012) High abundance of CD271(+) multipotential stromal cells (MSCs) in intramedullary cavities of long bones. Bone 50(2):510–517. https://doi.org/10.1016/j.bone.2011.07.016

Diaz-Laviada I, Ruiz-Llorente L (2005) Signal transduction activated by cannabinoid receptors. Mini Rev Med Chem 5(7):619–630

Felten SY, Felten DL, Bellinger DL, Olschowka JA (1992) Noradrenergic and peptidergic innervation of lymphoid organs. Chem Immunol 52:25–48

Fitch SR, Kimber GM, Wilson NK, Parker A, Mirshekar-Syahkal B, Gottgens B, Medvinsky A, Dzierzak E, Ottersbach K (2012) Signaling from the sympathetic nervous system regulates hematopoietic stem cell emergence during embryogenesis. Cell Stem Cell 11 (4):554–566. https://doi.org/10.1016/j.stem.2012.07.002

Fonseca-Pereira D, Arroz-Madeira S, Rodrigues-Campos-M, Barbosa IA, Domingues RG, Bento T, Almeida AR, Ribeiro H, Potocnik AJ, Enomoto H, Veiga-Fernandes H (2014) The neurotrophic factor receptor RET drives haematopoietic stem cell survival and function. Nature 514(7520):98–101. https://doi.org/10.1038/nature13498

Garcia-Garcia A, Korn C, Garcia-Fernandez M, Domingues O, Villadiego J, Martin-Perez D, Isern J, Bejarano-Garcia JA, Zimmer J, Perez-Simon JA, Toledo-Aral JJ, Michel T, Airaksinen MS, Mendez-Ferrer S (2019) Dual cholinergic signals regulate daily migration of hematopoietic stem cells and leukocytes. Blood 133(3):224–236. https://doi.org/10.1182/blood-2018-08-867648

Gerard NP, Bao L, Xiao-Ping H, Gerard C (1993) Molecular aspects of the tachykinin receptors. Regul Pept 43 (1–2):21–35

Gol'dberg ED, Zakharova O, Dygai AM, Simanina EV, Agafonov VI (1987) Modulating effect of enkephalins on hemopoieis under stress. Biull Eksp Biol Med 103 (5):589–590

Gowran A, McKayed K, Campbell VA (2013) The cannabinoid receptor type 1 is essential for mesenchymal stem cell survival and differentiation: implications for bone health. Stem Cells Int 2013:796715. https://doi.org/10.1155/2013/796715

Greco SJ, Corcoran KE, Cho KJ, Rameshwar P (2004) Tachykinins in the emerging immune system: relevance to bone marrow homeostasis and maintenance of hematopoietic stem cells. Front Biosci 9:1782–1793

Han D, Li X, Fan WS, Chen JW, Gou TT, Su T, Fan MM, Xu MQ, Wang YB, Ma S, Qiu Y, Cao F (2017) Activation of cannabinoid receptor type II by AM1241 protects adipose-derived mesenchymal stem cells from oxidative damage and enhances their therapeutic efficacy in myocardial infarction mice via Stat3 activation. Oncotarget 8(39):64853–64866. https://doi.org/10.18632/oncotarget.17614

Hempstead BL, Martin-Zanca D, Kaplan DR, Parada LF, Chao MV (1991) High-affinity NGF binding requires coexpression of the trk proto-oncogene and the low-affinity NGF receptor. Nature 350 (6320):678–683. https://doi.org/10.1038/350678a0

Herzog H, Hort YJ, Shine J, Selbie LA (1993) Molecular cloning, characterization, and localization of the human homolog to the reported bovine NPY Y3 receptor: lack of NPY binding and activation. DNA Cell Biol 12(6):465–471. https://doi.org/10.1089/dna.1993.12.465

Higuchi S, Ii M, Zhu P, Ashraf M (2012) Delta-opioid receptor activation promotes mesenchymal stem cell survival via PKC/STAT3 signaling pathway. Circ J 76 (1):204–212

Hirabayashi A, Nishiwaki K, Taki K, Shimada Y, Ishikawa N (1994) Effects of neuropeptide Y on lung vascular permeability in the pulmonary circulation of rats. Eur J Pharmacol 256(2):227–230

Holzer P (1988) Local effector functions of capsaicin-sensitive sensory nerve endings: involvement of tachykinins, calcitonin gene-related peptide and other neuropeptides. Neuroscience 24(3):739–768

Idris AI, van 't Hof RJ, Greig IR, Ridge SA, Baker D, Ross RA, Ralston SH (2005) Regulation of bone mass, bone loss and osteoclast activity by cannabinoid receptors. Nat Med 11(7):774–779. https://doi.org/10.1038/nm1255

Idris AI, Sophocleous A, Landao-Bassonga E, van't Hof RJ, Ralston SH (2008) Regulation of bone mass, osteoclast function, and ovariectomy-induced bone loss by the type 2 cannabinoid receptor. Endocrinology 149 (11):5619–5626. https://doi.org/10.1210/en.2008-0150

Idris AI, Sophocleous A, Landao-Bassonga E, Canals M, Milligan G, Baker D, van't Hof RJ, Ralston SH (2009) Cannabinoid receptor type 1 protects against age-related osteoporosis by regulating osteoblast and adipocyte differentiation in marrow stromal cells. Cell Metab 10(2):139–147. https://doi.org/10.1016/j.cmet.2009.07.006

Igura K, Haider H, Ahmed RP, Sheriff S, Ashraf M (2011) Neuropeptide y and neuropeptide y y5 receptor interaction restores impaired growth potential of aging bone marrow stromal cells. Rejuvenation Res 14 (4):393–403. https://doi.org/10.1089/rej.2010.1129

Ip NY (1998) The neurotrophins and neuropoietic cytokines: two families of growth factors acting on neural and hematopoietic cells. Ann N Y Acad Sci 840:97–106

Ip NY, Stitt TN, Tapley P, Klein R, Glass DJ, Fandl J, Greene LA, Barbacid M, Yancopoulos GD (1993) Similarities and differences in the way neurotrophins interact with the Trk receptors in neuronal and nonneuronal cells. Neuron 10(2):137–149

Jiang S, Alberich-Jorda M, Zagozdzon R, Parmar K, Fu Y, Mauch P, Banu N, Makriyannis A, Tenen DG, Avraham S, Groopman JE, Avraham HK (2011a) Cannabinoid receptor 2 and its agonists mediate hematopoiesis and hematopoietic stem and progenitor cell mobilization. Blood 117(3):827–838. https://doi.org/10.1182/blood-2010-01-265082

Jiang S, Fu Y, Avraham HK (2011b) Regulation of hematopoietic stem cell trafficking and mobilization by the endocannabinoid system. Transfusion 51 (Suppl 4):65S–71S. https://doi.org/10.1111/j.1537-2995.2011.03368.x

Jones E, English A, Churchman SM, Kouroupis D, Boxall SA, Kinsey S, Giannoudis PG, Emery P, McGonagle D (2010) Large-scale extraction and characterization of CD271+ multipotential stromal cells from trabecular bone in health and osteoarthritis: implications for bone regeneration strategies based on uncultured or minimally cultured multipotential stromal cells. Arthritis Rheum 62(7):1944–1954. https://doi.org/10.1002/art. 27451

Jorda MA, Verbakel SE, Valk PJ, Vankan-Berkhoudt YV, Maccarrone M, Finazzi-Agro A, Lowenberg B, Delwel R (2002) Hematopoietic cells expressing the peripheral cannabinoid receptor migrate in response to the endocannabinoid 2-arachidonoylglycerol. Blood 99 (8):2786–2793

Joshi DD, Dang A, Yadav P, Qian J, Bandari PS, Chen K, Donnelly R, Castro T, Gascon P, Haider A, Rameshwar P (2001) Negative feedback on the effects of stem cell factor on hematopoiesis is partly mediated through neutral endopeptidase activity on substance P: a combined functional and proteomic study. Blood 98 (9):2697–2706

Jung WC, Levesque JP, Ruitenberg MJ (2017) It takes nerve to fight back: The significance of neural innervation of the bone marrow and spleen for immune function. Semin Cell Dev Biol 61:60–70. https://doi.org/10. 1016/j.semcdb.2016.08.010

Kalinkovich A, Spiegel A, Shivtiel S, Kollet O, Jordaney N, Piacibello W, Lapidot T (2009) Blood-forming stem cells are nervous: direct and indirect regulation of immature human CD34+ cells by the nervous system. Brain Behav Immun 23(8):1059–1065. https://doi.org/10. 1016/j.bbi.2009.03.008

Kang HS, Trzaska KA, Corcoran K, Chang VT, Rameshwar P (2004) Neurokinin receptors: relevance to the emerging immune system. Arch Immunol Ther Exp 52(5):338–347

Katayama Y, Battista M, Kao WM, Hidalgo A, Peired AJ, Thomas SA, Frenette PS (2006) Signals from the sympathetic nervous system regulate hematopoietic stem cell egress from bone marrow. Cell 124(2):407–421. https://doi.org/10.1016/j.cell.2005.10.041

Kieffer BL, Evans CJ (2009) Opioid receptors: from binding sites to visible molecules in vivo. Neuropharmacology 56(Suppl 1):205–212. https://doi.org/10.1016/ j.neuropharm.2008.07.033

Klassert TE, Pinto F, Hernandez M, Candenas ML, Hernandez MC, Abreu J, Almeida TA (2008) Differential expression of neurokinin B and hemokinin-1 in human immune cells. J Neuroimmunol 196 (1–2):27–34. https://doi.org/10.1016/j.jneuroim.2008. 02.010

Klassert TE, Patel SA, Rameshwar P (2010) Tachykinins and Neurokinin Receptors in Bone Marrow Functions: Neural-Hematopoietic Link. J Receptor Ligand Channel Res 2010(3):51–61

Knight JM, Szabo A, Zhao S, Lyness JM, Sahler OJ, Liesveld JL, Sander T, Rizzo JD, Hillard CJ, Moynihan JA (2015) Circulating endocannabinoids during hematopoietic stem cell transplantation: a pilot study. Neurobiol Stress 2:44–50. https://doi.org/10. 1016/j.ynstr.2015.05.001

Kose S, Aerts-Kaya F, Kopru CZ, Nemutlu E, Kuskonmaz B, Karaosmanoglu B, Taskiran EZ, Altun B, Uckan Cetinkaya D, Korkusuz P (2018) Human bone marrow mesenchymal stem cells secrete endocannabinoids that stimulate in vitro hematopoietic stem cell migration effectively comparable to beta-adrenergic stimulation. Exp Hematol 57:30–41 e31. https://doi.org/10.1016/j.exphem.2017.09.009

Krizanac-Bengez LJ, Breljak D, Boranic M (1996) Suppressive effect of met-enkephalin on bone marrow cell proliferation in vitro shows circadian pattern and depends on the presence of adherent accessory cells. Biomed Pharmacother 50(2):85–91

Kuci S, Kuci Z, Kreyenberg H, Deak E, Putsch K, Huenecke S, Amara C, Koller S, Rettinger E, Grez M, Koehl U, Latifi-Pupovci H, Henschler R, Tonn T, von Laer D, Klingebiel T, Bader P (2010) CD271 antigen defines a subset of multipotent stromal cells with immunosuppressive and lymphohematopoietic engraftment-promoting properties. Haematologica 95(4):651–659. https://doi.org/10. 3324/haematol.2009.015065

Kulkarni-Narla A, Walcheck B, Brown DR (2001) Opioid receptors on bone marrow neutrophils modulate chemotaxis and CD11b/CD18 expression. Eur J Pharmacol 414(2–3):289–294

Kuntz A, Richins CA (1945) Innervation of the bone marrow. J Comp Neurol 83:213–222

Kuo LE, Abe K, Zukowska Z (2007) Stress, NPY and vascular remodeling: Implications for stress-related diseases. Peptides 28(2):435–440. https://doi.org/10. 1016/j.peptides.2006.08.035

Lapidot T, Kollet O (2010) The brain-bone-blood triad: traffic lights for stem-cell homing and mobilization. Hematology Am Soc Hematol Educ Program 2010:1–6. https://doi.org/10.1182/asheducation-2010.1.1

Lee NJ, Herzog H (2009) NPY regulation of bone remodelling. Neuropeptides 43(6):457–463. https:// doi.org/10.1016/j.npep.2009.08.006

Lee NJ, Doyle KL, Sainsbury A, Enriquez RF, Hort YJ, Riepler SJ, Baldock PA, Herzog H (2010) Critical role for Y1 receptors in mesenchymal progenitor cell differentiation and osteoblast activity. J Bone Miner Res 25(8):1736–1747. https://doi.org/10.1002/jbmr.61

Levi-Montalcini R, Angeletti PU (1963) Essential role of the nerve growth factor in the survival and maintenance of dissociated sensory and sympathetic embryonic nerve cells in vitro. Dev Biol 6:653–659

Lin S, Boey D, Herzog H (2004) NPY and Y receptors: lessons from transgenic and knockout models. Neuropeptides 38(4):189–200. https://doi.org/10. 1016/j.npep.2004.05.005

Liu K, Castillo MD, Murthy RG, Patel N, Rameshwar P (2007) Tachykinins and hematopoiesis. Clin Chim Acta 385(1–2):28–34. https://doi.org/10.1016/j.cca. 2007.07.008

Liu S, Jin D, Wu JQ, Xu ZY, Fu S, Mei G, Zou ZL, Ma SH (2016) Neuropeptide Y stimulates osteoblastic differentiation and VEGF expression of bone marrow mesenchymal stem cells related to canonical Wnt signaling activating in vitro. Neuropeptides 56:105–113. https://doi.org/10.1016/j.npep.2015.12.008

Lu B, Pang PT, Woo NH (2005) The yin and yang of neurotrophin action. Nat Rev Neurosci 6(8):603–614. https://doi.org/10.1038/nrn1726

Lucas D, Scheiermann C, Chow A, Kunisaki Y, Bruns I, Barrick C, Tessarollo L, Frenette PS (2013) Chemotherapy-induced bone marrow nerve injury impairs hematopoietic regeneration. Nat Med 19 (6):695–703. https://doi.org/10.1038/nm.3155

Luder E, Ramer R, Peters K, Hinz B (2017) Decisive role of P42/44 mitogen-activated protein kinase in Delta (9)-tetrahydrocannabinol-induced migration of human mesenchymal stem cells. Oncotarget 8 (62):105984–105994. https://doi.org/10.18632/oncotarget.22517

Lundberg P, Allison SJ, Lee NJ, Baldock PA, Brouard N, Rost S, Enriquez RF, Sainsbury A, Lamghari M, Simmons P, Eisman JA, Gardiner EM, Herzog H (2007) Greater bone formation of Y2 knockout mice is associated with increased osteoprogenitor numbers and altered Y1 receptor expression. J Biol Chem 282 (26):19082–19091. https://doi.org/10.1074/jbc.M609629200

Maccarrone M, Bab I, Biro T, Cabral GA, Dey SK, Di Marzo V, Konje JC, Kunos G, Mechoulam R, Pacher P, Sharkey KA, Zimmer A (2015) Endocannabinoid signaling at the periphery: 50 years after THC. Trends Pharmacol Sci 36(5):277–296. https://doi.org/10.1016/j.tips.2015.02.008

Maestroni GJ (1998) kappa-Opioid receptors in marrow stroma mediate the hematopoietic effects of melatonin-induced opioid cytokines. Ann N Y Acad Sci 840:411–419

Maestroni GJ, Conti A (1994) Modulation of hematopoiesis via alpha 1-adrenergic receptors on bone marrow cells. Exp Hematol 22(3):313–320

Maestroni GJ, Conti A, Pedrinis E (1992) Effect of adrenergic agents on hematopoiesis after syngeneic bone marrow transplantation in mice. Blood 80 (5):1178–1182

Maestroni GJ, Zammaretti F, Pedrinis E (1999) Hematopoietic effect of melatonin involvement of type 1 kappa-opioid receptor on bone marrow macrophages and interleukin-1. J Pineal Res 27 (3):145–153

Marino F, Cosentino M, Bombelli R, Ferrari M, Maestroni GJ, Conti A, Lecchini S, Frigo G (1997) Measurement of catecholamines in mouse bone marrow by means of HPLC with electrochemical detection. Haematologica 82(4):392–394

Matsuda H, Coughlin MD, Bienenstock J, Denburg JA (1988) Nerve growth factor promotes human hemopoietic colony growth and differentiation. Proc Natl Acad Sci U S A 85(17):6508–6512. https://doi.org/10.1073/pnas.85.17.6508

McAllister AK (2001) Neurotrophins and neuronal differentiation in the central nervous system. Cell Mol Life Sci 58(8):1054–1060. https://doi.org/10.1007/PL00000920

Mendez-Ferrer S, Lucas D, Battista M, Frenette PS (2008) Haematopoietic stem cell release is regulated by circadian oscillations. Nature 452(7186):442–447. https://doi.org/10.1038/nature06685

Mendez-Ferrer S, Battista M, Frenette PS (2010) Cooperation of beta(2)- and beta(3)-adrenergic receptors in hematopoietic progenitor cell mobilization. Ann N Y Acad Sci 1192:139–144. https://doi.org/10.1111/j.1749-6632.2010.05390.x

Mo FM, Offertaler L, Kunos G (2004) Atypical cannabinoid stimulates endothelial cell migration via a Gi/Go-coupled receptor distinct from CB1, CB2 or EDG-1. Eur J Pharmacol 489(1–2):21–27. https://doi.org/10.1016/j.ejphar.2004.02.034

Morrison SJ, Scadden DT (2014) The bone marrow niche for haematopoietic stem cells. Nature 505 (7483):327–334. https://doi.org/10.1038/nature12984

Morteau O, Lu B, Gerard C, Gerard NP (2001) Hemokinin 1 is a full agonist at the substance P receptor. Nat Immunol 2(12):1088. https://doi.org/10.1038/ni1201-1088

Muthu K, Iyer S, He LK, Szilagyi A, Gamelli RL, Shankar R, Jones SB (2007) Murine hematopoietic stem cells and progenitors express adrenergic receptors. J Neuroimmunol 186(1–2):27–36. https://doi.org/10.1016/j.jneuroim.2007.02.007

Nan YS, Feng GG, Hotta Y, Nishiwaki K, Shimada Y, Ishikawa A, Kurimoto N, Shigei T, Ishikawa N (2004) Neuropeptide Y enhances permeability across a rat aortic endothelial cell monolayer. Am J Physiol Heart Circ Physiol 286(3):H1027–H1033. https://doi.org/10.1152/ajpheart.00630.2003

Nowicki M, Miskowiak B (2003) Substance P–a potent risk factor in childhood lymphoblastic leukaemia. Leukemia 17(6):1096–1099. https://doi.org/10.1038/sj.leu.2402920

Nowicki M, Miskowiak B, Ostalska-Nowicka D (2003) Detection of substance P and its mRNA in human blast cells in childhood lymphoblastic leukaemia using immunocytochemistry and in situ hybridisation. Folia Histochem Cytobiol 41(1):33–36

Nowicki M, Ostalska-Nowicka D, Kondraciuk B, Miskowiak B (2007) The significance of substance P in physiological and malignant haematopoiesis. J Clin Pathol 60(7):749–755. https://doi.org/10.1136/jcp.2006.041475

Nykjaer A, Willnow TE, Petersen CM (2005) p75NTR--live or let die. Curr Opin Neurobiol 15(1):49–57. https://doi.org/10.1016/j.conb.2005.01.004

Ofek O, Karsak M, Leclerc N, Fogel M, Frenkel B, Wright K, Tam J, Attar-Namdar M, Kram V, Shohami E, Mechoulam R, Zimmer A, Bab I (2006) Peripheral cannabinoid receptor, CB2, regulates bone mass. Proc Natl Acad Sci U S A 103(3):696–701. https://doi.org/10.1073/pnas.0504187103

Park MH, Jin HK, Min WK, Lee WW, Lee JE, Akiyama H, Herzog H, Enikolopov GN, Schuchman EH, Bae JS (2015a) Neuropeptide Y regulates the hematopoietic stem cell microenvironment and prevents nerve injury in the bone marrow. EMBO J 34(12):1648–1660. https://doi.org/10.15252/embj.201490174

Park MH, Min WK, Jin HK, Bae JS (2015b) Role of neuropeptide Y in the bone marrow hematopoietic stem cell microenvironment. BMB Rep 48 (12):645–646

Park MH, Lee JK, Kim N, Min WK, Lee JE, Kim KT, Akiyama H, Herzog H, Schuchman EH, Jin HK, Bae JS (2016) Neuropeptide Y Induces Hematopoietic Stem/Progenitor Cell Mobilization by Regulating Matrix Metalloproteinase-9 Activity Through Y1 Receptor in Osteoblasts. Stem Cells 34 (8):2145–2156. https://doi.org/10.1002/stem.2383

Patinkin D, Milman G, Breuer A, Fride E, Mechoulam R (2008) Endocannabinoids as positive or negative factors in hematopoietic cell migration and differentiation. Eur J Pharmacol 595(1–3):1–6. https://doi.org/10.1016/j.ejphar.2008.05.002

Pelletier L, Angonin R, Regnard J, Fellmann D, Charbord P (2002) Human bone marrow angiogenesis: in vitro modulation by substance P and neurokinin A. Br J Haematol 119(4):1083–1089

Peng S, Zhou YL, Song ZY, Lin S (2017) Effects of Neuropeptide Y on Stem Cells and Their Potential Applications in Disease Therapy. Stem Cells Int 2017:6823917. https://doi.org/10.1155/2017/6823917

Pennefather JN, Lecci A, Candenas ML, Patak E, Pinto FM, Maggi CA (2004) Tachykinins and tachykinin receptors: a growing family. Life Sci 74 (12):1445–1463

Petit I, Szyper-Kravitz M, Nagler A, Lahav M, Peled A, Habler L, Ponomaryov T, Taichman RS, Arenzana-Seisdedos F, Fujii N, Sandbank J, Zipori D, Lapidot T (2002) G-CSF induces stem cell mobilization by decreasing bone marrow SDF-1 and up-regulating CXCR4. Nat Immunol 3(7):687–694. https://doi.org/10.1038/ni813

Quirici N, Soligo D, Bossolasco P, Servida F, Lumini C, Deliliers GL (2002) Isolation of bone marrow mesenchymal stem cells by anti-nerve growth factor receptor antibodies. Exp Hematol 30(7):783–791

Rameshwar P, Gascon P (1995) Substance P (SP) mediates production of stem cell factor and interleukin-1 in bone marrow stroma: potential autoregulatory role for these cytokines in SP receptor expression and induction. Blood 86(2):482–490

Rameshwar P, Gascon P (1996) Induction of negative hematopoietic regulators by neurokinin-A in bone marrow stroma. Blood 88(1):98–106

Rameshwar P, Ganea D, Gascon P (1993a) In vitro stimulatory effect of substance P on hematopoiesis. Blood 81(2):391–398

Rameshwar P, Gascon P, Ganea D (1993b) Stimulation of IL-2 production in murine lymphocytes by substance P and related tachykinins. J Immunol 151(5):2484–2496

Rameshwar P, Poddar A, Gascon P (1997) Hematopoietic regulation mediated by interactions among the neurokinins and cytokines. Leuk Lymphoma 28 (1–2):1–10. https://doi.org/10.3109/10428199709058325

Rezaee F, Rellick SL, Piedimonte G, Akers SM, O'Leary HA, Martin K, Craig MD, Gibson LF (2010) Neurotrophins regulate bone marrow stromal cell IL-6 expression through the MAPK pathway. PLoS One 5(3):e9690. https://doi.org/10.1371/journal.pone.0009690

Rogers ML, Beare A, Zola H, Rush RA (2008) CD 271 (P75 neurotrophin receptor). J Biol Regul Homeost Agents 22(1):1–6

Rossi F, Bernardo ME, Bellini G, Luongo L, Conforti A, Manzo I, Guida F, Cristino L, Imperatore R, Petrosino S, Nobili B, Di Marzo V, Locatelli F, Maione S (2013) The cannabinoid receptor type 2 as mediator of mesenchymal stromal cell immunosuppressive properties. PLoS One 8(11):e80022. https://doi.org/10.1371/journal.pone.0080022

Rozenfeld-Granot G, Toren A, Amariglio N, Nagler A, Rosenthal E, Biniaminov M, Brok-Simoni F, Rechavi G (2002) MAP kinase activation by mu opioid receptor in cord blood CD34(+)CD38(−) cells. Exp Hematol 30 (5):473–480

Ruhl T, Kim BS, Beier JP (2018) Cannabidiol restores differentiation capacity of LPS exposed adipose tissue mesenchymal stromal cells. Exp Cell Res 370(2):653–662. https://doi.org/10.1016/j.yexcr.2018.07.030

Satoh M, Minami M (1995) Molecular pharmacology of the opioid receptors. Pharmacol Ther 68(3):343–364

Schmuhl E, Ramer R, Salamon A, Peters K, Hinz B (2014) Increase of mesenchymal stem cell migration by cannabidiol via activation of p42/44 MAPK. Biochem Pharmacol 87(3):489–501. https://doi.org/10.1016/j.bcp.2013.11.016

Scutt A, Williamson EM (2007) Cannabinoids stimulate fibroblastic colony formation by bone marrow cells indirectly via CB2 receptors. Calcif Tissue Int 80 (1):50–59. https://doi.org/10.1007/s00223-006-0171-7

Shahrokhi S, Ebtekar M, Alimoghaddam K, Sharifi Z, Ghaffari SH, Pourfathollah AA, Kheirandish M, Mohseni M, Ghavamzadeh A (2010) Communication of substance P, calcitonin-gene-related neuropeptides and chemokine receptor 4 (CXCR4) in cord blood hematopoietic stem cells. Neuropeptides 44(5):385–389. https://doi.org/10.1016/j.npep.2010.06.002

Sharp BM, Roy S, Bidlack JM (1998) Evidence for opioid receptors on cells involved in host defense and the immune system. J Neuroimmunol 83(1–2):45–56

Simone MD, De Santis S, Vigneti E, Papa G, Amadori S, Aloe L (1999) Nerve growth factor: a survey of activity on immune and hematopoietic cells. Hematol Oncol 17 (1):1–10

Singh P, Hoggatt J, Kamocka MM, Mohammad KS, Saunders MR, Li H, Speth J, Carlesso N, Guise TA, Pelus LM (2017) Neuropeptide Y regulates a vascular gateway for hematopoietic stem and progenitor cells. J Clin Invest 127(12):4527–4540. https://doi.org/10.1172/JCI94687

Smith M, Wilson R, O'Brien S, Tufarelli C, Anderson SI, O'Sullivan SE (2015) The Effects of the Endocannabinoids Anandamide and 2-Arachidonoylglycerol on Human Osteoblast Proliferation and Differentiation. PLoS One 10(9): e0136546. https://doi.org/10.1371/journal.pone.0136546

Sousa DM, Herzog H, Lamghari M (2009) NPY signalling pathway in bone homeostasis: Y1 receptor as a potential drug target. Curr Drug Targets 10(1):9–19

Sousa DM, Baldock PA, Enriquez RF, Zhang L, Sainsbury A, Lamghari M, Herzog H (2012) Neuropeptide Y Y1 receptor antagonism increases bone mass in mice. Bone 51(1):8–16. https://doi.org/10.1016/j.bone.2012.03.020

Spiegel A, Shivtiel S, Kalinkovich A, Ludin A, Netzer N, Goichberg P, Azaria Y, Resnick I, Hardan I, Ben-Hur H, Nagler A, Rubinstein M, Lapidot T (2007) Catecholaminergic neurotransmitters regulate migration and repopulation of immature human CD34 + cells through Wnt signaling. Nat Immunol 8 (10):1123–1131. https://doi.org/10.1038/ni1509

Stead RH, Bienenstock J, Stanisz AM (1987) Neuropeptide regulation of mucosal immunity. Immunol Rev 100:333–359

Steidl U, Bork S, Schaub S, Selbach O, Seres J, Aivado M, Schroeder T, Rohr UP, Fenk R, Kliszewski S, Maercker C, Neubert P, Bornstein SR, Haas HL, Kobbe G, Tenen DG, Haas R, Kronenwett R (2004) Primary human CD34+ hematopoietic stem and progenitor cells express functionally active receptors of neuromediators. Blood 104(1):81–88. https://doi.org/10.1182/blood-2004-01-0373

Tabarowski Z, Gibson-Berry K, Felten SY (1996) Noradrenergic and peptidergic innervation of the mouse femur bone marrow. Acta Histochem 98(4):453–457. https://doi.org/10.1016/S0065-1281(96)80013-4

Taga T (1996) Gp130, a shared signal transducing receptor component for hematopoietic and neuropoietic cytokines. J Neurochem 67(1):1–10

Teixeira L, Sousa DM, Nunes AF, Sousa MM, Herzog H, Lamghari M (2009) NPY revealed as a critical modulator of osteoblast function in vitro: new insights into the role of Y1 and Y2 receptors. J Cell Biochem 107(5):908–916. https://doi.org/10.1002/jcb.22194

Tian M, Broxmeyer HE, Fan Y, Lai Z, Zhang S, Aronica S, Cooper S, Bigsby RM, Steinmetz R, Engle SJ, Mestek A, Pollock JD, Lehman MN, Jansen HT, Ying M, Stambrook PJ, Tischfield JA, Yu L (1997) Altered hematopoiesis, behavior, and sexual function in mu opioid receptor-deficient mice. J Exp Med 185(8):1517–1522. https://doi.org/10.1084/jem.185.8.1517

Tsuda T, Wong D, Dolovich J, Bienenstock J, Marshall J, Denburg JA (1991) Synergistic effects of nerve growth factor and granulocyte-macrophage colony-stimulating factor on human basophilic cell differentiation. Blood 77(5):971–979

Ulum B (2019) Assessment of the role of Neuropeptide Y in the regulation of hematopoietic stem cells. Middle East Technical University, Ankara

Valk P, Verbakel S, Vankan Y, Hol S, Mancham S, Ploemacher R, Mayen A, Lowenberg B, Delwel R (1997) Anandamide, a natural ligand for the peripheral cannabinoid receptor is a novel synergistic growth factor for hematopoietic cells. Blood 90(4):1448–1457

Waldhoer M, Bartlett SE, Whistler JL (2004) Opioid receptors. Annu Rev Biochem 73:953–990. https://doi.org/10.1146/annurev.biochem.73.011303.073940

Wang Y, Zhang D, Ashraf M, Zhao T, Huang W, Ashraf A, Balasubramaniam A (2010) Combining neuropeptide Y and mesenchymal stem cells reverses remodeling after myocardial infarction. Am J Physiol Heart Circ Physiol 298(1):H275–H286. https://doi.org/10.1152/ajpheart.00765.2009

Wang L, Yang L, Tian L, Mai P, Jia S, Yang L, Li L (2017) Cannabinoid Receptor 1 Mediates Homing of Bone Marrow-Derived Mesenchymal Stem Cells Triggered by Chronic Liver Injury. J Cell Physiol 232 (1):110–121. https://doi.org/10.1002/jcp.25395

Wee NKY, Sinder BP, Novak S, Wang X, Stoddard C, Matthews BG, Kalajzic I (2019) Skeletal phenotype of the neuropeptide Y knockout mouse. Neuropeptides 73:78–88. https://doi.org/10.1016/j.npep.2018.11.009

Wheway J, Mackay CR, Newton RA, Sainsbury A, Boey D, Herzog H, Mackay F (2005) A fundamental bimodal role for neuropeptide Y1 receptor in the immune system. J Exp Med 202(11):1527–1538. https://doi.org/10.1084/jem.20051971

Whyte LS, Ryberg E, Sims NA, Ridge SA, Mackie K, Greasley PJ, Ross RA, Rogers MJ (2009) The putative cannabinoid receptor GPR55 affects osteoclast function in vitro and bone mass in vivo. Proc Natl Acad Sci U S A 106(38):16511–16516. https://doi.org/10.1073/pnas.0902743106

Wollank Y, Ramer R, Ivanov I, Salamon A, Peters K, Hinz B (2015) Inhibition of FAAH confers increased stem cell migration via PPARalpha. J Lipid Res 56 (10):1947–1960. https://doi.org/10.1194/jlr.M061473

Wu J, Liu S, Meng H, Qu T, Fu S, Wang Z, Yang J, Jin D, Yu B (2017) Neuropeptide Y enhances proliferation and prevents apoptosis in rat bone marrow stromal cells in association with activation of the Wnt/beta-catenin pathway in vitro. Stem Cell Res 21:74–84. https://doi.org/10.1016/j.scr.2017.04.001

Xie J, Xiao D, Xu Y, Zhao J, Jiang L, Hu X, Zhang Y, Yu L (2016) Up-regulation of immunomodulatory effects of mouse bone-marrow derived mesenchymal stem cells by tetrahydrocannabinol pre-treatment involving cannabinoid receptor CB2. Oncotarget 7 (6):6436–6447. https://doi.org/10.18632/oncotarget.7042

Yamaguchi M, Levy RM (2016) beta-Caryophyllene promotes osteoblastic mineralization, and suppresses osteoclastogenesis and adipogenesis in mouse bone marrow cultures in vitro. Exp Ther Med 12 (6):3602–3606. https://doi.org/10.3892/etm.2016. 3818

Yamazaki K, Allen TD (1990) Ultrastructural morphometric study of efferent nerve terminals on murine bone marrow stromal cells, and the recognition of a novel anatomical unit: the "neuro-reticular complex". Am J Anat 187(3):261–276. https://doi.org/10.1002/aja. 1001870306

Yamazaki S, Ema H, Karlsson G, Yamaguchi T, Miyoshi H, Shioda S, Taketo MM, Karlsson S,

Iwama A, Nakauchi H (2011) Nonmyelinating Schwann cells maintain hematopoietic stem cell hibernation in the bone marrow niche. Cell 147 (5):1146–1158. https://doi.org/10.1016/j.cell.2011.09. 053

Zhang Y, Paige CJ (2003) T-cell developmental blockage by tachykinin antagonists and the role of hemokinin 1 in T lymphopoiesis. Blood 102(6):2165–2172. https://doi.org/10.1182/blood-2002-11-3572

Zhang Y, Lu L, Furlonger C, Wu GE, Paige CJ (2000) Hemokinin is a hematopoietic-specific tachykinin that regulates B lymphopoiesis. Nat Immunol 1 (5):392–397. https://doi.org/10.1038/80826

Adv Exp Med Biol – Cell Biology and Translational Medicine (2019) 6: 155–178
https://doi.org/10.1007/5584_2019_349
© Springer Nature Switzerland AG 2019
Published online: 4 April 2019

Homeobox Genes and Homeodomain Proteins: New Insights into Cardiac Development, Degeneration and Regeneration

Rokas Miksiunas, Ali Mobasheri, and Daiva Bironaite

Abstract

Cardiovascular diseases are the most common cause of human death in the developing world. Extensive evidence indicates that various toxic environmental factors and unhealthy lifestyle choices contribute to the risk, incidence and severity of cardiovascular diseases. Alterations in the genetic level of myocardium affects normal heart development and initiates pathological processes leading to various types of cardiac diseases. Homeobox genes are a large and highly specialized family of closely related genes that direct the formation of body structure, including cardiac development. Homeobox genes encode homeodomain proteins that function as transcription factors with characteristic structures that allow them to bind to DNA, regulate gene expression and subsequently control the proper physiological function of cells, tissues and organs. Mutations in homeobox genes are rare and usually lethal with evident alterations in cardiac function at or soon after the birth. Our understanding of homeobox gene family expression and function has expanded significantly during the recent years. However, the involvement of homeobox genes in the development of human and animal cardiac tissue requires further investigation. The phenotype of human congenital heart defects unveils only some aspects of human heart development. Therefore, mouse models are often used to gain a better understanding of human heart function, pathology and regeneration. In this review, we have focused on the role of homeobox genes in the development and pathology of human heart as potential tools for the future development of targeted regenerative strategies for various heart malfunctions.

Keywords

Cardiac development · Cardiac regeneration · Heart disease · Homeobox genes

R. Miksiunas, A. Mobasheri, and D. Bironaite (✉)
Department of Regenerative Medicine, State Research Institute Centre for Innovative Medicine, Vilnius, Lithuania
e-mail: daibironai@gmail.com;
daiva.bironaite@imcentras.lt

Abbreviations

AMHC1	atrial myosin heavy chain-1
ANTP	Antennapedia
BMP	bone morphogenetic protein
Cdh2	cadherin 2
CDK	cyclin-dependent kinases
Cited2	Cbp/P300 interacting transactivator with Glu/Asp Rich Carboxy-Terminal Domain 2
CNS	central nerve system

ESC	embryonic stem cells
FGF	fibroblast growth factor
FHF	first heart field
Flk1	fetal liver kinase 1
GJA5	gap junction protein alpha 5
GSC	goosecoid
H3K27me3	histone H3 methylation on the amino (N) terminal tail
Hcn4	hyperpolarization-activated cyclic nucleotide-gated channel 4 gene
HOXL	homeobox transcription factor Hox-like
Irx	Iroquois family of homeobox genes
ISL1	LIM-homeodomain transcription factor islet 1/insulin gene enhancer protein ISL-1
JMJD3	JmjC domain-containing protein 3
MEF2C	myocyte-specific enhancer factor 2C
MESP1	mesoderm posterior BHLH transcription factor 1
MSCs	mesenchymal stem cells
Myocd	myocardin
NKL	NK-like
Nkx2-5	homeobox protein NK-2 homolog E
Nodal	nodal growth differentiation factor
Nppa	natriuretic peptide A
OFT	outflow tract
PCBP2	poly(rC)-binding protein 2
Pitx2	paired like homeodomain 2
Pitx2c	paired-like homeodomain transcription factor 2
PROS	prospero
RA	retinoic acid
SAN	sinoatrial node
SHF	second heart field
Shox2	short stature homeobox 2
SMAD	main signal transducers for receptors of the transforming growth factor beta (TGF-β) superfamily;
TALE	three-amino-acid loop extension
Tbx5	T-box transcription factor 5
TF	transcription factors

TGF-β	transforming growth factor beta;
VCS	ventricular conduction system
ZEB2	zinc finger E-box binding homeobox 2
ZF	zinc finger
Ziro	zebrafish iroquois homeobox genes
ZO-3	tight junction protein 3

1 Introduction

Homeobox genes are a large family of genes that direct the formation of body structures along the head-tail axis in multicellular animal species (Innis 1997; Shashikant et al. 1991). It is also known that homeobox genes (Hox genes), as an ancient class of transcription factors, are important for the body patterning during embryo development (Innis 1997; Shashikant et al. 1991). Many of the homeobox genes play very important part in the spatiotemporal development of human heart (Lage et al. 2010). Likewise, some of these genes shape the human heart and control its multistep developmental process from simple crescent cells to a fully functional organ. For example, homeobox genes like homeobox protein NK-2 homolog E (Nkx2-5), LIM-homeodomain transcription factor islet 1 (Isl1), paired like homeodomain 2 (Pitx2) are widely known to be important for the proper development of human heart (Akazawa and Komuro 2005; Luo et al. 2014; Franco et al. 2017). However, there are many more homeobox genes that play substantial roles in cardiac function but thus far there are less known and/or less investigated.

Specific inherited gene mutations cause congenital heart defects such as atrial or ventricle septal defects, abnormalities of outflow tract and etc. (Bao et al. 1999). Similarly, various pathological lifestyle factors like smoking, low physical activity, toxic and noxious agents and other environmental factors might also negatively affect cardiovascular function and promote heart failure (O'Toole et al. 2008; Nayor and Vasan

2015). Since it is impossible to exactly pinpoint how certain gene mutations influence development of human heart at the earliest stages, different mouse models have been created to better understand regulation of human heart development and its relation to various diseases (Camacho et al. 2016). Many of the genes studies in mouse models have similar vital roles in the development and function of human heart (Xu and Baldini 2007). Therefore, investigation of human disease and cues from mouse heart development models have revealed an important role of homeobox genes, including those that encode transcription factors.

Aside from already known homeobox genes, there are more homeobox genes that are essential for the formation of human and/or mouse myocardium. Some of these homeobox genes code transcription factors (TF), whereas other form a tight network regulating heart development and fate of heart progenitors. Several review articles have explored individual families of homeobox gene and their roles in embryo development. However, knowledge concerning the involvement of homeobox genes and homeodomain TF in the development of human heart referring mouse models are still lacking. Therefore, in this review we describe the role of more than 20 homeobox genes that are mainly involved in heart development and around 15 homeobox genes that are known to play minor or less investigated, but nonetheless important roles in cardiac development. Data summarized in this review will help to broaden the possible future applications of homeobox genes and their coded TF in targeted therapeutic strategies for cardiac regeneration and therapy.

2   Development of the Human Heart

Starting from the day first of fertilization, the zygote undergoes multiple cell divisions leading to the formation of third germ layer, known as the mesoderm (Moorman et al. 2003). Later

mesodermal cells migrate towards anterior part of embryo to form a distinct crescent-shaped epithelium, named the cardiac crescent (Buckingham et al. 2005). Cells situated in the distinct anterior-lateral territory within the cardiac crescent contribute to the formation of first heart field (FHF), distinguished by the expression of hyperpolarization-activated cyclic nucleotide-gated channel 4 gene (Hcn4) (Liang et al. 2013). Cardiac progenitor cells also develop into second heart field (SHF), which is located medially to the cardiac crescent and extend posteriorly (Cai et al. 2003). Formation of SHF is marked by the expression of LIM-homeodomain transcription factor islet 1 (ISL1) (Cai et al. 2003). Sometimes progenitors of FHF and SHF are called cardiogenic or cardiac mesoderm (Dupays et al. 2015; Liu et al. 2014; Kitajima et al. 2000). These distinct heart fields fuse to form heart tube, which eventually develops into functional heart (Fig. 1) (Moorman et al. 2003; Nemer 2008). During this time, the primitive cardiac conduction system, including sinoatrial node (SAN), ventricular conduction system and other, starts to form (van Weerd and Christoffels 2016). The FHF cells develop into the left ventricle, as well as into the atrioventricular canal and part of the atria, whereas SHF cells develop into the right ventricle and outflow tract, with contribution to the formation of atria and inflow vessels (Buckingham et al. 2005). Once the heart fields are formed, they fuse into heart tube and undergo process called heart looping. During this phase the whole heart tube twists in the rightward direction eventually forming clearly visible, but still primitive, heart chambers (Santini et al. 2016). Later on, the heart undergoes septation to fully separated left and right sides of the heart.

There are many factors regulating human and mouse heart development, however only some of them may be considered to be core regulators of cardiogenesis. One of the most important TF is GATA4 which orchestrates expression of multiple transcriptions including other major determinants of cardiomyogenesis like Nkx2-5, T-box transcription factor 5 (Tbx5), heart- and

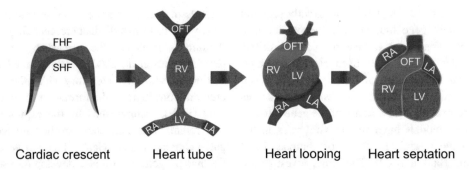

Cardiac crescent   Heart tube   Heart looping   Heart septation

Fig. 1 Schematic representation of heart development in human and mouse. *FGF* first heart field, *SHF* second heart field, *RV* right ventricle, *LV* left ventricle, *RA* right atrium, *LA* left atrium. (Scheme adapted from (Nemer, 2008))

neural crest derivatives-expressed protein 1 (HAND1/2) and others (Bruneau et al. 2001a; Belaguli et al. 2000; Sepulveda et al. 1998). GATA4 integrates bone morphogenetic proteins (BMP) and SMAD, the main signal transducers for receptors of the transforming growth factor beta (TGF-β) superfamily, to ensure cardiac cell survival and stable lineage during cardiac development (Benchabane and Wrana 2003). Of course, there are other factors that promote development of various structures within the heart. For example, it is know that Tbx5 controls atrial gene expression, whereas myocyte-specific enhancer factor 2C (MEF2C) promotes development of ventricle and vasculogenesis (Bruneau et al. 2001a; Lin et al. 1997). Altogether, the development of human heart is a carefully controlled multistep process involving many genes, intracellular and extracellular signalling factors leading to proper cardiac function. The miss-controlled heart development process leads to various inherited or acquired cardiac disorders. This review focuses mostly on the homeodomain proteins, as one of the most important group of transcription factors regulating heart development, function and impairment.

3   Homeobox Genes

Homeodomain proteins are one of the most important group of proteins/transcription factors regulating plan of body structure and organogenesis in eukaryotes including heart development and disorders. DNA binding proteins have been extensively studied, but even today there are no established rules for predicting the specificity of DNA sequence based upon the amino acid sequence of the proteins. Homeodomain proteins are characterized by specific 60 amino acid long helix-turn-helix DNA binding homeodomain motif (Seifert et al. 2015). The homeodomain is a very highly conserved structure and consists of three helical regions folded into a tight globular structure that binds a 5′-TAAT-3′ core motif. The high degree of conservation of homeodomain proteins is an ideal model to study specific protein-DNA interactions. The DNA sequence that encodes the homeodomain is called the "homeobox" and homeobox-containing genes are known as "hox" genes.

Most of the transcription factors belonging to this group are not only structurally but also evolutionary conserved and play crucial roles in embryonic patterning and differentiation (Pearson et al. 2005). The main role of homeodomain proteins *in vivo* is to control the genetic determination of development and implementation of the genetic body plan. There are 102 homeobox gene families that represent 235 active human homeodomain proteins, but only some homeodomain classes have close association with cardiac development and/or diseases (Bürglin and Affolter 2016). This review covers description of around 20 homeobox genes that up today are known to have a major

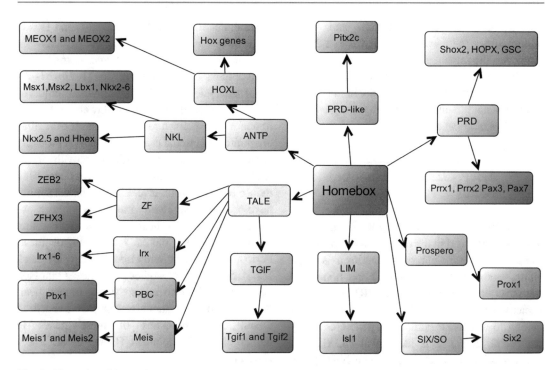

Fig. 2 Hierarchy of homeobox genes involved in heart development. Pink boxes indicate homeobox gene classes, yellow box indicates TALE gene superclass. Red boxes indicate genes with major involvement in heart development and diseases. Blue boxes indicate genes having less important role in heart development and diseases

impact in the regulation of heart development and functioning (Fig. 2).

4 Homeobox Genes in Mouse Heart Development and Human Disease

Humans have more than 67 genes that are important for cardiac hypertrophy and over 92 genes that control cardiovascular system, therefore it is reasonable to assume that some of the genes might be responsible for cardiac development and disease (van der Harst et al. 2016; Smith and Newton-Cheh 2015). Congenital heart disease (CHD) have structural heart anomaly, including atrial or ventricle septal defects, overriding aorta, right atrium isomerism and other structural changes in new born heart. Patients do usually display multiple symptoms, like, rapid breathing, bluish skin, poor weight gain and feeling tired in general (Sun et al. 2015). The main cause of these abnormalities is reduced blood oxygen levels in whole organism, which is a result of improper heart septation leading to the mixture of oxygenated and deoxygenated blood in systemic blood circulation. Additionally, CHD could be caused by genetical or environmental factors like infections during pregnancy such as Rubella, drugs and maternal illness (Sun et al. 2015). Since homeobox genes are important for the heart development, some mutation of the homeobox genes including *MEIS2*, *Nkx2-5* and others can potentially cause CHD (Zakariyah et al. 2017; Johansson et al. 2014). Of course, homeobox gene mutations is not a sole cause of heart defect present at human birth. For example, mutations in *GATA4* and *Tbx5* can also affect integrity of heart tissue (McCulley and Black 2012). In addition, impaired gene functioning might be also related to the other heart diseases like cardiomyopathy, hypertrophy, defects of the heart rhythm and other abnormalities (Kathiresan and Srivastava 2012). However, the impact of homeobox genes in heart disorders is still not clear and needs further investigation to clarify their role not only in the cardiac development but also in the physiological and pathophysiological conditions.

5 ANTP Class of Homeobox Genes

The Antennapedia (ANTP)-class of homeobox genes are involved in the determination of pattern formation along the anterior-posterior axis of the animal embryo. NK-like (NKL) and homeobox transcription factor Hox-like (HOXL) are an ancient subclasses of homeobox genes that belong to an ANTP homeobox gene class (Holland et al. 2007). It is likely that HOXL gene clusters originate from NKL, since NKL genes are widespread throughout the genome as tight clustered Hox genes (Bürglin and Affolter 2016). Both subclasses of the genes (NKL and HOXL) are evolutionarily conserved and play predetermined roles in heart patterning and disease.

6 Hox Gene Families of HOXL Subclass

Hox gene families belong to the HOXL subclass of ANTP class of homeobox. Hox genes code transcription factors which are important for whole body patterning and development (Pearson et al. 2005). It total, there are 39 human Hox genes grouped in HOXA, HOXB, HOXC and HOXD gene clusters. Hox genes are highly conserved, because they play a vital role in anterior-posterior formation of body axis (Pearson et al. 2005). The precise function of these genes is achieved by their specific temporal and spatial expression over the life course. During early mouse cardiac development, the retinoic acid (RA) might be responsible for the anterior-posterior patterning in SHF

(Bertrand et al. 2011). Hoxb1, Hoxa1, and Hoxa3 act as downstream targets of RA and participate in forming outflow tract (OFT) and normal SHF development (Bertrand et al. 2011). *Hoxb1−/−* or *Hoxa1−/−*, *Hoxb1+/−* mouse embryos develop shortened OFT and display abnormal proliferation and premature differentiation of cardiac progenitors (Bertrand et al. 2011). This is probably related to the altered fibroblast growth factor (FGF) and BMP signaling pathways in developing mouse embryo (Roux et al. 2015). Clinical studies have revealed that Hoxa1 mutations might cause congenital human heart defects and other abnormalities like, mental retardation, deafness, horizontal gaze restriction and etc. (Bosley et al. 2008). Several other studies have shown that Hox genes might be also related to the human heart diseases, however more research is needed to unveil exact functions of these genes in human heart development (Gong et al. 2005; Haas et al. 2013).

7 Nk4 Gene Family of NKL Subclass

There are multiple NKL genes in mouse and humans regulating various developmental processes, however only some of them contribute to the development of the heart (Larroux et al. 2007). The *Nkx2-5* and *Nkx2-6* genes are the members of the NK4 homeobox gene family of NKL subclass and are closely related to the Drosophila *tinman* gene (Bürglin and Affolter 2016; Harvey 1996). To our knowledge, only *Nkx2-6* and *Nkx2-5* relate to the mouse and human heart development and disease, whereas *Nkx2-3*, *Nkx2-7*, *Nkx2-8* and *Nkx2-10* might be important for the heart development of zebrafish, frog or chicken (Newman and Krieg 1998; Wang et al. 2014; Tu et al. 2009; Allen et al. 2006; Brand et al. 1997).

During the early stages of embryogenesis, Nkx2-5 is expressed in myocardium and pharyngeal endoderm, whereas Nkx2-6 can be found in sinus venosus, pharyngeal endoderm and myocardium of the outflow tract (Lints et al. 1993).

Moreover, during normal heart development, Nkx2-5 expression is essential for the looping of vertebrate embryonic heart, heart septation and formation of cardiac conduction system, whereas most of the *Nkx-2-5* mutations are related to human congenital heart disease and conduction defects (Tanaka et al. 1999). Inactivation of *Nkx2-5* arrested heart formation at the looping stage revealing its critical role in cardiac development (Lyons et al. 1995). However, targeted disruption of *Nkx-2.6* did not cause any abnormalities in the heart suggesting a possible compensatory function of Nkx-2.5 (Tanaka et al. 2000).

It is important to note that *Nkx2-5* mutations lead to an altered spatiotemporal development of human heart, improper heart septation and formation of cardiac conduction system (Dupays et al. 2015; McCulley and Black 2012; McElhinney et al. 2003). Analysis of human *Nkx2-5* mutants and gene truncations showed that most of the mutations affected *Nkx2-5* binding to DNA or its localization but not protein-protein interactions (McCulley and Black 2012; Reamon-Buettner et al. 2004). Several different studies of mice *Nkx2-5* knockout and human embryonic stem cells (ESC) revealed that *Nkx2-5* mutations might alter gene expression of specific transcription factors like SP1, SRY, JUND, STAT6, *MYCN*, *PRDM16*, *HEY2* and others (Anderson et al. 2018; Li et al. 2015). Also some studies support an idea that, *Nkx2-5* mutant proteins might alter space and time specific human cardiac development by dysregulating BMP, Notch and Wnt signalling pathways (Anderson et al. 2018; Wang et al. 2011; Luxán et al. 2016; Cambier et al. 2014). There is a possibility that Nkx2-5 modulates these pathways by interacting with multiple transcription factors in time-dependent mode. For example, in mouse heart Nkx2-5 interacts with Hand2 transcriptions factor to activate Irx4, which is necessary for the ventricular identity (Yamagishi et al. 2001). Conversely, Nkx2-5 expression is also timely regulated since Nkx2-5 overexpression leads to an improper SAN formation in early mouse development (Roux et al. 2015). Mammalian

heart development is also regulated by the combination of cardiac transcription factors having specific DNA motifs in their centrally located DNA binding domains. It was also shown that Nkx2-5, GATA4 and Tbx5 can physically interact and synergistically regulate targeted genes (Hiroi et al. 2001; Pradhan et al. 2016). Since these genes are the master regulators of heart development, functional mutations in these genes are linked to various types of congenital heart diseases (Benson 2002; Hatcher et al. 2003). Taken together, Nkx2-5 and other transcription factors like Isl1, GATA4, Tbx5, Hand2, MEF2C, Irx4 form a core of transcription factors essential for the heart development and congenital heart disease (Fig. 3) (McCulley and Black 2012).

8 HHEX Gene Family of NKL Subclass

Proline rich homeodomain protein or homeobox protein (PRH/HHEX) expressed by hematopoietic system is a transcription factor belonging to the family of NKL subclass gene (Bedford et al. 1993). As the name implies, it is important for the development of hematopoietic cell, but not less is essential for the development of other systems, including heart (Bedford et al. 1993). Mouse double *HHex* mutants have multiple developmental issues, including defective vasculogenesis, hypoplasia of the right ventricle, aberrant development of the compact myocardium and other complications related to forebrain, thyroid and liver developmental disorders (Hallaq et al. 1998). Additional studies have revealed that HHex plays distinct role in mouse cardiac mesoderm specification and development. HHex expression is controlled by Sox17 transcription factor, which is known to be essential for the formation of mouse cardiac mesoderm (Liu et al. 2014). Several studies of human population have shown that common variants of *HHex* gene (rs7923837 and rs1111875) may also be associated with diabetes (Karns et al. 2013; Kelliny et al. 2009; Pechlivanis et al. 2010).

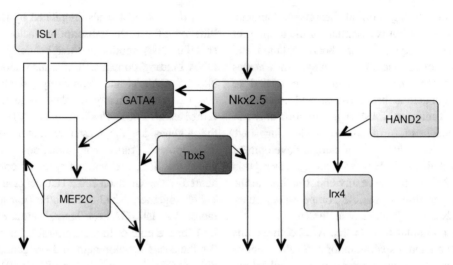

Fig. 3 Core transcription factors important for the heart development and congenital heart disease. (Scheme adapted from (McCulley and Black, 2012))

9   PRD and PRD-Like Homeobox Class

The PRD class is the second largest of the homeobox gene classes in animal genomes and, like the ANTP class, these genes have been found only in animals. The PRD class derives its name from the *Paired* (*Prd*) gene of *Drosophila*. Multiple gene families belong to the PRD homeobox class, including *Shox2*, *Hopx*, *GSC*, *Pitx2* and others, which are important for the heart development and disease.

10   Shox Gene Family of PRD Class

Short stature homeobox 2 (Shox2) is a homeobox gene belonging to the PRD class of homeobox. Shox2 is an essential for the development of limb and cardiac conduction systems, including formation of sinoatrial node (SAN) in mice and humans (Gu et al. 2008; Blaschke et al. 2007; Liu et al. 2011). Studies of Shox2 function during the mouse development revealed several cues how this homeodomain transcription factor in particular controls formation of SAN (Blaschke et al. 2007). Shox2 mice null mutants displayed severe

cardiac conduction defects, such as low heart rhythm rate and drastically reduced cell proliferation (Espinoza-Lewis et al. 2009). This phenotype is probably related to the downregulation of *HCN4*, *Tbx3* and the upregulation of natriuretic peptide A (*Nppa*), gap junction protein alpha 5 (*GJA5*) and *Nkx2-5* gene expressions (Espinoza-Lewis et al. 2009). It is also known that HCN channels play a vital role in autonomic control of heart rate, so it is no surprise why Shox2 null mutants do not develop SAN (Alig et al. 2009). During the normal development of mouse cardiac expression of Shox2 is also tightly controlled by several transcription factors. For example, transcription factor Tbx5 activate Shox2 expression, however transcription factors like Pitx2c and NKX2-5 potentially silence Shox2 expression (Espinoza-Lewis et al. 2011; Puskaric et al. 2010). Paired-like homeodomain transcription factor 2 (Pitx2c) also can potentially inhibit left-sided pacemaker specification by suppressing Shox2 expression in left atrium, therefore SAN develops only in the region of right atrium (Wang et al. 2010). All these results indicate that Shox2 is essential for the maintaining pacemaker cell program during the heart development. On the other hand, in adulthood Nkx2-5

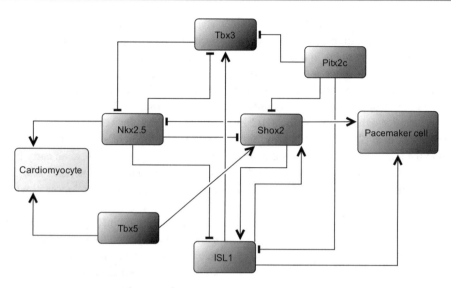

Fig. 4 Signaling networks governing pacemaker and atrial differentiation of cardiomyocyte in developing mouse heart. (Scheme adapted from (Liang et al. 2017))

antagonizes Shox2 and promotes cardiomyocyte formation (Fig. 4) (Liang et al. 2017).

Taken together, these studies indicate that Shox2 might be a good candidate to develop biological pacemaker. Results from mouse ESCs and canine mesenchymal stem cells (MSCs) have shown that cells overexpressing Shox2 induce expression of SAN markers such as HCN4, Cx45 and Tbx3 (Ionta et al. 2015; Feng et al. 2016). In addition, mouse embryonic bodies overexpressing Shox2 showed better contractile phenotype compared to the control group of embryonic bodies (Ionta et al. 2015). Moreover, human patients with an early-onset atrial fibrillation had significantly downregulated expression of *Shox2* gene (Hoffmann et al. 2016). These results are promising for patients suffering from heart rhythm defects, however, more studies are needed to test functions of biological pacemaker in order to treat human arrhythmias.

11 HOPX Gene Family of PRD Class

HOPX is another PRD-class homeobox gene family important for the multiple organ development and tissue homeostasis in adults (Chen et al. 2015; Schneider et al. 2015; Mariotto et al. 2016). Since it lacks a DNA binding domain, HOPX can only modulate gene expression by forming complexes with other regulatory proteins (Kook et al. 2006). In general HOPX acts as a cell proliferation inhibitor in humans cancer cells, however its function in mouse cardiac cell differentiation is not entirely clear (Chen et al. 2015; Waraya et al. 2012; Yap et al. 2016). Studies of mouse development have shown that HOPX plays a critical function in early formation of cardiomyocyte progenitors. HOPX integrates BMP and WNT signalling in developing mouse heart by interacting with SMAD proteins and inhibiting WNT signalling pathway leading to the formation and differentiation of cardiomyocyte progenitors (Jain et al. 2015).

On the other hand, there are some cues that HOPX might act as a negative regulator of cardiac differentiation in mice. HOPX interacts with HDAC2, thus reducing GATA4 transcriptional activity by deacetylation (Trivedi et al. 2010). These findings are consistent with previous reports that overexpression of HDAC2 inhibits the development of cardiomyocytes by down-regulating the expression of *GATA4* and *Nkx2-5* genes (Kawamura et al. 2005; Karamboulas et al. 2006). All these results highlight the complex nature of HOPX and its partners in heart development. HOPX undeniably plays a critical role in early heart development, because some mouse mutants cannot develop a functional myocardium and display cardiac conduction defects (Chen et al. 2002; Ismat et al. 2005). HOPX also might be related to the human heart failure, since HOPX is downregulated in patients having cardiac hypertrophy (Güleç et al. 2014; Trivedi et al. 2011).

12 Goosecoid Gene Family of PRD Class

Goosecoid (GSC) is another protein that belongs to the bicoid related paired (PRD) homeobox class of genes. Goosecoid is often associated with limb, skeletal and craniofacial development, although, it might be important for cardiac mesoderm formation, since its expression is controlled by Mesp1 (Zhu et al. 1998). Mesoderm posterior BHLH transcription factor 1 (MESP1) preferentially binds to two variations of E-box sequences and activates critical mesoderm modulators, including Gata4, mix paired-like homeobox (Mixl1) and GSC homeobox (Soibam et al. 2015). In addition, mesoderm formation can be induced with l-proline and trans-4-hydroxy-l-proline resulting in increased expression of Mixl1 and GSC (Date et al. 2013). GSC also is important for the cell migration in early embryonic development, therefore the overexpression of goosecoid enhances oncogenic cell growth and metastasis (Kang et al. 2014).

13 Pitx Gene Family of PRD-Like Class

Paired like homeodomain 2 *(Pitx2)* is a PRD-like homeobox class gene which is important for the establishment of the left-right axis and for the asymmetrical development of the mouse and probably human heart, lung, and spleen, twisting of the gut and stomach, as well as the development of the eyes (Campione et al. 1999; Shiratori et al. 2006; Evans and Gage, 2005). There are several alternative *Pitx2* transcripts, however only Pitx2c isoform plays determined role in the asymmetric development of mouse heart (Liu et al. 2002). Higher vertebrates, at an early heart development stage and after the heart tube formation, undergo embryonic heart looping, which is the first visual evidence of embryo asymmetry (Harvey 2002). Transcription factors like nodal growth differentiation factor (Nodal) and Cbp/P300 interacting transactivator with Glu/Asp Rich Carboxy-Terminal Domain 2 (Cited2) activate Pitx2 transcription leading to the rightward twist of the heart tube and forming prospective embryonic atrial and ventricular chambers. Deletion of *Pitx2c* in mouse caused drastic alteration of looping process leading to various heart defects including the isomerism of right atrium and ventricle (Lin et al. 1999; Yu et al. 2001).

Humans with Pitx2c mutations develop various heart abnormalities, including an improper formation of ventricle and atrial chambers septa, atrial fibrillation and others. It is quite likely that septation defects are caused by the downregulation of transcription factors downstream of Pitx2, since certain *Pitx2* mutants displayed reduced cardiac transcriptional activity in human patients (Wang et al. 2013; Wei et al. 2014). Surprisingly, the overexpression of Pitx2c in mouse R1-embryonic stem cells results in elevated gene expression of essential cardiac transcription factors like GATA4, MEF2C, Nkx2-5 and others (Lozano-Velasco et al. 2011). Consequently, *Pitx2c* might be a good candidate for heart regeneration, since it positively regulates multiple transcription factors important for cardiac development. Recently it was shown that mouse embryonic stem cells overexpressing Pitx2c could restore mouse heart function after a myocardium infarct through the multiple mechanisms including efficient terminal differentiation, regulation of action potentials of cardiomyocytes and positive paracrine effects (Guddati et al. 2009). However, more studies need to be done to determine the utility of Pitx2c in human heart regeneration strategies.

14 TALE Homeobox Superclass

Three-amino-acid loop extension (TALE) is another superclass of homeobox genes, which codes for highly conserved transcription regulators essential for various developmental programs. These genes encode proteins with atypical homeodomain structure, defined by having three additional amino acids in homeodomain. TALE homeobox gene superclass includes the main zinc finger (ZF), PBC and Meis homeobox 1 (Meis) classes. Out of 20 human homeodomains only Meis1, Meis2 and Iroquois homeobox proteins 1-6 (Irx1-6) have their clearly defined function in heart development and disease.

15 Meis Genes Family of Meis Class

Meis1 encodes the TALE superclass homeobox transcription factor implicated in cardiac, hematopoietic and neural development (Mariotto et al. 2013; Azcoitia et al. 2005; Hisa et al. 2004). *Meis1* deficient mice have malformed cardiac outflow tracts with overriding aorta and ventricular septal defect (Stankunas et al. 2008). Downregulation of Meis1 leads to cardiac hypertrophy in humans and mice. Meis1 binds poly (rC)-binding protein 2 (*PCBP2*) gene promoter and activates its expression in order to suppress human or mouse heart hypertrophy (Zhang et al. 2016). In turn, PCBP2 represses angiotensin II, which enhances hypertrophic human or mouse cardiac growth (Zhang et al. 2015). There are around 79 cardiac specific genes that have Meis1 and NKX2-5 binding sites in developing mouse heart, some of them are associated with

cell signaling and cardiac progenitor differentiation, like Tbx20, myocardin, cadherin 2 (Cdh2), Wnt11, and Wnt2 (Dupays et al. 2015). Adult mouse cardiomyocytes with mutant *MEIS1* exhibit increased proliferation and progression of the cell cycle (Mariotto et al. 2013). This function is emphasized in adult mouse hearts since Meis1 activates inhibitors of cyclin-dependent kinases (CDK) like p15, p16 and p21 (Mariotto et al. 2013). In humans non-synonymous *Meis1* gene variants might be associated with congenital heart defects, whereas patients carrying 2p14 microdeletions show symptoms of deafness and cardiomyopathy (Mathieu et al. 2017; Arrington et al. 2012). It is likely that Meis1 is required for the control of spatiotemporal cell proliferation in early developing heart to prevent hypertrophy, however more studies need to be done to fully understand the role of Meis1 in cardiac development and disease.

Meis2 encodes TALE homeobox superclass transcription factor essential for the development of mouse cranial and cardiac neural crest (Machon et al. 2015). Recent findings indicate that Meis2 might be an important factor for the proliferation of fetal human cardiomyocyte cells (Wu et al. 2015). Reduction of *Meis2* gene expression by miR-134 results in slowed progression of human cardiomyocyte progenitor cell cycle (Wu et al. 2015). A clinical and genetic study also revealed that small *Meis2* deletion can negatively affect several developmental processes: human patients with small *Meis2* non-frame shift deletion (c.998_1000del:p.Arg333del) had serious cleft palate and cardiac septal defects (Louw et al. 2015). It is known that Meis2 interacts with DNA and forms multimeric complexes with Hox and Pbx proteins (Louw et al. 2015). Single deletion of arginine residue affects the ability of Meis2 to bind DNA leading to serious developmental problems of human heart (Louw et al. 2015). Clinical studies have shown that patients having only one functioning Meis2 gene copy survive, however they have similar phenotype such as clefting and ventricular septal defects leading to delayed motor development and learning disability (Johansson et al. 2014).

16   Irx Gene Family of IRX Class

Iroquois homeobox genes and their coded homeodomain proteins are another class of transcription factors belonging to TALE superclass of homeobox genes. Iroquois-class homeodomain TF (Irx) defining feature is atypical homeodomain structure and specific Iroquois (IRO) homeodomain family sequence motif, which is important for the recognition of DNA sequence (Gómez-Skarmeta and Modolell 2002; Cavodeassi et al. 2001). Humans and mice have six Irx proteins, which are important for the development of lung, nervous system, eye, pancreas, female gonad, early limb and, of course, heart patterning (Cavodeassi et al. 2001; Cheng et al. 2005; Schwab et al. 2006; van Tuyl et al. 2006; Ragvin et al. 2010; Jorgensen and Gao 2005; McDonald et al. 2010). Irx1 and Irx2 are expressed in interventricular septum from E14.5 onward, however, mouse *Irx2* mutants are viable and display no notable phenotype defects in the developing heart (Christoffels et al. 2000; Lebel et al. 2003). *Irx1* gene variants might be related to the congenital heart disease in humans (Guo et al. 2017).

Irx3 gene in mice seems to be very important for the ventricular conduction system (VCS) (Christoffels et al. 2000). Various studies suggest that Irx3 is required to maintain rapid electric conduction through the VCS for proper ventricular activation, via antithetical regulation of Cx40 and Cx43 expression (Zhang et al. 2011; Kasahara et al. 2003). Clinical studies have revealed that defects of *Irx3* gene can cause lethal cardiac arrhythmias in human patients (Koizumi et al. 2016). Irx3 function appears to be evolutionary conserved, since expression of Ziro3a, a Irx3 homologue in zebrafish, is detected in developing fish heart (Zhang et al. 2011).

Irx4 is associated with the formation of ventricular myocardium in mouse and humans (Christoffels et al. 2000; Cheng et al. 2011).

Data from mouse and chicken indicate that Irx4 suppresses atrial gene expression by down regulating atrial myosin heavy chain-1 (AMHC1) (Bao et al. 1999; Bruneau et al. 2001b). Several *Irx4* mutations have been identified that might be associated with human congenital heart disease, particularly ventricular septal defect (Cheng et al. 2011).

Irx5 is expressed in adult mouse heart and maintains proper action potentials, particularly regulates T-wave seen in ECG (Costantini et al. 2005). Mice lacking Irx5 develop properly without any structural abnormalities in the heart (Costantini et al. 2005). This indicates that *Irx5* is not required for cardiac development or that other *Irx* genes can compensate for the loss of Irx5.

Irx6 is detectable in mouse developing heart, however its expression is relatively weak compared to other *Irx* genes (Christoffels et al. 2000).

17   Zeb Gene Family of ZF Class

ZEB2 or zinc finger E-box binding homeobox 2 is a gene coding transcription factor belonging to class of ZF homeobox gene and homeodomain class of ZN proteins (Bürglin and Affolter, 2016). It has multiple functional domains (E-box, Zinc finger, homeobox), so naturally it can control gene expression with a variety of transcription factors (Gheldof et al. 2012). The complex nature of Zeb2 shows that it drives multiple processes including the development of heart and neural systems, however, it usually acts as a transcription repressor rather than activator (Hegarty et al. 2015). Systematic study of mouse and human ESC transcriptome differentiation profiles revealed that Zeb2 might play important role in cardiac specialization. Human ESC with silenced *Zeb2* gene proliferate more slowly and fail to differentiate into mature cardiomyocytes compared to the wild cells (Busser et al. 2015). In addition, cardiomyocytes with silenced *Zeb2* do not show any contractile properties, although cardiac differentiation program is activated. More detailed analysis has revealed that silencing of *Zeb2* gene negatively affects human striated muscle contraction program, including genes related to calmodulin pathway, HCN and potassium channels (Busser et al. 2015). Targeted regulation of *Zeb2* gene expression improves cardiomyogenic processes and heart regeneration.

Zeb2 mutation is also often associated with Mowat-Wilson syndrome (Garavelli and Mainardi 2007). Major signs of this disorder frequently include distinctive facial features, intellectual disability, delayed development, an intestinal disorder called Hirschsprung disease, Congenital Heart Disease and other types of birth defects (Garavelli and Mainardi 2007). All mentioned disorders are related to the improper heart development caused by the Zeb2 defective heart cells. In addition, Zeb2 repress epithelial genes (claudins, tight junction protein 3 (ZO-3), connexins, E-cadherin, plakophilin 2, desmoplakin, and crumbs3) in order to induce epithelial to mesenchymal transition (EMT), which is crucial for the developmental processes such as gastrulation, neural crest formation, heart morphogenesis, formation of the musculoskeletal system, and craniofacial structures (Vandewalle et al. 2009; Garavelli et al. 2017).

18   LIM Homeobox Class

LIM homeobox class genes encode two Lim domains and one homeodomain. Lim domain is a 50–60 amino acid length zinc finger motif, which is primarily involved in protein-protein interactions, so naturally LIM transcription factors can interact with multiple proteins in cell, thus regulating its phenotype.

19   Isl Gene Family of LIM Homeobox Class

Isl1 is a LIM homeobox class member that encodes a homeodomain transcription factor important for cell differentiation, fate determination and generation of cell diversity in multiple mouse and human tissues including central nerve system (CNS), pancreas and heart (Zhuang et al. 2013). During early cardiac development, Isl1, Nkx2-5 and fetal liver kinase 1 (Flk1) support the formation of SHF, which gives rise to the right ventricle, outflow tract and part of the atria (Dyer and Kirby, 2009). Isl1 promotes expansion, migration and proliferation of SHF progenitor cells during the development of the mouse heart (Witzel et al. 2012). Additionally, Isl1+ mouse heart cells have potential to differentiate into multiple cell types within the heart, including cardiomyocytes, smooth muscle, pacemaker and endothelial cells (Laugwitz et al. 2007).

There are multiple mechanisms explaining how Isl1 can promote expression of target genes, which suggests the expression of Isl1 is tightly controlled during the mouse heart development. For example, Nkx2-5 homeodomain transcription factor downregulates Isl1 expression in order to promote ventricular development in mouse heart (Witzel et al. 2012; Prall et al. 2007). The newest studies indicate, that Isl1 may repress development of mouse heart ventricle in order to promote the development of SAN (Dorn et al. 2015). Mouse embryos overexpressing Isl1 develop SAN-like cells instead of ventricle myocardium (Dorn et al. 2015). It is likely that the expression of Isl1 activates Nkx2-5 expression in SHF progenitor cells, however, in later staged of heart development Nkx2-5 shuts down ISL1 expression to promote ventricular development (Dorn et al. 2015). Isl1 orchestrates the expression of hundreds of potential genes implicated in cardiac differentiation, mainly through epigenetic mechanisms (Wang et al. 2016). Isl1 in mouse ESCs acts together with JmjC domain-containing protein 3 (JMJD3) histone demethylase to promote the demethylation or tri-methylation of core histone H3 on the amino (N) terminal tail (H3K27me3) at the enhancer's place of key downstream target genes, such as myocardin (*Myocd*), *MEF2C* and others (Wang et al. 2016). In addition, Isl1 may reduce histone methylation near *GATA4* and *Nkx2-5* genes after the expression of Isl1 lentiviral gene, and can also recruit p300 histone acetyltransferase to the promoter of

MEF2C gene in order to promote Mef2c expression in developing mouse embryo (Yu et al. 2013). Other data suggest that lentiviral-induced overexpression of *Isl1* gene promotes not only *MEF2C* gene acetylation, but also *GATA4* and *Nkx2-5* in C3H10T1/2 mouse cell line (Xu et al. 2016). The tight control of *Ils1* gene expression is required, since it acts as a positive cardiomyogenic gene regulator reducing methylation and increasing acetylation levels of genes and histones by direct and indirect methods. Most of the published data concerning Isl1 function have come from the studies of mouse development, however there are some studies that link *Isl1* gene expression with the susceptibility to human congenital heart disease (Luo et al. 2014; Stevens et al. 2010). Development of mechanisms that could control expression of Isl1 might be important target in further regulation of heart regeneration.

20 PROS Homeobox Class

Homeobox prospero (PROS) genes code atypical C terminal prospero domain and belongs to a distinctive class of Prospero homeodomain proteins (Yousef and Matthews, 2005). The PROS domain is a DNA binding domain of approximately 100 amino acids. In addition, PROS homeobox genes code additional three amino acids in their HD domain (Yousef and Matthews 2005).

21 PROX Gene Family of PROS Homeobox Class

Prospero homeobox 1 or Prox1 is a gene coding a transcription factor that plays important role in the development of mouse heart, CNS, eye, liver and lymphatic system (Elsir et al. 2012). Firstly, it was discovered in Drosophila as an important player in the development of central nervous system in insects. However, later Prox1 homologues were found in vertebrates and mammals (Elsir

et al. 2012). In mouse heart development of Prox1 is important for the sarcomere formation and muscle contraction (Risebro et al. 2009). Mouse Prox1 conditional mutants show increased number of fast twitch fibers compared to slow twitch fibers. *Prox1* mutant mice develop fatal dilated cardiomyopathy and die around 7–14th week (Petchey et al. 2014). It was shown that in mice Prox1 acts as a transcriptional repressor of genes like *Tnnt3*, *Tnni2* and *Myl1* that are essential for the formation of fast twitch fibbers (Petchey et al. 2014). Prox1 might also be important for the maintenance of cardiac conduction system in adult mice. It was also shown that uncontrolled Nkx2-5 expression led to cardiac conduction defects, surprisingly suggesting that Prox1 might act as a direct upstream modifier of *Nkx2-5* gene expression (Risebro et al. 2012). In humans dysregulation of *Prox1* gene expression might also lead to congenital heart disease, like, hypoplastic left heart (Gill et al. 2009). Thus, the close connection of Prox1 with Nkx2-5 and other heart development and diseases regulating genes makes it an attractive target in cardiac regeneration field.

22 Role of Homeobox Genes in Cardiomyogenesis

The summarized and reviewed data of estimated involvement of homeobox genes in the heart development, diseases and/or regeneration processes suggest that some homeobox genes play more important role than the other. Data summarized in Table 1 show the homeobox genes that have been most commonly investigated with important roles in cardiomyogenesis.

It is quite evident that dozens of homeobox genes are required for early cardiomyogenesis, heart septation, formation of pacemaker cell, cardiomyocyte and etc. Some of the homeobox genes are directly related to the development of CHD, atrial fibrillations and other cardiac pathologies. However, there are much more

Table 1 Homeobox genes with major involvement in heart development and diseases

Gene	Development	Disease	Reference
Hox	Hoxa1, Hoxb2 and Hoxb2 is important for anterior-posterior patterning in SHF. Integrating FGF and BMP signalling.	Hoxa1 mutations might cause CHD.	Pearson et al. (2005), Bertrand et al. (2011), Bosley et al. (2008), Gong et al. (2005) and Haas et al. (2013)
		HOXB13, and HOXC5 mutations might be related to heart disease.	
Nkx2–5	Heart looping, heart septation and cardiac conduction system formation. Integrates BMP, notch and WNT signaling during development.	Multiple gene variants and truncations are related to CHD.	McCulley and Black (2012), Tanaka et al. (1999), McElhinney et al. (2003), Anderson et al. (2018), Wang et al. (2011), Luxán et al. (2016) and Cambier et al. (2014)
Hhex	Cardiac mesoderm specification.	HHex gene variants might be associated with diabetes.	Liu et al. (2014), Karns et al. (2013), Kelliny et al. (2009) and Pechlivanis et al. (2010)
Shox2	Cardiac conduction system development.	Downregulation during early-onset atrial fibrillation.	Blaschke et al. (2007) and Hoffmann et al. (2016)
Hopx	Cardiomyocyte progenitor formation in mouse early heart. Negative regulator of GATA4 expression.	Downregulated in patients having cardiac hypertrophy.	Jain et al. (2015), Trivedi et al. (2010) and Trivedi et al. (2011)
GSC	Cardiac mesoderm specification.		Zhu et al. (1998)
Pitx2c	Establishment of the left-right axis in heart development. Heart looping and chamber septation.	Mutations cause improper ventricle and atrial chambers septa formation, atrial fibrillation.	Liu et al. (2002), Wang et al. (2013) and Wei et al. (2014)
Meis1 and Meis2	Meis1 and Meis2 control of cell cycle progression during heart development.	Meis1 and Meis2 gene variants might be associated with CHD.	Mariotto et al. (2013), Arrington et al. (2012), Wu et al. (2015) and Louw et al. (2015)
Irx1-6	Irx3 very important for ventricular conduction system.	Irx1 gene variants might be associated with CHD.	Christoffels et al. (2000), Guo et al. (2017), Koizumi et al. (2016) and Cheng et al. (2011)
	Irx4 is associated with the formation of ventricular myocardium in mouse and humans.	Irx3 gene defects can cause lethal cardiac arrhythmias in human patients.	
ZEB2	Controls striated muscle development and contraction.	Gene variants cause Mowat-Wilson syndrome. Patients display CHD and other defects.	Busser et al. (2015) and Garavelli and Mainardi (2007)
Islet1	Cell expansion, migration and proliferation. Marks formation of SHF. Repress ventricular fate in order to promote sinoatrial node development. Positive gene regulator, which reduces gene and histone methylation levels and increase acetylation.	Gene variants might be related to CHD.	Witzel et al. (2012), Dorn et al. (2015), Wang et al. (2016) and Stevens et al. (2010)
Prox1	Important for sarcomere formation and muscle contraction.	Dysregulation of Prox1 might lead to CHD.	Elsir et al. (2012) and Petchey et al. (2014)

homeobox genes related to the heart development, that so far have been less investigated or in one or another model system showed less direct involvement in cardiomyogenic processes (Table 2). Data summarized in Table 2 also highlight the fact that many more studies are needed to understand regulation of homeobox genes and their role in cardiomyogenic processes.

Table 2 Homeobox genes having less important role in heart development and diseases

Class	Subclass	Gene	Development and disease	Reference
ANTP	HOXL	*MEOX1* and *MEOX2*	Control of vascular endothelial cells proliferation in mice. Dysregulation might be associated with heart disease in mouse.	Lu et al. (2018), Douville et al. (2011)
ANTP	NKL	*Msx1* and *Msx2*	Regulate survival of secondary heart field precursors and post-migratory proliferation of cardiac neural crest in the outflow tract	Chen et al. (2007)
ANTP	NKL	*Lbx1*	Specification of a subpopulation of cardiac neural crest necessary for normal heart development.	Schäfer et al. (2003)
ANTP	NKL	*Nkx2-6*	NKX2-6 mutation predisposes to familial atrial fibrillation.	Wang et al. (2014)
PRD		*Prrx1* and *Prrx2*	Formation of cardiovascular system and connective tissues of the heart and in the great arteries and veins.	Bergwerff et al. (2000)
PRD		*Pax3, Pax7*	Involved in neural crest and cardiac development	JA (1996)
ZF (TALE)		*ZFHX3*	Genetic polymorphisms in are associated with atrial fibrillation in a Chinese Han population.	Liu et al. (2014)
PBC (TALE)		*Pbx1*	Patterning of the great arteries and cardiac outflow tract.	Stankunas et al. (2008) and Arrington et al. (2012), Chang et al. (2008)
			Pbx acts with Hand2 in early myocardial differentiation in zebrafish. Non-synonymous variants in PBX genes are associated with congenital heart defects.	
TGIF (TALE)		*Tgif1* and *Tgif2*	Left-right asymmetry formation and embryonic heart looping.	Powers et al. (2010)
SIX/SO		*Six2*	Six2 marks a dynamic subset of second heart field progenitors.	Zhou et al. (2017)

23 Concluding Remarks

Heart development is a complex process requiring strict spatiotemporal development to form a healthy organ providing properly functioning organism. Multiple transcription factors, signalling pathways, morphogens and other stimuli govern the heart development process. However, it is possible to state that multiple homeobox genes and their coded transcription factors come into the heart developmental stages when their function is needed (Tables 1 and 2). None of these homeodomain transcription factors can be separated from each other, since their ability to bind DNA affects patterns of multiple gene thus resulting in changed transcriptome level of multiple cells.

Homeodomain proteins are not the only transcription factors important for cardiac development. The transcription factors of other gene families also significantly contribute heart development. For example, GATA, Tbx, HAND, Mef2c and other accompany homeodomain factors like Nkx2-5, Isl1, etc. (Hiroi et al. 2001; Gao et al. 2011; Maves et al. 2009; Skerjanc et al. 1998). Most of these homeodomain transcription factors are conserved and display coexistence and codependence in heart development of human as well as simple invertebrates like fruit fly or ascidians (Jensen et al. 2013a; Olson 2006). Only birds and mammals display fully separated heart, however reptilians still have no septum between right and left ventricles (Jensen et al. 2013b). Deeper further insights into septum formation of lower vertebrates like snakes, lizards and turtles could also help to understand signalling networks of human congenital heart diseases. Maybe in the future will be possible to engineer a reptile with four chambered heart, thus leading to better understanding of cardiac regeneration process and allowing to develop new therapeutic strategies for human cardiac congenital and other types of diseases.

Acknowledgements The study is funded by the Lithuanian Research council, project No. S-MIP-17-13.

Ethics Approval and Consent to Participate Not applicable.

Consent for Publication All authors agree to the publication of this manuscript.

Availability of Data and Material Not applicable.

Competing Interests The authors declare that they have no competing interests.

Authors' Contributions RM wrote the manuscript draft. DB revised the manuscript. AM read, corrected and approved the final manuscript.

Funding The study is funded by the Lithuanian Research council, project No. S-MIP-17-13.

References

Akazawa H, Komuro I (2005) Cardiac transcription factor Csx/Nkx2-5: its role in cardiac development and diseases. Pharmacol Ther 107:252–268. https://doi.org/10.1016/j.pharmthera.2005.03.005

Alig J, Marger L, Mesirca P, Ehmke H, Mangoni ME, Isbrandt D (2009) Control of heart rate by cAMP sensitivity of HCN channels. Proc Natl Acad Sci U S A 106:12189–12194. https://doi.org/10.1073/pnas.0810332106

Allen BG, Allen-Brady K, Weeks DL (2006) Reduction of XNkx2-10 expression leads to anterior defects and malformation of the embryonic heart. Mech Dev 123:719–729. https://doi.org/10.1016/j.mod.2006.07.008

Anderson DJ, Kaplan DI, Bell KM, Koutsis K, Haynes JM, Mills RJ, Phelan DG, Qian EL, Leitoguinho AR, Arasaratnam D, Labonne T, Ng ES, Davis RP, Casini S, Passier R, Hudson JE, Porrello ER, Costa MW, Rafii A, Curl CL, Delbridge LM et al (2018) NKX2-5 regulates human cardiomyogenesis via a HEY2 dependent transcriptional network. Nat Commun 9:1–13. https://doi.org/10.1038/s41467-018-03714-x

Arrington CB, Dowse BR, Bleyl SB, Bowles NE (2012) Non-synonymous variants in pre-B cell leukemia homeobox (PBX) genes are associated with congenital heart defects. Eur J Med Genet 55:235–237. https://doi.org/10.1016/j.ejmg.2012.02.002.Non-synonymous

Azcoitia V, Aracil M, Martínez-A C, Torres M (2005) The homeodomain protein Meis1 is essential for definitive hematopoiesis and vascular patterning in the mouse embryo. Dev Biol 280:307–320. https://doi.org/10.1016/j.ydbio.2005.01.004

Bao ZZ, Bruneau BG, Seidman JG, Seidman CE, Cepko CL (1999) Regulation of chamber-specific gene expression in the developing heart by irx4. Science 283:1161–1164

Bedford FK, Ashworth A, Enver T, Wiedemann LM (1993) HEX: a novel homeobox gene expressed during haematopoiesis and conserved between mouse and human. Nucleic Acids Res 21:1245–1249

Belaguli NS, Sepulveda JL, Nigam V, Charron F, Nemer M, Schwartz RJ (2000) Cardiac tissue enriched factors serum response factor and GATA-4 are mutual coregulators. Mol Cell Biol 20:7550–7558

Benchabane H, Wrana JL (2003) GATA- and Smad1-dependent enhancers in the Smad7 gene differentially interpret bone morphogenetic protein concentrations. Mol Cell Biol 23:6646–6661

Benson DW (2002) The genetics of congenital heart disease: a point in the revolution. Cardiol Clin 20:385–394

Bergwerff M, Gittenberger-de Groot AC, Wisse LJ, DeRuiter MC, Wessels A, Martin JF, Olson EN, Kern MJ (2000) Loss of function of the Prx1 and Prx2 homeobox genes alters architecture of the great elastic arteries and ductus arteriosus. Virchows Arch 436:12–19

Bertrand N, Roux M, Ryckebüsch L, Niederreither K, Dollé P, Moon A, Capecchi M, Zaffran S (2011) Hox genes define distinct progenitor sub-domains within the second heart field. Dev Biol 353:266–274. https://doi.org/10.1016/j.ydbio.2011.02.029

Blaschke RJ, Hahurij ND, Kuijper S, Just S, Wisse LJ, Deissler K, Maxelon T, Anastassiadis K, Spitzer J, Hardt SE, Schöler H, Feitsma H, Rottbauer W, Blum M, Meijlink F, Rappold G, Gittenberger-de Groot AC (2007) Targeted mutation reveals essential functions of the homeodomain transcription factor Shox2 in sinoatrial and pacemaking development. Circulation 115:1830–1838. https://doi.org/10.1161/CIRCULATIONAHA.106.637819

Bosley TM, Alorainy IA, Salih MA, Aldhalaan HM, Abu-Amero KK, Oystreck DT, Tischfield MA, Engle EC, Erickson RP (2008) The clinical spectrum of homozygous HOXA1 mutations. Am J Med Genet A 146:1235–1240. https://doi.org/10.1002/ajmg.a.32262

Brand T, Andrée B, Schneider A, Buchberger A, Arnold H (1997) Chicken NKx2-8, a novel homeobox gene expressed during early heart and foregut development. Mech Dev 64:53–59

Bruneau BG, Nemer G, Schmitt JP, Charron F, Robitaille L, Caron S, Conner DA, Gessler M, Nemer M, Seidman CE, Seidman JG (2001a) A murine model of Holt-Oram syndrome defines roles of the

T-box transcription factor Tbx5 in cardiogenesis and disease. Cell 106:709–721

Bruneau BG, Bao ZZ, Fatkin D, Xavier-Neto JGD, Maguire CT, Berul CI, Kass DA, Kuroski-de Bold ML, de Bold AJ, Conner DA, Rosenthal N, Cepko CL, Seidman CE, Seidman JG (2001b) Cardiomyopathy in irx4-deficient mice is preceded by abnormal ventricular gene expression. Mol Cell Biol 21:1730–1736. https://doi.org/10.1128/MCB.21.5.1730-1736.2001

Buckingham M, Meilhac S, Zaffran S (2005) Building the mammalian heart from two sources of myocardial cells. Nat Rev Genet 6:826–837. https://doi.org/10.1038/nrg1710

Bürglin TR, Affolter M (2016) Homeodomain proteins: an update. Chromosoma 125:497–521. https://doi.org/10.1007/s00412-015-0543-8

Busser BW, Lin Y, Yang Y et al (2015) An orthologous epigenetic gene expression signature derived from differentiating embryonic stem cells identifies regulators of cardiogenesis. PLoS One 10:e0141066. https://doi.org/10.1371/journal.pone.0141066

Cai CL, Liang X, Shi Y, Chu PH, Pfaff SL, Chen J, Evans S (2003) Isl1 identifies a cardiac progenitor population that proliferates prior to differentiation and contributes a majority of cells to the heart. Dev Cell 5:877–889

Camacho P, Fan H, Liu Z, He J (2016) Small mammalian animal models of heart disease. Am J Cardiovasc Dis 6:70–80

Cambier L, Plate M, Sucov HM, Pashmforoush M (2014) Nkx2-5 regulates cardiac growth through modulation of Wnt signaling by R-spondin3. Development 141:2959–2971. https://doi.org/10.1242/dev.103416

Campione M, Steinbeisser H, Schweickert A, Deissler K, van Bebber F, Lowe LA, Nowotschin S, Viebahn C, Haffter P, Kuehn MR, Blum M (1999) The homeobox gene Pitx2: mediator of asymmetric left-right signaling in vertebrate heart and gut looping. Development 126:1225–1234

Cavodeassi F, Modolell J, Gómez-Skarmeta JL (2001) The Iroquois family of genes: from body building to neural patterning. Development 128:2847–2855

Chang CP, Stankunas K, Shang C, Kao SC, Twu KY, Cleary ML (2008) Pbx1 functions in distinct regulatory networks to pattern the great arteries and cardiac outflow tract. Development 135:3577–3586. https://doi.org/10.1242/dev.022350

Chen F, Kook H, Milewski R, Gitler AD, Lu MM, Li J, Nazarian R, Schnepp R, Jen K, Biben C, Runke G, Mackay JP, Novotny J, Schwartz RJ, Harvey RP, Mullins MC, Epstein JA (2002) Hop is an unusual homeobox gene that modulates cardiac development. Cell 110:713–723

Chen YH, Ishii M, Sun J, Sucov HM, Maxson RE Jr (2007) Msx1 and Msx2 regulate survival of secondary heart field precursors and post-migratory proliferation of cardiac neural crest in the outflow tract. Dev Biol 308:421–437. https://doi.org/10.1016/j.ydbio.2007.05.037

Chen Y, Yang L, Cui T, Pacyna-Gengelbach M, Petersen I (2015) Hopx is methylated and exerts tumour-suppressive function through ras-induced senescence in human lung cancer. J Pathol 235:397–407. https://doi.org/10.1002/path.4469

Cheng CW, Chow RL, Lebel M, Sakuma R, Cheung HO, Thanabalasingham V, Zhang X, Bruneau BG, Birch DG, Hui CC, McInnes RR, Cheng S (2005) The Iroquois homeobox gene, Irx5, is required for retinal cone bipolar cell development. Dev Biol 287:48–60. https://doi.org/10.1016/j.ydbio.2005.08.029

Cheng Z, Wang J, Su D, Pan H, Huang G, Li X, Li Z, Shen A, Xie X, Wang B, Ma X (2011) Two novel mutations of the IRX4 gene in patients with congenital heart disease. Hum Genet 130:657–662. https://doi.org/10.1007/s00439-011-0996-7

Christoffels VM, Keijser AG, Houweling AC, Clout DE, Moorman AFM (2000) Patterning the embryonic heart: identification of five mouse Iroquois homeobox genes in the developing heart. Dev Biol 224:263–274. https://doi.org/10.1006/dbio.2000.9801

Costantini DL, Arruda EP, Agarwal P, Kim KH, Zhu Y, Zhu W, Lebel M, Cheng CW, Park CY, Pierce SA, Guerchicoff A, Pollevick GD, Chan TY, Kabir MG, Cheng SH, Husain M, Antzelevitch C, Srivastava D, Gross GJ, Hui CC, Backx PH, Bruneau BG (2005) The homeodomain transcription factor Irx5 establishes the mouse cardiac ventricular repolarization gradient. Cell 123:347–358. https://doi.org/10.1016/j.cell.2005.08.004

Date Y, Hasegawa S, Yamada T, Inoue Y, Mizutani H, Nakata S, Akamatsu H (2013) Major amino acids in collagen hydrolysate regulate the differentiation of mouse embryoid bodies. J Biosci Bioeng 116:386–390. https://doi.org/10.1016/j.jbiosc.2013.03.014

Dorn T, Goedel A, Lam JT, Haas J, Tian Q, Herrmann F, Bundschu K, Dobreva G, Schiemann M, Dirschinger R, Guo Y, Kühl SJ, Sinnecker D, Lipp P, Laugwitz K-L, Kühl M, Moretti A (2015) Direct Nkx2-5 transcriptional repression of isl1 controls cardiomyocyte subtype identity. Stem Cells 33:1113–1129. https://doi.org/10.1002/stem.1923

Douville JM, Cheung DY, Herbert KL, Moffatt T, Wigle JT (2011) Mechanisms of MEOX1 and MEOX2 regulation of the cyclin dependent kinase inhibitors p21 and p16 in vascular endothelial cells. PLoS One 6:e29099. https://doi.org/10.1371/journal.pone.0029099

Dupays L, Shang C, Wilson R, Kotecha S, Wood S, Towers N, Mohun T (2015) Sequential binding of MEIS1 and NKX2-5 on the Popdc2 gene: a mechanism for spatiotemporal regulation of enhancers during cardiogenesis. Cell Rep 13:183–195. https://doi.org/10.1016/j.celrep.2015.08.065

Dyer LA, Kirby ML (2009) The role of secondary heart field in cardiac development. Dev Biol 336:137–144. https://doi.org/10.1016/j.ydbio.2009.10.009

Elsir T, Smits A, Lindström MS, Nister M (2012) Transcription factor PROX1: its role in development and cancer. Cancer Metastasis Rev 31:793–805. https://doi.org/10.1007/s10555-012-9390-8

Espinoza-Lewis RA, Yu L, He F, Liu H, Tang R, Shi J, Sun X, Martin JF, Wang D, Yang J, Chen Y (2009) Shox2 is essential for the differentiation of cardiac pacemaker cells by repressing Nkx2-5. Dev Biol 327:376–385. https://doi.org/10.1021/nl061786n. Core-Shell

Espinoza-Lewis RA, Liu H, Sun C, Chen C, Jiao K, Chen Y (2011) Ectopic expression of Nkx2.5 suppresses the formation of the sinoatrial node in mice. Dev Biol 356:359–369. https://doi.org/10.1016/j.ydbio.2011.05.663

Epstein JA (1996) Pax3, neural crest and cardiovascular development. Trends Cardiovasc Med 6:255–260. https://doi.org/10.1016/S1050-1738(96)00110-7

Evans AL, Gage PJ (2005) Expression of the homeobox gene Pitx2 in neural crest is required for optic stalk and ocular anterior segment development. Hum Mol Genet 14:3347–3359. https://doi.org/10.1093/hmg/ddi365

Feng Y, Yang P, Luo S, Zhang Z, Li H, Zhu P, Song Z (2016) Shox2 influences mesenchymal stem cell fate in a co-culture model in vitro. Mol Med Rep 14:637–642. https://doi.org/10.3892/mmr.2016.5306

Franco D, Sedmera D, Lozano-Velasco E (2017) Multiple roles of Pitx2 in cardiac development and disease. J Cardiovasc Dev Dis 4:16. https://doi.org/10.3390/jcdd4040016

Gao XR, Tan YZ, Wang HJ (2011) Overexpression of Csx/Nkx2.5 and GATA-4 enhances the efficacy of mesenchymal stem cell transplantation after myocardial infarction. Circ J 75:2683–2691. https://doi.org/10.1253/circj.CJ-11-0238

Garavelli L, Mainardi PC (2007) Mowat-Wilson syndrome. Orphanet J Rare Dis 12:1–12. https://doi.org/10.1186/1750-1172-2-42

Garavelli L, Ivanovski I, Caraffi SG, Santodirocco D, Pollazzon M, Cordelli DM, Abdalla E, Accorsi P, Adam MP, Baldo C, Bayat A, Belligni E, Bonvicini F, Breckpot J, Callewaert B, Cocchi G, Cuturilo G, Devriendt K, Dinulos MB, Djuric O et al (2017) Neuroimaging findings in Mowat – Wilson syndrome: a study of 54 patients. Genet Med 19:691–700. https://doi.org/10.1038/gim.2016.176

Gheldof A, Hulpiau P, van Roy F, De Craene B, Berx G (2012) Evolutionary functional analysis and molecular regulation of the ZEB transcription factors. Cell Mol Life Sci 69:2527–2541. https://doi.org/10.1007/s00018-012-0935-3

Gill HK, Parsons SR, Spalluto C, Davies AF, Knorz VJ, Burlinson CE, Ng B, Carter NP, Ogilvie CM, Wilson DI, Roberts RG (2009) Separation of the PROX1 gene from upstream conserved elements in a complex inversion/translocation patient with hypoplastic left heart. Eur J Hum Genet 17:1423–1431. https://doi.org/10.1038/ejhg.2009.91

Gómez-Skarmeta JL, Modolell J (2002) Iroquois genes: genomic organization and function in vertebrate neural development. Curr Opin Genet Dev 12:403–408

Gong LG, Qiu GR, Jiang H, Xu XY, Zhu HY, Sun KL (2005) Analysis of single nucleotide polymorphisms and haplotypes in HOXC gene cluster within susceptible region 12q13 of simple congenital heart disease. Zhonghua Yi Xue Yi Chuan Xue Za Zhi 22:497–501

Gu S, Wei N, Yu L, Fei J, Chen Y (2008) Shox2-deficiency leads to dysplasia and ankylosis of the temporomandibular joint in mice. Mech Dev 125:729–742. https://doi.org/10.1016/j.mod.2008.04.003

Guddati AK, Otero JJ, Kessler E, Aistrup G, Wasserstrom JA, Han X, Lomasney JW, Kessler JA (2009) Embryonic stem cells overexpressing Pitx2 engraft in infarcted myocardium and improve cardiac function. Int Heart J 50:783–799

Güleç Ç, Abacı N, Bayrak F, Kömürcü Bayrak E, Kahveci G, Güven C, Ünaltuna NE (2014) Association between non-coding polymorphisms of HOPX gene and syncope in hypertrophic cardiomyopathy. Anadolu Kardiyol Derg 14:617–624. https://doi.org/10.5152/akd.2014.4972

Guo C, Wang Q, Wang Y, Yang L, Luo H, Cao XF, An L, Qiu Y, Du M, Ma X, Hui L, Lu C (2017) Exome sequencing reveals novel IRXI mutation in congenital heart disease. Mol Med Rep 15:3193–3197. https://doi.org/10.3892/mmr.2017.6410

Haas J, Frese KS, Park YJ, Keller A, Vogel B, Lindroth AM, Weichenhan D, Franke J, Fischer S, Bauer A, Marquart S, Sedaghat-Hamedani FKE, Köhler D, Wolf NM, Hassel S, Nietsch R, Wieland T, Ehlermann P, Schultz JH, Dösch A, Mereles D, Hardt S, Backs J, Hoheisel JD, Plass C, Katus HA, Meder B (2013) Alterations in cardiac DNA methylation in human dilated cardiomyopathy. EMBO Mol Med 5:413–429. https://doi.org/10.1002/emmm.201201553

Hallaq H, Pinter E, Enciso J et al (1998) A null mutation of Hhex results in abnormal cardiac development, defective vasculogenesis and elevated Vegfa levels. Development 131:5197–5209. https://doi.org/10.1242/dev.01393

Harvey RP (1996) NK-2 homeobox genes and heart development. Dev Biol 178:203–216

Harvey RP (2002) Patterning the vertebrate heart. Nat Rev Genet 3:544–556. https://doi.org/10.1038/nrg843

Hatcher CJ, Diman NY, McDermott DA, Basson CT (2003) Transcription factor cascades in congenital heart malformation. Trends Mol Med 9:512–515

Hegarty SV, Sullivan AM, Keeffe GWO (2015) Progress in Neurobiology Zeb2: a multifunctional regulator of nervous system development. Prog Neurobiol 132:81–95. https://doi.org/10.1016/j.pneurobio.2015.07.001

Hiroi Y, Kudoh S, Monzen K, Ikeda Y, Yazaki Y, Nagai R, Komuro I (2001) Tbx5 associates with Nkx2-5 and synergistically promotes cardiomyocyte differentiation. Nat Genet 28:276–280. https://doi.org/10.1038/90123

Hisa T, Spence SE, Rachel RA, Fujita M, Nakamura T, Ward JM, Devor-Henneman DE, Saiki Y, Kutsuna H, Tessarollo L, Jenkins NA, Copeland NG (2004) Hematopoietic, angiogenic and eye defects in Meis1 mutant animals. EMBO J 23:450–459. https://doi.org/10.1038/sj.emboj.7600038

Hoffmann S, Clauss S, Berger IM, Weiß B, Montalbano A, Röth R, Bucher M, Klier I, Wakili R, Seitz H, Schulze-Bahr E, Katus HA, Flachsbart F, Nebel A, Guenther SP, Bagaev E, Rottbauer W, Kääb S, Just S, Rappold G (2016) Coding and non-coding variants in the SHOX2 gene in patients with early-onset atrial fibrillation. Basic Res Cardiol 111:36. https://doi.org/10.1007/s00395-016-0557-2

Holland PWH, Booth HAF, Bruford EA (2007) Classification and nomenclature of all human homeobox genes. BMC Biol 5:1–29. https://doi.org/10.1186/1741-7007-5-47

Innis JW (1997) Role of HOX genes in human development. Curr Opin Pediatr 9:617–622

Ionta V, Liang W, Kim EH, Rafie R, Giacomello A, Marbán E, Cho C (2015) SHOX2 overexpression favors differentiation of embryonic stem cells into cardiac pacemaker cells, improving biological pacing ability. Stem Cell Reports 4:129–142. https://doi.org/10.1016/j.stemcr.2014.11.004

Ismat FA, Zhang M, Kook H, Huang B, Zhou R, Ferrari VA, Epstein JA, Patel VV (2005) Homeobox protein Hop functions in the adult cardiac conduction system. Circ Res 96:898–903. https://doi.org/10.1161/01.RES.0000163108.47258.f3

JA E (1996) Pax3, neural crest and cardiovascular development. Trends Cardiovasc Med 6:255–60. https://doi.org/10.1016/S1050-1738(96)00110-7

Jain R, Li D, Gupta M, Manderfield LJ, Ifkovits JL, Wang Q, Liu F, Liu Y, Poleshko A, Padmanabhan A, Raum JC, Li L, Morrisey EE, Lu MM, Won KJ, Epstein JA (2015) Integration of Bmp and Wnt signaling by Hopx specifies commitment of cardiomyoblasts. Science 348:aaa6071. https://doi.org/10.1016/j.joca.2015.05.020.Osteoarthritic

Jensen B, Wang T, Christoffels VM, Moorman AFM (2013a) Evolution and development of the building plan of the vertebrate heart. Biochim Biophys Acta 1833:783–794. https://doi.org/10.1016/j.bbamcr.2012.10.004

Jensen B, van den Berg G, van den Doel R, Oostra RJ, Wang T, Moorman AFM (2013b) Development of the hearts of lizards and snakes and perspectives to cardiac evolution. PLoS One 8:e63651. https://doi.org/10.1371/journal.pone.0063651

Johansson S, Berland S, Gradek GA, Bongers E, de Leeuw N, Pfundt R, Fannemel M, Rødningen O, Brendehaug A, Haukanes BI, Hovland R, Helland G, Houge G (2014) Haploinsufficiency of MEIS2 is associated with orofacial clefting and learning disability. Am J Med Genet A 164A:1622–1626. https://doi.org/10.1002/ajmg.a.36498

Jorgensen JS, Gao L (2005) Irx3 is differentially up-regulated in female gonads during sex determination. Gene Expr Pattern 5:756–762. https://doi.org/10.1016/j.modgep.2005.04.011

Kang KW, Lee MJ, Song JA, Jeong JY, Kim YK, Lee C, Kim TH, Kwak KB, Kim OJ, An HJ (2014) Overexpression of goosecoid homeobox is associated with chemoresistance and poor prognosis in ovarian carcinoma. Oncol Rep 32:189–198. https://doi.org/10.3892/or.2014.3203

Karamboulas C, Swedani A, Ward C, Al-Madhoun AS, Wilton S, Boisvenue S, Ridgeway AG, Skerjanc IS (2006) HDAC activity regulates entry of mesoderm cells into the cardiac muscle lineage. J Cell Sci 119:4305–4314. https://doi.org/10.1242/jcs.03185

Karns R, Succop P, Zhang G, Sun G, Indugula SR, Havas-Augustin D, Novokmet N, Durakovic Z, Milanovic SM, Missoni S, Vuletic S, Chakraborty R, Rudan P, Deka R (2013) Modeling metabolic syndrome through structural equations of metabolic traits, comorbid diseases, and GWAS variants. Obesity 21:745–754. https://doi.org/10.1002/oby.20445

Kasahara H, Ueyama T, Wakimoto H, Liu MK, Maguire CT, Converso KL, Kang PM, Manning WJ, Lawitts J, Paul DL, Berul CI, Izumo S (2003) Nkx2.5 homeoprotein regulates expression of gap junction protein connexin 43 and sarcomere organization in postnatal cardiomyocytes. J Mol Cell Cardiol 35:243–256. https://doi.org/10.1016/S0022-2828(03)00002-6

Kathiresan S, Srivastava D (2012) Genetics of human cardiovascular disease. Cell 148:1242–1257. https://doi.org/10.1016/j.cell.2012.03.001

Kawamura T, Ono K, Morimoto T et al (2005) Acetylation of GATA-4 is involved in the differentiation of embryonic stem cells into cardiac myocytes. J Biol Chem 280:19682–19688. https://doi.org/10.1074/jbc.M412428200

Kelliny C, Ekelund U, Andersen LB, Brage S, Loos RJ, Wareham NJ, Langenberg C (2009) Common genetic determinants of glucose homeostasis in healthy children: the European Youth Heart Study. Diabetes 58:2939–2945. https://doi.org/10.2337/db09-0374

Kitajima S, Takagi A, Inoue T, Saga Y (2000) MesP1 and MesP2 are essential for the development of cardiac mesoderm. Development 127:3215–3226

Koizumi A, Sasano T, Kimura W, Miyamoto Y, Aiba T, Ishikawa T, Nogami A, Fukamizu S, Sakurada H, Takahashi Y, Nakamura H, Ishikura T, Koseki H, Arimura T, Kimura A, Hirao K, Isobe M, Shimizu W, Miura N, Furukawa T (2016) Genetic defects in a His-Purkinje system transcription factor, IRX3, cause lethal cardiac arrhythmias. Eur Heart J 37:1469–1475. https://doi.org/10.1093/eurheartj/ehv449

Kook H, Yung WW, Simpson RJ, Kee HJ, Shin S, Lowry JA, Loughlin FE, Yin Z, Epstein JA, Mackay J (2006) Analysis of the structure and function of the transcriptional coregulator HOP. Biochemistry 45:10584–10590. https://doi.org/10.1021/bi060641s

Lage K, Møllgård K, Greenway S, Wakimoto H, Gorham JM, Workman CT, Bendsen E, Hansen NT, Rigina O, Roque FS, Wiese C, Christoffels VM, Roberts AE, Smoot LB, Pu WT, Donahoe PK, Tommerup N, Brunak S, Seidman CE, Seidman JG, Larsen LA (2010) Dissecting spatio-temporal protein networks driving human heart development and related disorders. Mol Syst Biol 6:381. https://doi.org/10.1038/msb.2010.36

Larroux C, Fahey B, Degnan SM, Adamski M, Rokhsar DS, Degnan BM (2007) The NK homeobox gene cluster predates the origin of Hox genes. Curr Biol 17:706–710. https://doi.org/10.1016/j.cub.2007.03.008

Laugwitz KL, Moretti A, Caron L, Nakano A, Chien KR (2007) Islet1 cardiovascular progenitors: a single source for heart lineages. Development 135:193–205. https://doi.org/10.1242/dev.001883

Lebel M, Agarwal P, Cheng CW, Kabir MG, Chan TY, Thanabalasingham V, Zhang X, Cohen DR, Husain M, Cheng SH, Bruneau BG, Cheng SH (2003) The Iroquois homeobox gene irx2 is not essential for normal development of the heart and midbrain-hindbrain boundary in mice. Mol Cell Biol 23:8216–8225

Li J, Cao Y, Wu Y, Chen W, Yuan Y, Ma X, Huang G (2015) The expression profile analysis of NKX2-5 knock-out embryonic mice to explore the pathogenesis of congenital heart disease. J Cardiol 66:527–531. https://doi.org/10.1016/j.jjcc.2014.12.022

Liang X, Wang G, Lin L, Lowe J, Zhang Q, Bu L, Chen Y, Chen J, Sun Y, Evans SM (2013) HCN4 dynamically marks the first heart field and conduction system precursors. Circ Res 113:399–407. https://doi.org/10.1161/CIRCRESAHA.113.301588

Liang X, Evans SM, Sun Y (2017) Development of the cardiac pacemaker. Cell Mol Life Sci 74:1247–1259. https://doi.org/10.1007/s00018-016-2400-1

Lin Q, Schwarz J, Bucana C, Olson EN (1997) Control of mouse cardiac morphogenesis and myogenesis by transcription factor MEF2C. Science 276:1404–1407

Lin CR, Kioussi C, O'Connell S et al (1999) Pitx2 regulates lung asymmetry, cardiac positioning and pituitary and tooth morphogenesis. Nature 401:279–282. https://doi.org/10.1038/45803

Lints TJ, Parsons LM, Hartley L, Lyons I, Harvey RP (1993) Nkx-2.5: a novel murine homeobox gene expressed in early heart progenitor cells and their myogenic descendants. Development 119:419–431

Liu CC, Liu WW, Palie JJ et al (2002) Pitx2c patterns anterior myocardium and aortic arch vessels and is required for local cell movement into atrioventricular cushions. Development 129:5081–5091. https://doi.org/10.1242/dev.00173

Liu H, Chen CH, Espinoza-Lewis RA, Jiao Z, Sheu I, Hu X, Lin M, Zhang Y, Chen Y (2011) Functional redundancy between human SHOX and mouse Shox2 genes in the regulation of sinoatrial node formation and Pacemaking function *. J Biol Chem

286:17029–17038. https://doi.org/10.1074/jbc.M111.234252

Liu Y, Kaneda R, Leja TW, Subkhankulova T, Tolmachov O, Minchiotti G, Schwartz RJ, Barahona M, Schneider M (2014) Hhex and Cer1 mediate the Sox17 pathway for cardiac mesoderm formation in embryonic stem. Stem Cells 32:1515–1526. https://doi.org/10.1002/stem.1695

Liu Y, Ni B, Lin Y, Chen XG, Fang Z, Zhao L, Hu Z, Zhang F, Ai X (2014) Genetic polymorphisms in ZFHX3 are associated with atrial fibrillation in a Chinese Han population. PLoS One 9:e101318. https://doi.org/10.1371/journal.pone.0101318

Louw JJ, Corveleyn A, Jia Y, Hens G, Gewillig M, Devriendt K (2015) MEIS2 involvement in cardiac development, cleft palate, and intellectual disability. Am J Med Genet A 167A:1142–1146. https://doi.org/10.1002/ajmg.a.36989

Lozano-Velasco E, Chinchilla A, Martínez-Fernández S et al (2011) Pitx2c modulates cardiac-specific transcription factors networks in differentiating cardiomyocytes from murine embryonic stem cells. Cells Tissues Organs 194:349–362. https://doi.org/10.1159/000323533

Lu D, Wang J, Li J, Guan F, Zhang X, Dong W, Liu N, Gao S, Zhang L (2018) Meox1 accelerates myocardial hypertrophic decompensation through Gata4. Cardiovasc Res 114:300–311. https://doi.org/10.1093/cvr/cvx222

Luo ZL, Sun H, Yang ZQ, Ma YH, Gu Y, He YQ, Wei D, Xia LB, Yang BH, Guo T (2014) Genetic variations of ISL1 associated with human congenital heart disease in Chinese Han people. Genet Mol Res 13:1329–1338. https://doi.org/10.4238/2014.February.28.5

Luxán G, D'Amato G, de la Pompa JL (2016) Endocardial notch signaling in cardiac development and disease. Circ Res 118:1–18. https://doi.org/10.1161/CIRCRESAHA.115.305350

Lyons I, Parsons LM, Hartley L, Li R, Andrews JE, Robb L, Harvey RP (1995) Myogenic and morphogenetic defects in the heart tubes of murine embryos lacking the homeo box gene Nkx2–5. Genes Dev 9:1654–1666

Machon O, Masek J, Machonova O, Krauss S, Kozmik Z (2015) Meis2 is essential for cranial and cardiac neural crest development. BMC Dev Biol 15:40. https://doi.org/10.1186/s12861-015-0093-6

Mariotto A, Pavlova O, Park HS, Huber M, Sadek HA (2013) Meis1 regulates postnatal cardiomyocyte cell cycle arrest. Nature 497:249–253. https://doi.org/10.1038/jid.2014.371

Mariotto A, Pavlova O, Park HS et al (2016) HOPX: the unusual homeodomain-containing protein. J Invest Dermatol 136:905–911. https://doi.org/10.1016/j.jid.2016.01.032

Mathieu ML, Demily C, Chantot-Bastaraud S, Afenjar A, Mignot C, Andrieux J, Gerard M, Catala-Mora J, Jouk PS, Labalme A, Edery P, Sanlaville D, Rossi M (2017) Clinical and molecular cytogenetic characterization of four unrelated patients carrying 2p14 microdeletions.

Am J Med Genet A 173:2268–2274. https://doi.org/10.1002/ajmg.a.38307

Maves L, Tyler A, Moens CB, Tapscott SJ (2009) Pbx acts with Hand2 in early myocardial differentiation. Dev Biol 333:409–418. https://doi.org/10.1016/j.ydbio.2009.07.004

McCulley DJ, Black BL (2012) Transcription factor pathways and congenital heart disease. Curr Top Dev Biol 100:253–277. https://doi.org/10.1016/B978-0-12-387786-4.00008-7.Transcription

McDonald LA, Gerrelli D, Fok Y, Hurst LD, Tickle C (2010) Comparison of Iroquois gene expression in limbs/fins of vertebrate embryos. J Anat 216:683–691. https://doi.org/10.1111/j.1469-7580.2010.01233.x

McElhinney DB, Geiger E, Blinder J, Benson DW, Goldmuntz E (2003) NKX2.5 mutations in patients with congenital heart disease. J Am Coll Cardiol 42:1650–1655

Moorman A, Webb S, Brown NA et al (2003) Development of the heart: (1) formation of the cardiac chambers and arterial trunks. Heart 89:806–814

Nayor M, Vasan RS (2015) Preventing heart failure: the role of physical activity. Curr Opin Cardiol 30:543–550. https://doi.org/10.1097/HCO.0000000000000206

Nemer M (2008) Genetic insights into normal and abnormal heart development. Cardiovasc Pathol 17:48–54. https://doi.org/10.1016/j.carpath.2007.06.005

Newman CS, Krieg PA (1998) Tinman-related genes expressed during heart development in Xenopus. Dev Genet 22:230–238. https://doi.org/10.1002/(SICI)1520-6408(1998)22:3<230::AID-DVG5>3.0.CO;2-7

O'Toole TE, Conklin DJ, Bhatnagar A (2008) Environmental risk factors for heart disease. Rev Environ Health 23:167–202. https://doi.org/10.1038/nrcardio.2015.152

Olson EN (2006) Gene regulatory networks in the evolution and development of the heart. Science 313:1922–1927. https://doi.org/10.1126/science.1132292

Pearson JC, Lemons D, McGinnis W (2005) Modulating Hox gene functions during animal body patterning. Nat Rev Genet 6:893–904. https://doi.org/10.1038/nrg1726

Pechlivanis S, Scherag A, Mühleisen TW, Möhlenkamp S, Horsthemke B, Boes T, Bröcker-Preuss M, Mann K, Erbel R, Jöckel KH, Nöthen MM, Moebus S (2010) Coronary artery calcification and its relationship to validated genetic variants for diabetes mellitus assessed in the Heinz Nixdorf recall cohort. Arterioscler Thromb Vasc Biol 30:1867–1872. https://doi.org/10.1161/ATVBAHA.110.208496

Petchey LK, Risebro CA, Vieira JM, Roberts T, Bryson JB, Greensmith L, Lythgoe MF, Riley PR (2014) Loss of Prox1 in striated muscle causes slow to fast skeletal muscle fiber conversion and dilated cardiomyopathy. Proc Natl Acad Sci U S A 111:9515–9520. https://doi.org/10.1073/pnas.1406191111

Powers SE, Taniguchi K, Yen W, Melhuish TA, Shen J, Walsh CA, Sutherland AE, Wotton D (2010) Tgif1 and Tgif2 regulate Nodal signaling and are required for gastrulation. Development 137:249–259. https://doi.org/10.1242/dev.040782

Pradhan L, Gopal S, Li S, Ashur S, Suryanarayanan S, Kasahara H, Nam H (2016) Intermolecular interactions of cardiac transcription factors NKX2.5 and TBX5. Biochemistry 55:1702–1710. https://doi.org/10.1021/acs.biochem.6b00171

Prall OW, Menon MK, Solloway MJ, Watanabe Y, Zaffran S, Bajolle F, Biben C, McBride JJ, Robertson BR, Chaulet H, Stennard FA, Wise N, Schaft D, Wolstein O, Furtado MB, Shiratori H, Chien KR, Hamada H, Black BL, Saga Y, Robertson EJ, Buckingham ME, Harvey RP (2007) An Nkx2-5/Bmp2/Smad1 negative feedback loop controls heart progenitor specification and proliferation. Cell 128:947–959

Puskaric S, Schmitteckert S, Mori AD, Glaser A, Schneider KU, Bruneau BG, Blaschke RJ, Steinbeisser H, Rappold G (2010) Shox2 mediates Tbx5 activity by regulating Bmp4 in the pacemaker region of the developing heart. Hum Mol Genet 19:4625–4633. https://doi.org/10.1093/hmg/ddq393

Ragvin A, Moro E, Fredman D, Navratilova P, Drivenes Ø, Engström PG, Alonso ME, de la Calle Mustienes E, Gómez Skarmeta JL, Tavares MJ, Casares F, Manzanares M, van Heyningen V, Molven A, Njølstad PR, Argenton F, Lenhard B, Becker TS (2010) Long-range gene regulation links genomic type 2 diabetes and obesity risk regions to HHEX, SOX4, and IRX3. Proc Natl Acad Sci U S A 107:775–780. https://doi.org/10.1073/pnas.0911591107

Reamon-Buettner SM, Hecker H, Spanel-Borowski K, Craatz S, Kuenzel E, Borlak J (2004) Novel NKX2-5 mutations in diseased heart tissues of patients with cardiac malformations. Am J Pathol 164:2117–2125. https://doi.org/10.1016/S0002-9440(10)63770-4

Risebro CA, Searles RG, Melville A, Athalie AD et al (2009) Prox1 maintains muscle structure and growth in the developing heart. Development 136:495–505. https://doi.org/10.1242/dev.030007

Risebro CA, Petchey LK, Smart N et al (2012) Epistatic rescue of Nkx2.5 adult cardiac conduction disease phenotypes by prospero-related homeobox protein 1 and HDAC3. Circ Res 111:e19-31. https://doi.org/10.1161/CIRCRESAHA.111.260695

Roux M, Laforest B, Capecchi M, Bertrand N, Zaffran S (2015) Hoxb1 regulates proliferation and differentiation of second heart field progenitors in pharyngeal mesoderm and genetically interacts with Hoxa1 during cardiac outflow tract development. Dev Biol 406:247–258. https://doi.org/10.1016/j.ydbio.2015.08.015

Santini MP, Forte E, Harvey RP, Kovacic JC (2016) Developmental origin and lineage plasticity of endogenous cardiac stem cells. Development 4:1242–1258. https://doi.org/10.1242/dev.111591

Schäfer K, Neuhaus P, Kruse J, Braun T (2003) The homeobox gene Lbx1 specifies a subpopulation of cardiac neural crest necessary for normal heart development. Circ Res 92:73–80

Schneider MD, Baker AH, Riley P (2015) Hopx and the cardiomyocyte parentage. Mol Ther 23:1420–1422. https://doi.org/10.1038/mt.2015.140

Schwab K, Hartman HA, Liang HC, Aronow BJ, Patterson LT, Potter SS (2006) Comprehensive microarray analysis of hoxa11/hoxd11 mutant kidney development. Dev Biol 293:540–554. https://doi.org/10.1016/j.ydbio.2006.02.023

Seifert A, Werheid DF, Knapp SM, Tobiasch E (2015) Role of Hox genes in stem cell differentiation. World J Stem Cells 7:583–595. https://doi.org/10.4252/wjsc.v7.i3.583

Sepulveda JL, Belaguli N, Nigam V, Chen CY, Nemer M, Schwartz RJ (1998) GATA-4 and Nkx-2.5 coactivate Nkx-2 DNA binding targets: role for regulating early cardiac gene expression. Mol Cell Biol 18:3405–3415

Shashikant CS, Utset MF, Violette SM, Wise TL, Einat P, Einat MPJ, Schughart K, Ruddle FH (1991) Homeobox genes in mouse development. Crit Rev Eukaryot Gene Expr 1:207–245

Shiratori H, Yashiro K, Shen MM, Hamada H (2006) Conserved regulation and role of Pitx2 in situs-specific morphogenesis of visceral organs. Development 133:3015–3025. https://doi.org/10.1242/dev.02470

Skerjanc IS, Petropoulos H, Ridgeway AG, Wilton S (1998) Myocyte enhancer factor 2C and Nkx2-5 up-regulate each other's expression and initiate cardiomyogenesis in P19 cells. J Biol Chem 273:34904–34910. https://doi.org/10.1074/jbc.273.52.34904

Smith JG, Newton-Cheh C (2015) Genome-wide association studies of late-onset cardiovascular disease. J Mol Cell Cardiol 83:131–141. https://doi.org/10.1016/j.yjmcc.2015.04.004

Soibam B, Benham A, Kim J, Weng KC, Yang L, Xu X, Robertson M, Azares A, Cooney AJ, Schwartz RJ, Liu Y (2015) Genome-wide identification of MESP1 targets demonstrates primary regulation over. Stem Cells 33:3254–3265. https://doi.org/10.1002/stem.2111

Stankunas K, Shang C, Twu KY, Kao SC, Jenkins NA, Copeland NG, Sanyal M, Selleri L, Cleary ML, Chang C (2008) Pbx/Meis deficiencies demonstrate multigenetic origins of congenital heart disease. Circ Res 103:702–709. https://doi.org/10.1161/CIRCRESAHA.108.175489

Stevens KN, Hakonarson H, Kim CE, Doevendans PA, Koeleman BP, Mital S, Raue J, Glessner JT, Coles JG, Moreno V, Granger A, Gruber SB, Gruber PJ (2010) Common variation in ISL1 confers genetic susceptibility for human congenital heart disease. PLoS One 26:e10855. https://doi.org/10.1371/journal.pone.0010855

Sun R, Liu M, Lu L, Zheng Y, Zhang P (2015) Congenital heart disease: causes, diagnosis, symptoms, and treatments. Cell Biochem Biophys 72:857–860. https://doi.org/10.1007/s12013-015-0551-6

Tanaka M, Chen Z, Bartunkova S, Yamasaki N, Izumo S (1999) The cardiac homeobox gene Csx/Nkx2.5 lies genetically upstream of multiple genes essential for heart development. Development 126:1269–1280

Tanaka M, Yamasaki N, Izumo S (2000) Phenotypic characterization of the murine Nkx2.6 homeobox gene by gene targeting. Mol Cell Biol 20:2874–2879

Trivedi CM, Zhu W, Wang Q et al (2010) Hopx and Hdac2 interact to modulate Gata4 acetylation and embryonic cardiac myocyte proliferation. Dev Cell 19:450–459. https://doi.org/10.1016/j.devcel.2010.08.012

Trivedi CM, Cappola TP, Margulies KB, Epstein JA (2011) Homeodomain only protein x is downregulated in human heart failure. J Mol Cell Cardiol 50:1056–1058. https://doi.org/10.1021/nl061786n. Core-Shell

Tu CT, Yang TC, Tsai HJ (2009) Nkx2.7 and Nkx2.5 function redundantly and are required for cardiac morphogenesis of zebrafish embryos. PLoS One 4:e4249. https://doi.org/10.1371/journal.pone.0004249

van der Harst P, van Setten J, Verweij N, Vogler G, Franke L, Maurano MT, Wang X, Mateo Leach I, Eijgelsheim M, Sotoodehnia N, Hayward C, Sorice R, Meirelles O, Lyytikäinen LP, Polašek O, Tanaka T, Arking DE, Ulivi S, Trompet S et al (2016) 52 genetic loci influencing myocardial mass. J Am Coll Cardiol 68:1435–1448. https://doi.org/10.1016/j.jacc.2016.07.729

van Tuyl M, Liu J, Groenman F, Ridsdale R, Han RN, Venkatesh V, Tibboel D, Post M (2006) Iroquois genes influence proximo-distal morphogenesis during rat lung development. Am J Physiol Lung Cell Mol Physiol 290:L777–L789. https://doi.org/10.1152/ajplung.00293.2005

van Weerd JH, Christoffels VM (2016) The formation and function of the cardiac conduction system. Development 143:197–210. https://doi.org/10.1242/dev.124883

Vandewalle C, Van Roy F, Berx G (2009) The role of the ZEB family of transcription factors in development and disease. Cell Mol Life Sci 66:773–787. https://doi.org/10.1007/s00018-008-8465-8

Wang J, Klysik E, Sood S, Johnson RL, Wehrens XH, Martin JF (2010) Pitx2 prevents susceptibility to atrial arrhythmias by inhibiting left-sided pacemaker specification. Proc Natl Acad Sci U S A 107:9753–9758. https://doi.org/10.1073/pnas.0912585107

Wang J, Greene SB, Martin JF (2011) BMP signaling in congenital heart disease: new developments and future directions. Birth Defects Res A Clin Mol Teratol 91:441–448. https://doi.org/10.1002/bdra.20785

Wang J, Xin YF, Xu WJ, Liu ZM, Qiu XB, Qu XK, Xu L, Li X, Yang Y (2013) Prevalence and spectrum of PITX2c mutations associated with congenital heart disease. DNA Cell Biol 32:708–716. https://doi.org/10.1089/dna.2013.2185

Wang J, Zhang DF, Sun YM, Li RG, Qiu XB, Qu XK, Liu X, Fang WY, Yang YQ (2014) NKX2-6 mutation predisposes to familial atrial fibrillation. Int J Mol Med

34:1581–1590. https://doi.org/10.3892/ijmm.2014. 1971

Wang Y, Li Y, Guo C, Lu Q, Wang W, Jia Z, Chen P, Ma K, Reinberg D, Zhou C (2016) ISL1 and JMJD3 synergistically control cardiac differentiation of embryonic stem cells. Nucl Acids Res 44:6741–6755. https://doi.org/10.1093/nar/gkw301

Waraya M, Yamashita K, Katoh H, Ooki A, Kawamata H, Nishimiya HNK, Ema A, Watanabe M (2012) Cancer specific promoter CpG Islands hypermethylation of HOP homeobox (HOPX) gene and its potential tumor suppressive role in pancreatic carcinogenesis. BMC Cancer 12:397. https://doi.org/10.1186/1471-2407-12-397

Wei D, Gong XH, Qiu G, Wang J, Yang Y (2014) Novel PITX2c loss-of-function mutations associated with complex congenital heart disease. Int J Mol Med 33:1201–1208. https://doi.org/10.3892/ijmm.2014. 1689

Witzel HR, Jungblut B, Choe CP, Crump JG, Braun T, Crump JG (2012) The LIM protein Ajuba restricts the second heart field progenitor pool by regulating Isl1 activity. Dev Cell 23:58–70. https://doi.org/10.1016/j.devcel.2012.06.005

Wu YH, Zhao H, Zhou LP, Zhao CX, Wu YF, Zhen LX, Li J, Ge DX, Xu L, Lin L, Liu Y, Liang DD, Chen Y (2015) miR-134 modulates the proliferation of human cardiomyocyte progenitor cells by targeting Meis2. Int J Mol Sci 6224:25199–25213. https://doi.org/10.3390/ijms161025199

Xu H, Baldini A (2007) Genetic pathways to mammalian heart development: recent progress from manipulation of the mouse genome. Semin Cell Dev Biol 18:77–83. https://doi.org/10.1016/j.semcdb.2006.11.011

Xu H, Yi Q, Yang C, Wang Y, Tian J, Zhu J (2016) Histone modifications interact with DNA methylation at the GATA4 promoter during differentiation of mesenchymal stem cells into cardiomyocyte-like cells. Cell Prolif 49:315–329. https://doi.org/10.1111/cpr.12253

Yamagishi H, Yamagishi C, Nakagawa O, Harvey RP, Olson EN, Srivastava D (2001) The combinatorial activities of Nkx2.5 and dHAND are essential for cardiac ventricle formation. Dev Biol 239:190–203. https://doi.org/10.1006/dbio.2001.0417

Yap LF, Lai SL, Patmanathan SN et al (2016) HOPX functions as a tumour suppressor in head and neck cancer. Sci Rep 6:38758. https://doi.org/10.1038/srep38758

Yousef MS, Matthews BW (2005) Structural basis of Prospero-DNA interaction: implications for transcription regulation in developing cells. Structure 13:601–607. https://doi.org/10.1016/j.str.2005.01.023

Yu X, St Amand TR, Wang S, Li G, Zhang Y, Hu YP, Nguyen L, Qiu MS, Chen Y (2001) Differential expression and functional analysis of Pitx2 isoforms in regulation of heart looping in the chick. Development 1013:1005–1013

Yu Z, Kong J, Pan B, Sun H, Lv T, Zhu J, Huang G, Tian J (2013) Islet-1 may function as an assistant factor for histone acetylation and regulation of cardiac development-related transcription factor Mef2c expression. PLoS One 8:e77690. https://doi.org/10.1371/journal.pone.0077690

Zakariyah AF, Rajgara RF, Veinot JP, Skerjanc IS, Burgon PG (2017) Congenital heart defect causing mutation in Nkx2.5 displays in vivo functional deficit. J Mol Cell Cardiol 105:89–98. https://doi.org/10.1016/j.yjmcc.2017.03.003

Zhang SS, Kim KH, Rosen A, Smyth JW, Sakuma R, Delgado-Olguín P, Davis M, Chi NC, Puviindran V, Gaborit N, Sukonnik T, Wylie JN, Brand-Arzamendi-K, Farman GP, Kim J, Rose RA, Marsden PA, Zhu Y, Zhou YQ, Miquerol L, Henkelman RM, Stainier DY, Shaw RM, Hui CC, Bruneau BG, Backx PH (2011) Iroquois homeobox gene 3 establishes fast conduction in the cardiac His-Purkinje network. Proc Natl Acad Sci U S A 108:13576–13581. https://doi.org/10.1073/pnas.1106911108

Zhang Y, Si Y, Ma N, Mei J (2015) The RNA-binding protein PCBP2 inhibits Ang II-induced hypertrophy of cardiomyocytes though promoting GPR56 mRNA degeneration. Biochem Biophys Res Commun 464:679–684. https://doi.org/10.1016/j.bbrc.2015.06.139

Zhang Y, Si Y, Ma N (2016) Meis1 promotes poly (rC)-binding protein 2 expression and inhibits angiotensin II-induced cardiomyocyte hypertrophy. IUBMB Life 68:13–22. https://doi.org/10.1002/iub.1456

Zhou Z, Wang J, Guo C, Chang W, Zhuang J, Zhu P, Li X (2017) Temporally distinct Six2-positive second heart field progenitors regulate mammalian heart development and disease. Cell Rep 18:1019–1032. https://doi.org/10.1016/j.celrep.2017.01.002

Zhu CC, Yamada G, Nakamura S, Terashi T, Schweickert A, Blum M (1998) Malformation of trachea and pelvic region in goosecoid mutant mice. Dev Dyn 211:374–381. https://doi.org/10.1002/(SICI)1097-0177(199804)211:4<374::AID-AJA8>3.0.CO;2-E

Zhuang S, Zhang Q, Zhuang T, Evans SM, Liang X, Sun Y (2013) Expression of Isl1 during mouse development. Gene Expr Patterns 13:407–412. https://doi.org/10.1016/j.gep.2013.07.001.Expression

Adv Exp Med Biol – Cell Biology and Translational Medicine (2019) 6: 179–220
https://doi.org/10.1007/5584_2019_340
© Springer Nature Switzerland AG 2019
Published online: 26 April 2019

Generation of Human Stem Cell-Derived Pancreatic Organoids (POs) for Regenerative Medicine

Victor Navarro-Tableros, Yonathan Gomez, Maria Felice Brizzi, and Giovanni Camussi

Abstract

Insulin-dependent diabetes mellitus or type 1 diabetes mellitus (T1DM) is an auto-immune condition characterized by the loss of pancreatic β-cells. The curative approach for highly selected patients is the pancreas or the pancreatic islet transplantation. Nevertheless, these options are limited by a growing shortage of donor organs and by the requirement of immunosuppression.

Xenotransplantation of porcine islets has been extensively investigated. Nevertheless, the strong xenoimmunity and the risk of transmission of porcine endogenous retroviruses, have limited their application in clinic. Generation of β-like cells from stem cells is one of the most promising strategies in regenerative medicine. Embryonic, and more recently, adult stem cells are currently the most promising cell sources exploited to generate functional β-cells in vitro. A number of studies demonstrated that stem cells could generate functional pancreatic organoids (POs), able to restore normoglycemia when implanted in different preclinical diabetic models. Nevertheless, a gradual loss of function and cell dead are commonly detected when POs are transplanted in immunocompetent animals. So far, the main issue to be solved is the post-transplanted islet loss, due to the host immune attack. To avoid this hurdle, nano-technology has provided a number of polymers currently under investigation for islet micro and macro-encapsulation. These new approaches, besides conferring PO immune protection, are able to supply oxygen and nutrients and to preserve PO morphology and long-term viability.

Herein, we summarize the current knowledge on bioengineered POs and the stem cell differentiation platforms. We also discuss the in vitro strategies used to generate functional POs, and the protocols currently used to confer immune-protection against the host immune attack (micro- and macro-encapsulation). In addition, the most relevant ongoing clinical trials, and the most relevant hurdles met to move towards clinical application are revised.

V. Navarro-Tableros
2i3T Società per la gestione dell'incubatore di imprese e per il trasferimento tecnologico Scarl, University of Turin, Turin, Italy

Y. Gomez and M. F. Brizzi
Department of Medical Sciences, University of Turin, Turin, Italy

G. Camussi (✉)
Department of Medical Sciences, University of Turin, Turin, Italy

Fondazione per la Ricerca Biomedica-ONLUS, Turin, Italy
e-mail: giovanni.camussi@unito.it

Keywords

Cell differentiation · Innovative therapies ·
Pancreatic organoids · Stem cells · Type 1
diabetes mellitus

Abbreviations

2D	Bi-dimensional
3D	Three-dimensional
A. ECM	Artificial ECM
AHFBRs	Alginate-filled hollow fiber bioreactors
ALDHhi	High aldehyde dehydrogenase activity
BAPs	Bioartificial pancreas
BCD	β-cell-derived cells
BM-MSC	Bone marrow MSC
CaO2	Calcium peroxide
Cas9	DNA endonuclease Cas9
CPO	Calcium peroxide
CRISPR	Clustered Regularly Interspaced Short Palindromic Repeats
Dex	Dexamethasone
ECM	Extracellular matrix
ED	Eudragit
EPCs	Endothelial progenitor cells
FBM	Functional β cell mass
FICCs	Fetal porcine islet-like cell clusters
FTY720	Fingolimod
GHRH	Growth hormone releasing hormone
Glut-2	Glucose transporter 2
HARV	High aspect ratio vessels
Hep-PEG	Glycosaminaglycans and heparin enriched capsules
hESCs	Human embryonic stem cells
hHPCs	Human hepatic progenitor cells
hiPSC	Human induced pluripotent stem cells
HLSC	Human liver stem cells
HLSC	Adult human liver stem-like cells
HMGB1	High mobility group box 1
huPI-MSC	Adult human pancreatic islet mesenchymal stromal cells
IBMIR	Instant blood-mediated inflammatory reaction
IL-2	Interleukine-2
IPCs	Insulin producing cells
iPSCs	Inducible pluripotent cells
KCl	Potassium chloride
LbL	Layer-by-layer
MGC	Methacrylated glycol chitosan
MMP	Matrix metalloproteinase
MP	Microparticles
mPEG	Methoxy polyethylene glycol
MSC	Mesenchymal stem cells
NeuroD1	Neuronal Differentiation 1 protein
NICC	Neonatal islet-like cell clusters
NICHE	Neovascularized implantable cell homing and encapsulation
NK	Natural Killer T cells
NPCCs	Neonatal pancreatic cell-clusters
NOD	Non-obese diabetic
PANC-1	Human pancreatic ductal-cells
PB	Probucol
PB-CDCA	PB-TCA alginate-microencapsulation
PCL/PVA	polyvinyl alcohol/polyvinyl alcohol
PDMS	Polydimethylsiloxane
PDX1	Pancreatic and duodenal homeobox 1
PEG	Poly(ethylene glycol)
PEGCol	Poly(ethylene glycol) hydrogels containing collagen type I
PEG-VS	Vinyl sulfone-terminated polyethylene glycol
PEOT/ PBT	Poly(ethylene oxide terephthalate)-poly(butylene terephthalate)
PERV	Porcine retrovirus
PES/PVP	Polyethersulfone/ polyvinylpyrrolidone
PFTBA	Perfluorotributylamine
PGA	Polyglycolides
PLA	Polylactic acid
PLGA	Poly(lactide-co-glycolide)
PLLA/ PVA	Poly-L-lactic acid-polyvinyl alcohol
PLO	Poly-L-ornithine
PMAA-Na	Sodium polymethacrylate
POs	Pancreatic organoids
PSCs	Pancreatic stem cells
Ptf1a	Pancreas Associated Transcription Factor 1a
PTFE	Polytetra-fluoroethylene

PTX	Pentoxifylline
RWV	Rotating wall vessels
SF	Silk fibroin macroporous
SIPN	Semi-interpenetrating polymer network
SP	Saccharide-peptide
SPO	Sodium percarbonate
starPEG	Star-shaped polyethylene glycol
T1DM	Type 1 diabetes mellitus
T2DM	Type 2 diabetes mellitus
TCA	Taurocholic acid (bile acid)
Th	T-helper
TRAFFIC	Thread-reinforced alginate fiber for islets encapsulation
TNF-α	Tumor necrosis factor-alpha
VEGF	Vascular endothelial growth factor
VEGFR2	Vascular endothelial growth factor receptor 2

1   Introduction

Type 1 diabetes mellitus (T1DM) develops when the endocrine pancreas fails to produce proper insulin concentrations to maintain the glucose homeostasis. The current approach to treat T1DM patients consists in the administration of exogenous insulin to supply the deficiency of pancreas secretion. This implies that pancreas transplantation would represent the finest therapeutic approach in these patients.

The protocols applied to isolate and purificate adult human pancreatic islets have been standardized and considerably improved over the last years (Brennan et al. 2016). Advances in peritransplant anti-inflammatory strategies, immunomodulation, and immunosuppression, have also contributed to the improvement of current clinical results (Ricordi et al. 2016). Nevertheless, only ~50% of recipients remain insulin-independent 5 years after pancreas transplantation. This mainly depends on the related inflammatory reaction, hypoxia and hypoxia–reoxygenation injury (Pepper et al. 2018).

Transplant of pancreatic islets is currently considered an approach to be used in highly selected patients with severe hypoglycemia or instable T1DM (Liu et al. 2015). For decades, the clinical application of islet transplantation has been limited by the inadequacy of the immunosuppressive therapy and the number of donor tissues (Liu et al. 2015). Once transplanted, only nearby 60% of the T1DM patients become insulin independent and maintain the glycemic control (Liu et al. 2015). In addition, most of these patients gradually lost insulin independence 2 years after transplantation (Shapiro et al. 2006). Beneficial long-term outcomes consisting in a long-term insulin independence (≥ 5 years) depends on an efficient immunosuppressive therapy able to improve islet mass engraftment and prevent recurrent autoimmunity (Bellin et al. 2012). However, long-term immunosuppressive therapy is associated with a significant morbidity, including the increased risk of cancer and infections (Shapiro et al. 2017). Moreover, as extensively demonstrated, nutrients cannot effectively reach the core of large-size transplanted islets resulting in cell death and islet loss. Such diffusional problems could be solved by breaking up islets into single islet cells. However, dissociated cells are less efficient than intact islets. Based in this concept, Li et al. (2017) established a novel islet engineering approach by encapsulating dissociated cells from human islets to generate newly formed islet-like organoids, similar in size and gene expression profile (Isl-1, Gcg, and insulin-1) to native islets. Furthermore, by limiting the diameter of these engineered islet cell clusters, to a maximum of 100 μm, cell viability and insulin secretion were improved (Li et al. 2017). Nevertheless, this approach faced the same problem as pancreas and islet transplantation, the shortage of donor organs (Matsumoto 2010).

Additionally, the lost of post-transplanted islets, due to the adverse immune and non-immune reactions (Croon et al. 2003), is associated with the failure of long-term insulin independence after intrahepatic islet transplantation (Bruni et al. 2014). This implies that to increase the number of transplantable T1DM patients, it is mandatory to obtain reliable and standardized sources of human transplantable islets, and avoid immunosuppression. To cover these needs, several efforts have been devoted to develop new potential strategies to produce functional pancreatic organoids (POs) (Zhou and Melton 2018). However, similar to pancreas or islet transplantation, transplanted POs activate the host immune response (Szot et al. 2015) which targets POs and leads to their loss of function (Szot et al. 2015).

Xenotransplantation using porcine islets, has been considered as an alternative source of human islets, and extensively investigated (Cardona et al. 2006; Hering et al. 2006). The easy and less expensive isolation procedure of fetal porcine islet-like cell clusters (FICCs) and neonatal islet-like cell clusters (NICCs) represents the main advantage at using xenotransplantation instead of allogenic/heterogenic allografts (Liu et al. 2017). Significant improvement in glycaemic control, has been provided by xenotransplantation (Liu et al. 2017). However, after transplantation, islet xenografts gradually lose efficacy (Safley et al. 2018a). Furhermore, the strong xenoimmunity and the risk of porcine endogenous retroviruses transmission, have limitated their application in clinic (Liu et al. 2017; Van Der Laan et al. 2000; Samy et al. 2014).

In the last decades the research has been focused on the identification of alternative cell sources from which functional insulin-secreting cells could be generated (Matsumoto et al. 2016). Cell lines are currently considered as potential sources able to generate new and mature insulin-secreting cells (Ravassard et al. 2011; Boss et al. 2017). For a long time, the introduction of the insulin gene along with components of the stimulus–secretion coupling pathways have been applied to induce the commitment of immature cells to insulin-producing cells. However, these cells failed to effectively store and release fully processed insulin (Ravassard et al. 2011; Clark et al. 1997; Scharfmann et al. 2014; Xie et al. 2016). Stem cells, isolated from embryonic and more recently, from adult tissues, are currently the most promising approaches (Balboa et al. 2018). Despite the great efforts and the promising results, so far, cell products covering all the morphological and functional profiles recapitulating the mature pancreatic islets are still missing.

Pancreatic islets are three-dimensional structures (3D) composed by different endocrine cells. The cellular ineractions resulting from this 3D conformation play an important role in the modulation of hormone secretion (Navarro-tableros et al. 2018; Kim et al. 2016a). Assembly of stem cells in 3D structures facilitates cellular interactions and promotes morphogenesis and tissue organization mimicking native pancreatic islets (Navarro-tableros et al. 2018; Kim et al. 2016a). Both natural and artificial materials can be used to create 3D synthetic organic and inorganic porosous scaffolds (Hinderer et al. 2016). Nanotechnology have provided new opportunities to design distinctive materials with specific physicochemical properties (Liu et al. 2018). Thus, generation of POs similar, in terms of morphology and function to native pancreatic islets, not only may enhance long-term transplant viability in clinical settings, but could also be applied for drug screening purposes (Scavuzzo et al. 2018). Hydrogels are extensively used for drug delivery and tissue regeneration due to their similarity in terms of structure and biocompatibility to the native extracellular matrix (ECM) (Dimatteo et al. 2018; Steffens et al. 2018). Biomaterials are classified in three major groups: (i) naturally-derived materials; (ii) synthetic polymers; and (iii) decellularized organ- or tissue-derived scaffolds.

The size of organoids should be also considered to generate encapsulated POs. In general, it is widely accepted that smaller capsules provide better β-cell oxygenation (Buchwald et al. 2018). However, concerns related to their removal still remains if compared to macroencapsulation devices (Scharp and Marchetti 2014). Preliminar studies have suggested that islet cell encapsulation within meter-long microfibers might overcomes this issue (Dolgin 2016). In this review, all 3D isolated structures (islets) or artificialy generated, either encapsulated or not, and independently of their origin (human and other species) will be referred as POs.

In summary, key challenges for human translation still include: (1) generation of biocompatible materials, (2) generation of 3D POs with functional insulin secreting cells, (3) improvemnt of PO survival, (4) protection of POs from the immune attack, and (5) the availability of nutrient supply (Tomei et al. 2015). Thererfore, the ideal encapsulation method, should allow long-term survival and functional capability of implanted POs (Ryan et al. 2017). Encapsulated islets or newly generated POs are typically delivered in extrahepatical sides such as intraperitoneally, subcutaneously or into the omentum (Berman et al. 2016). However, the host immune response and the activation of the fibrotic process occurring after transplantation, still remain a challenge (Safley et al. 2018a).

In this review, the new strategies to generate *in vitro* POs (stem cells-based, gene editing and alternative 3D culture methods), immunoisolation approaches, such as macro- and micro-encapsulation, and polymers used for these purposes will be discussed. Moreover, the ongoing clinical trials, and the hurdles to move towards their clinical application, will be also overviewed.

2 The Relevance of 3D Structure

Islet architecture, including arrangement and interaction between endocrine cells is relevant for their precise function (Orci et al. 1975; Kitasato et al. 1996; Kanno et al. 2002). More importantly, the homologous and heterologous intercellular contacts are relevant to provide a finest regulation of insulin secretion (Wojtusciszyn et al. 2008). Furthermore, the paracrine and autocrine interaction between α- and *β-cells*, contributes to the firm dynamic of hormone secretion, on which depends the effective control of glucose homeostasis (Jo et al. 2009). Additionally, gap-junction-mediated cell-cell interactions protect pancreatic cells against apoptosis (Klee et al. 2011), and correlate with a correct developmental acquisition of a mature insulin secretory profile (Santos-Silva et al. 2012). Therefore, the 3D architecture and the intercellular regulatory mechanisms involving ions (electrical coupling) and hormones, particularly involved in the cross-talk among α-, δ-, and *β-cells* within the islets, are crucial to ensure a correct response to paracrine signals (Rorsman and Ashcroft 2018). Modification or loss of these properties results in abnormal glucose control (Rutter and Hodson 2015; Kilimnik et al. 2011). Upon islet isolation, insulin-producing cells undergo multiple cell death processes including apoptosis, anoikis, and necrosis, which have been attributed to the loss of critical interactions between cells, ECM and the vasculature (Paraskevas et al. 2000; Irving-Rodgers et al. 2014; De Vos et al. 2016). Modification of the islet architecture also occurs during isolation and culture of cadaveric human islets, prior to transplantation (Lavallard et al. 2016) or after infusion into the portal vein (Henriksnäs et al. 2012). This results in the failure to obtain a fine glycemic metabolic control (Morini et al. 2006). Therefore, the preservation of 3D cellular architecture of human islets or POs, derived from different cell sources, have a functional relevance for islet engraftment (Navarro-tableros et al. 2018). Improved protocols/devices for PO *in vitro* culture may thus offer crucial indications for appropriate growth and differentiation of stem/progenitor cells (S/PCs), allowing safeguarding of specific characteristics, such as multi-lineage differentiation capabilities and paracrine activity.

3 Microencapsulation

Native pancreatic islets are heterogeneous structures formed by almost 5 different endocrine cells: α, β, γ, δ, ε, and pancreatic polypeptide (PP) cells which by interacting in a complex and synchronized manner allow a fine autocrine and paracrine regulation. Generation of POs, reminiscent of native pancreatic islets in structure and function, should be pursued to provide a more efficient treatment. Indeed, generation of pancreatic endoderm and endocrine cells in 3D microenvironment has been extensively documented. Wang et al. (2017) have generated human embryonic stem cells (hESCs)-derived POs expressing high levels of *β-cell* markers such as Pdx1, Ngn3, Insulin, MafA, and the Glucose transporter 2 (Glut-2). More importantly, they demonstrated that these POs contained pancreatic α, β, δ, and PP producing cells. Insulin biosynthesis was confirmed by the high C-peptide expression and by the presence of insulin-secretory granules. The improvement of insulin secretion response to high glucose concentration has finally demonstrated their functional maturation (Wang et al. 2017).

3.1 Nanotopographies as Drivers of Cell Differentiation

A number of studies strongly provided evidences that engraftment, proliferation and differentiation of stem cells into a specialized phenotype are modulated by their structural organzation (Guilak et al. 2009). The interactions between extracellular binding sites and cytoskeletal elements strongly affect *in vitro* cell morphology, adhesion, and consequently gene expression (Bettinger et al. 2009).

Nanotopographies are involved in stem cell adhesion, and regulate their functions by mechanotransduction signals through the cell membrane (Yim et al. 2010). On the other hand, nanotopographies inducing lower cell adhesion cues, allow cells to form POs by promoting cell−cell interaction (Shen et al. 2015). Evidences have been also provided that differentiating hESCs and iPSCs are able to perceive nanotopographical signals, and efficiently differentiate into pancreatic cells (Kim et al. 2016b). It has been also described that compared to traditional 2D cultures, human iPSCs (hiPSCs) cultured in nanofibrous scaffolds and driven to differentiate into POs are formed by insulin-producing cells, and exhibit a morphological and functional profile similar to mature pancreatic *β-cells* (Nassiri Mansour et al. 2018).

The establishment and preservation of a functional *β-cell* mass (FBM) are replacement. FBM is dogged by a direct sensing of its key components, i.e. the number and function of *β-cells* (Pipeleers et al. 2008). Contrasting results

have been obtained by De Mesmaeker et al. (2018). They demonstrated that a sustained FBM could be obtained using immature proliferative porcine *β-cells*, but not human adult islet cells. These observations suggest that differatiation depends not only on the employed methodology, but also on the cell of origin.

Thereby, cell-based therapy requires simple and safe encapsulation methods to produce POs of uniform size and shape. Indeed, encapsulation promotes not only a quick cell aggregation of mesenchymal stem cells (MSC) in dense cell clusters, but also increases the expression of insulin and Pdx-1 mRNA (Barati et al. 2018). These concepts are currently exploited and widely applied in biomedical areas (Lee et al. 2011).

Encapsulation has been also pursued to provide immuno-isolation biocompatibility, optimized nutrient diffusion and insulin release by POs (Orive et al. 2015). In line with these needs, a number of new polymers have been generated (Gálvez-Martín et al. 2017). Besides the characteristics and configuration of microcapsules, the nature of the cells should be also considered. A number of different cell sources has been investigated for cell therapy including stem cells (autologous and allogenic), mature somatic cells, modified human cells, xenogenic cells and others (Kang et al. 2014). Alginate-based microcapsules have been reported not only to enable the proliferation and differentiation of hESCs into definitive endoderm-derived cells, but also to enhance their viability and proliferation, and to promote their aggregation into POs (Chayosumrit et al. 2010).

Independently of the origin, single cell within microcapsules, tend to spontaneously self-aggregate to form POs (Chayosumrit et al. 2010). These POs tend to mimic the structure and the features of the original in vivo-like cytoarchitectures (Wang et al. 2006; Zhang et al. 2006). Moreover, it is important to keep in mind that spheroid size also influences microcapsule efficacy (Perignon et al. 2015). POs with small diameters and with a low cell content, could not provide suitable conditions for cell aggregation due to inadequate number of adhering cells, while large POs can undergo oxygen depletion with consequent hypoxia (Huang et al. 2012).

One of the most commonly used islet micro-encapsulation biomaterial is the alginate. Microcapsules have an ideal surface/volume ratio, which allows a better exchange of nutrients, insulin and glucose. Furthermore, islets could individually be included in an single capsule (Ryan et al. 2017; Desai and Shea 2017). Alginate is an anionic polysaccharide composed by unbranched polymers of 1,4-linked β-D-mannuronic and α-L-guluronic acid residues which forms a gel in the presence of multivalent cations such as Ca^{2+} or Ba^{2+} and provides a 3D biomimetic environment to the cells that resembles the in vivo conditions (Datar et al. 2015). An important issue of first alginate products relied on their chemical instability during long-term implantation that hindered feasibility of cell therapy. It has been reported that alginate-Ca^{2+} hydrogels tend to degrade, thus resulting in the contact of incapsulated structures with the host's immune system (Scharp and Marchetti 2014). Clinical and pre-clinical studies have noted that reducing the volume of encapsulated materials and the corrisponding diffusional restrictions are critical for the engraftment (Scharp and Marchetti 2014). Experts in biomaterials have improved the stability of alginate hydrogels by modifying their biomechanical properties by a process called "click" crosslinking (Breger et al. 2015). These modifications confer to the capsules a superior stability, but a higher permeability to small size "diffusates". The alginate matrix can stabilize the cell cluster size and can allow a more homogeneous cell morphology, leading to a long-term culture of these clusters compared to non-encapsulated ones (Formo et al. 2015). Moreover, alginate-encapsulated human stem cell-derived POs could be effectively protected from the immune reaction when intraperitoneally implanted in mice (Vegas et al. 2016). In fact, micro-encapsulation is the most investigated approach (Zimmermann et al. 2007), and a number of pre-clinical studies in non-human primate models (Sun et al. 1996; Elliott et al. 2005), and human trials (Soon-Shiong et al. 1994; Calafiore et al. 2006; Tuch

et al. 2009; Jacobs-Tulleneers-Thevissen et al. 2013) based on this technology are ongoing.

Polymers applied to produce microcapsules should also have chemo-mechanical stability, and an easy to handle and appropriate pore size to allow a bi-directional diffusion of molecules in the semipermeable membrane (Kang et al. 2014). Currently, microcapsules have been obtained in many different forms, sizes, compositions, and with different permeability (De Vos et al. 2014). Both synthetic and natural polymers have been used for encapsulation purposes. Nevertheless, only alginate has been largely studied and it is currently certified as safe for human application (De Vos et al. 2014). As other polymers, alginate tends to be largely contaminated by original and additional contaminants such as polyphenols, endotoxins, and proteins introduced during the industrial extraction processes (Vos et al. 2006). In fact, a number of medicated alginates has been generated to solve this issue.

In general, it is accepted that smaller capsules (0.5 mm) better engraft functional POs (Buchwald et al. 2018). However, recent investigations suggested that the biocompatibility of alginate-encapsulated POs can be significantly enhanced by the use of larger capsules. In particular, it has been reported that larger capsules (1–1.5 mm) could generate a reduced immune reaction and fibrosis and avoid altered pattern of glucose-induced insulin secretion than smaller spheres (Veiseh et al. 2015). In contrast, other studies indicated that delayed and blunted responses to glucose and KCl depolarization, are present in larger capsules. As a consequence, larger capsules provide a reduced insulin response and a sustained and slow release of insulin in response to glucose (Buchwald et al. 2018).

A number of new microcapsules enriched with different combination of factors, has been developed. Recently, new capsules generated by a combination of antihyperlipidemic drug (probucol; PB), bile acid (taurocholic acid; TCA) and alginate-microencapsulation (PB-Alginate) called PB-CDCA capsules have been described. They are able to improve PO function and to reduce cell apoptosis driven by the hyperglycemic setting (Mooranian et al. 2016). Additional effects include improved viability under the hyperglycemic environment, increased insulin production, and reduced TNF-α release by pancreatic *β-cells* (Mooranian et al. 2018). Furthermore, the PB-CDCA capsules, offer the mechanical stability, the buoyancy, the PB release, the thermal stability, and contain antioxidants (Mooranian et al. 2018). Similar results were obtained when capsules were enriched with absorption-enhancer chenodeoxycholic acid and Eudragit (ED) polymers (Mooranian et al. 2018) or ECM proteins (Mooranian et al. 2018). Moreover, it has been demonstrated that the addition of ECM proteins or trophic factors to alginate-encapsulated POs stabilizes the cluster size, cell morphology, and improves the oxygen consumption rate and PO survival in long-term culture (Formo et al. 2015). A newly ECM-based encapsulation system, named meter-long core shell alginate-hydrogel microfibers, has been shown to allow the formation of a core of pancreatic islet cells surrounded by ECM proteins and is considered a promising approach which can facilitate PO implantation and removing (Onoe et al. 2013).

The microencapsulated POs composed by pig islets seeded in human decellularized collagen matrix has been recently implanted in non-human primates (Dufrane et al. 2010). Long-term PO survival and insulin release were associated with improved glycemic control for 6 months, in the absence of immunosuppressive medications (Dufrane et al. 2010). However, the risk of porcine retrovirus (PERV) transfer, even if reduced, was still detected (Dufrane et al. 2010), possibly due to the damage of the alginate barrier (Crossan et al. 2018). Thus, additional quality control is mandatory before moving toward their clinical application.

By investigating alginate-encapsulated adult porcine islet transplants, it has become evident that despite long-term glucose normalization, a massive fibrosis occurred in harvested capsules (Vaithilingam et al. 2011). Nevertheless, improvement in avoiding fibrosis was obtained by using microcapsules enriched in photo-crosslinked methacrylated glycol chitosan (MGC) (Hillberg et al. 2015) or Rampamycin-PEG (Park et al. 2017). Although relevant data have been obtained long-term and large

preclinical studies are still required (Park et al. 2017).

3.2 Immunomodulation

Even if the donor and organ managements, the surgical procedure, and the recipient management have been improved, the immune rejection still remains the most relevant issue to be solved (Schuetz et al. 2018). In order to safeguard graft survival, robust immunosuppressive treatments are required, some of which are toxic for *β-cells* (Nir et al. 2007). It has been demonstrated that the addition of multilayers composed bypolymers, such as poly-L-ornithine (PLO) or activated methoxy polyethylene glycol (mPEG) to the surface of alginate microcapsules could allow immune protection and decrease interleukine-2 (IL-2) secretion (Nabavimanesh et al. 2015).

As above discussed, encapsulation is an effective approach to protect transplanted cells from rejection by the host immune response by using a semi-permeable artificial membrane. Microencapsulation has the advantage to allow implantation of allogeneic and xenogeneic cells, and free bidirectional diffusion of nutrients, oxygen, and other molecules like insulin. Moreover, it is also able to prevent the host immune response (Steele et al. 2014). Thus, effcient encapsulation approaches should be pursued to reduce or abolish the requirement of pharmacological immunosuppression. For example, double-layer alginate/dextran–spermine microcapsules combined with pentoxifylline (PTX) have been shown to be effective in preventing the immune attack over standard alginate-based encapsulation (Azadi et al. 2016). PTX could be easily released from the porous microcapsules, however, extended destruction of the membrane capsules has been documented (Azadi et al. 2016).

Alternative approaches aimed to improve immune-protection consist on the blockage of one of the most expressed proteins on the islets: the high mobility group box 1 (HMGB1) which is known to induce inflammation (Wang et al. 2016). Therefore, the regulation of HMGB1-mediated inflammation and particularly of the HMGB1 A box (Itoh et al. 2011) is actually considered a new target for immnuno-protection (Sama et al. 2004). In fact, recently, an anti-HMGB1 receptor-enriched encapsulation approach has been developed (Jo et al. 2015). Jo et al. (2015) have demonstrated that HMGB1 A box offers a protective effect on islet transplantation by decreasing the amount of TNF-α secreted by macrophages. It was also demonstrated that the HMGB-enriched-encapsulated method strogly improves PO survival rate after intraperitoneally xenotransplants into diabetic mice (Jo et al. 2015).

3.3   Encapsulated Xenografts

The possibility to use encapsulated porcine-islets to treat diabetes has been also evaluated. Promising results indicate that encapsulated porcine islets are more efficient than non-encapsulated porcine islets in restoring normoglycemia in diabetic monkeys (Sun et al. 1996). Co-encapsulation of pancreatic islets with MSC futher improved vascularization and oxygenation of the PO graft (Vériter et al. 2014). However, despite a glycemic control was detected up to 32 weeks, not substantial improvement in the xenograft function was observed (Vériter et al. 2014).

A first phase I/IIa clinical trial aimed to test the microbiological safety of porcine islet xenotransplantation has been opened by recruiting nonimmunosuppressed subjects with unstable T1DM (Wynyard et al. 2014) transplanted with microencapsulated neonatal porcine islets (Wynyard et al. 2014). A reduction in unaware hypoglycemia events, the HbA1c levels, and a decrease of daily insulin requirement have been reported (Matsumoto et al. 2014). Pig-islet-based PO biosafety was further demonstrated in a phase IIa efficacy trial in humans (Morozov et al. 2017). Interestingly, plasma porcine C-peptide level has been applied to demonstrate the function of alginate-encapsulated porcine *β-cells* implanted in large and small animals (Montanucci et al. 2013; Chen et al. 2015). However, insight on the relationship between the initial *β-cell* dose, the viability of the implanted *β-cells* and their

function in term of metabolic control within time, are still missing (Dufrane et al. 2006; Foster et al. 2007).

It is important to remind that notwithstanding the relevant progress in xenotransplantation, significant limitations to clinical application persist:

1. No clinical applicable immunosuppressive protocols for preventing xenograft rejection are avalaible (Samy et al. 2014).
2. High number of designated pathogen-free donor pancreata are required to manufacture neonatal porcine (Thompson et al. 2011) and adult pig islets for each patient (Rogers et al. 2011).
3. Manufacturing of NICC is still challenging and costly (Korbutt et al. 1996).

4   Macroencapsulation

Human pancreatic islets are usually transplanted in the intrahepatic site, via the portal vein. However, this environment directly exposes the islets to the blood flow, thus eliciting the "instant blood-mediated inflammatory reaction" (IBMIR) and leads to islet damage and loss (Moberg et al. 2002; Nilsson et al. 2011). In addition, the concomitant reduced oxygen tension and the high intrahepatic concentrations of immunosuppressants further interfere with survival and engraftment of islets (Carlsson et al. 2001; Olsson et al. 2011; Desai et al. 2003). Unlike liver, the subcutaneous space has been proposed as an ideal implantation site, since it possesses a high grade of vascularization and can offer mechanical protections to the implanted islets (Pepper et al. 2015).

Even if transplantation of pancreatic islets is currently considered a promising therapeutically approach to cure diabetes, islet availability is limited by the number of donors and a large number of transplanted patients still requires insulin therapy to obtain the glycemic control (Shapiro et al. 2006). Additionally, due to the high oxygen demand, a highly vascularization is necessary to allow the long-term survival and function of transplanted islets. Furthermore, accessible and minimally invasive sites are fundamental to allow implantation, replenishment and graft retrieval (Pagliuca and Melton 2013).

Recent advances in regenerative medicine and tissue engineering have preferred new strategies for the generation of 3D scaffolds. Thus, a number of biodegradable and biocompatible synthetic polymers have been used to produce nanofibers with the aim to improve maturation and function of the newly generated POs (De Vos et al. 2014). These materials rapidly undergo adaptation to the human body and do not stimulate the immune system (Ellis et al. 2017). These platforms provide a 3D environment that improves stem cell differentiation into β-like cells (Nadri et al. 2017; Abazari et al. 2018). It appears that these 3D nanofiber scaffolds are able to support an efficient cell-ECM interaction and cell-cell contact, by mimicking the in vivo condition (Mahboudi et al. 2018).

Macroencapsulation is an old technology also applied in diabetes research to provide a better survival of POs after transplantation (Desai and Shea 2017). Macroencapsulation devices are heterogeneous in geometry and in materials (Song and Roy 2016). A number of studies has shown the ability of pancreatic progenitor cells or endocrine cells to persist and function within subcutaneously implanted devices (Agulnick et al. 2015).

4.1 Differentiation of Macroencapsulated Cells

Compared to 2D cultures, PCL/PVA 3D scaffolds could efficiently improve differentiation of iPSCs into β-like cells by inducing the expression of endocrine markers such as Glucagon, Insulin, Pdx1, Ngn3 and Glut-2 genes (Enderami et al. 2017). Differentiation properties of Poly(lactide-co-glycolide) (PLGA)-based microporous scaffolds have been also documented (Blomeier et al. 2006; Salvay et al. 2008; Mao et al. 2009). Poly(ethylene glycol) hydrogels containing collagen type I (PEGCol) have been also reported to promote aggregation and differentiation of hESC-derived progenitors in glucose-sensitive POs, also displaying enhancement of long-term viability

and morphology preservation (Mason et al. 2009; Amer et al. 2015). One of the most promising materials currently used for the generation of encapsulated POs is the poly-L-lactic acid-polyvinyl alcohol (PLLA/PVA) polymer, which provides a better microenvironment than 2D cultures (Mobarra et al. 2018). It has been reported that PLLA/PVA polymer is able to promote differentiation of iPSCs into insulin producing cells (iPSCs-IPCs) by inducing the expression of pancreatic-specific transcription factors such as Pdx1, insulin, glucagon and Ngn3 (Mobarra et al. 2018).

4.2 Macroencapsulation and Immunoprotection

Bioscaffolds used as cell or drug carriers are usually constituted of non-degradable or degradable biomaterials. Current methodologies use ECM and ECM-like materials, or ECM-synthetic polymer hybrids (Hinderer et al. 2016). Hydrogels have been exploited for regenerative applications due to their biocompatibility and similarity in structure to the native extracellular matrix (Dimatteo et al. 2018). Several aliphatic polyesters, such as polymers of lactic acid [polylactides (PLA)], glycolic acid [polyglycolides (PGA)], and their copolymers [polylactoglycolides (PLGA)] have been extensivly selected as biocompatible and bioresorbable matrices to reconstitute soft and hard synthetic engineered tissues (Mironov et al. 2017).

Polycaprolactone and polyvinyl alcohol (PCL/PVA)-based scaffolds display biological characteristics which improve differentiation of pancreatic β-like cells and organization into islet-like structures. PCL is a hydrophobic polyester which is biocompatible and biodegradable and has high mechanical stability. It is therefore considered a good candidate to produce 3D culture (Zarekhalili et al. 2017). Recent studies have demonstrated that the low-immunogenic polyethylene glycol (PEG)-based hydrogels used for islet transplantation have the ability to support islet engraftment and function (Jeong et al. 2013; Rengifo et al. 2014) and allow long-term function and restoration of normoglycemia in diabetic mice

(Rios et al. 2016). Some studies demonstrated that using polymer films and electrospun meshes made with poly(ethylene oxide terephthalate)-poly (butylene terephthalate) (PEOT/PBT) blocks copolymer, provides a protective environment to preserve islet morphology by preventing their aggregation in implanted islets (Buitinga et al. 2013). However, even if these scaffolds can allow a higher nutrient diffusion (glucose flux), and maintain the insulin and glucagon expression, a decrease of *β-cell* density in the islet core was observed (Buitinga et al. 2013). Further studies demonstrated that cell death, resulting from non homogeneous vascularization whithin the core and the outer shell of the scaffold-seeded islets (Buitinga et al. 2013), could be avoided by optimizing the pore size of the PEOT/PBT-based scaffolds (Buitinga et al. 2017).

In addition, it has been shown that microporous scaffolds, made with non-degradable polyethylene glycol (PEG)-based hydrogels, protect islets from the host immune response and show comparable results to the initial engraftment and function of non-encapsulated islets. Furthermore, unlike encapsulated islets which lose vascular connections to the host tissue, the microporous scaffolds allow islet revascularization after transplantation (Rios et al. 2018).

Recent progresses in nanotechnology and materials used for encapsulation, as well as in immunomodulatory strategies have significantly improved vascularization. Meanwhile, differentiation strategies of ESC and iPSC into islet-like structures have achieved an upscalable production for potential clinical applications. However, these cells can potentially drive tumour development (Päth et al. 2019). New strategies based on micro- and nano-materials (i.e., PEG, PLGA, chitosan, liposomes and silica) alone or enriched with trophic factors, have been recently developed to achieve a better stability of the capsules and, more importantly, to avoid loss of *β-cell* mass and PO function (Hinderer et al. 2016). Bioengineers are focusing on the development of new synthetic materials able to maintain tissue- and organ-specific differentiation and morphogenesis. Indeed, the newly generated bio-products have important advantages such as (1) a great surface area for oxygenation and nutrient/cathabolite transport, (2) a porous structure allowing infiltration of cells and blood vessels, (3) a satisfactory mechanical strength supporting cell attachment, (4) could be easily implanted, and (5) eventually, undergo degradation over time (Lutolf and Hubbell 2005).

To promote vascularization of transplantable devices housing pancreatic islets is the use of TheraCyte™ system. TheraCyte™ device is made by a bilayered polytetra-fluoroethylene (PTFE) where either free or microencapsulated islets are placed in the membrane to obtain a planar, bilaminar membranous pouch (Sörenby et al. 2008). A first clinical study has demonstrated that one-year after transplantation, TheraCyte device was biologically inert and lacked adverse effects, when transplanted in humans. Nevertheless, marked fibroblast overgrowth occurred and almost all the tissues were fibrotic (Tibell et al. 2001). A further study has reported that an immortalized cell line derived from human islets (betalox5), and induced to differentiate, formed long-term survival and functional POs also immune-protected when allocated in TheraCyte devices (Itkin-Ansari et al. 2003). Afterthat, some preclinical studies were carried on to test the TheraCyte device housing rat islets (Sörenby et al. 2008), human islets and human fetal pancreatic POs (Lee et al. 2009). Sörenby et al. (Sörenby et al. 2008) improved the efficacy of the TheraCyte system by trasplanting rat islets in preimplanted TheraCyte devices. Briefly, empty and capped devices were first subcutaneously pre-implantated on the back of athymic mice before diabetes induction. Subsequently, the islets were transplanted in the already implanted and newly pre-vascularized device. Following these modifications, an improvement of PO efficacy was obtained. Moreover, the preimplantation procedure significantly reduced the number of macroencapsulated islets required to restore normoglycemia (Sörenby et al. 2008), without a detectable T-cell response (Lee et al. 2009). It is important to remark that *β-cells* at this stage of maturation may avoid the immune T-cell response (Lee et al. 2009).

In vivo differentiation into predominantly *β-cells* was also obtained when huESC-derived endoderm cells were seeded into the TheraCyte device (Motte et al. 2014). Biosafety was supported by the absence of increased biomass or hESC cell escape for up to 150 days (Kirk et al. 2014). However, an accumulation of CD8[+] T cells surrounding the membrane was observed in some of these devices (Boettler et al. 2016), indicating a certain grade of immune reaction against the graft.

In vitro PEGylation maintains viability and insulin secretory capabilities of transplanted islets (Lee et al. 2002), but also protects them from cytokines secreted by immune cells (Lee et al. 2004). Moreover, PEGylated-macroencapsulation allows a stable blood glucose level of the allotransplants after one year (Lee et al. 2006a). Since PEGylation makes transplanted POs protected against the recipient immune system, a reduction of cyclosporine A dosage was requested (Lee et al. 2006b), and a reduced graft immune infiltration was detected (Lee et al. 2006c). The use of a combination of immunossuppresive drugs was associated with better outcomes (Im et al. 2013). Coated thin PEG-layers on islet surface named "Layer-by-layer (LbL) PEGylation" have been shown to allow a further reduction of islet volume per unit, which facilitates the exploitment of the portal vein for transplantation (Wilson et al. 2008).

4.3 Drug-Charged Devices/Scaffolds

The use of dexamethasone-charged macroporous scaffolds has been proposed to inhibit the host immunoreaction. This original approach was developed to accelerate islet engraftment by promoting the expansion of the anti-inflammatory M2 macrophages (Jiang et al. 2017). To generate favorable host responses and to improve the overall outcomes of the transplant polydimethylsiloxane (PDMS)-based 3D scaffold platform, a local and controlled delivery of dexamethasone (Dex) was used (Jiang et al. 2017). In particular, Dex-scaffold accelerated islet engraftment in a diabetic mouse model, improving glucose control early after transplantation. Remarkably, it was demonstrated that lower doses of Dex (0.1% or 0.25%) were able to induce a M2 phenotype of macrophages interfering with inflammation during the first post-implantation week, whereas higher doses of Dex (0.5% and 1%) significantly delayed the engraftment and function of islets (Jiang et al. 2017).

4.4 Vascularization-Enhanced Macroencapsulation

The use of immune-isolated macrodevices designed to islet delivery into extrahepatic transplant site is not limited to synthetic PEG-based hydrogel macrodevices. In fact, a two-component synthetic PEG hydrogel macrodevice system has been generated. These PEG-devices consist in a hydrogel core cross-linked with a non-degradable PEG-dithiol and a proteolytically sensitive vasculogenic outer layer to allow matrix degradation and to enhance vessel infiltration (Weaver et al. 2018). These PEG-dithiol devices promoted engraftment and overall graft efficacy, resulting in enhanced vascular density and in the improvement of islet viability when transplanted in diabetic rats (Weaver et al. 2018).

Through 3D printing technology, Farine et al. (2017) designed an innovative and refillable encapsulation system transcutaneously implated. The PLA-system generated a prompt and efficient PO vasacularization, and supported the long-term graft survival (Farina et al. 2017). Nevertheless, in this first study the in vivo experiments were performed in immunodeficient mice, which did not allow the evaluation of their efficacy in preventing the host immune attack (Farina et al. 2017). More recently, the same group developed a new PLA encapsulation system named neovascularized implantable cell homing and encapsulation (NICHE) that seems to be a promising alternative approach due to their ability to maintain PO viability and a robust hormone secretion (Farina et al. 2018). Despite NICHE system has provided promises to improve long-term viability and function using Leydig cells (Farina et al. 2018), its feasibility in POs still requires validation.

The pro-angiogenic and immunomodulatory effects of the local delivery of the immunomodulating drug fingolimod (FTY720) allow a better oxygenated and tolerant environment for islet engrafment (Bowers et al. 2018). FTY720 is a small molecule that activates sphingosine-1-phosphate receptors and is involved in the regulation of immunomodulatory and pro-angiogenic signals. Based on this concept, Bowers et al. (Bowers et al. 2018) developed a new encapsulation membrane for islet transplantation outside the portal circulation. The FTY720-releasing nanofiber-based semi-permeable membrane increased PO revascularization, and was also able to block the immune response by reducing the number of macrophages and their released cytokines (Bowers et al. 2018).

A similar strategy which can facilitate PO transplantation and preserve their function in extrahepatic sites is represented by the encapsulation of pancreatic islets in synthetic saccharide-peptide (SP) hydrogels (Liao et al. 2013). The SP hydrogel is constituted by natural blocks of amino acids and saccharides, nontoxic and entirely biodegradable (Chawla et al. 2011). The SP hydrogel also displays several exclusive properties, such as ease of handling, crosslinkable at mild physiological conditions, biocompatibility, and in situ polymerization after injection (Chawla et al. 2011). The SP hydrogel allows PO survival and function of normal islet structure, but also promotes a rapid vascularization (Liao et al. 2013). Furthermore, it has been demonstrated that SP hydrogels are also able to induce minimal inflammatory cell infiltration. However, they tend to degenerate (Liao et al. 2013).

4.5   Removal/Retrievability

Despite the huge efforts in improving the encapsulation technology, the major problem still relies on the removal (Scharp and Marchetti 2014). This raises significant concerns since this can be associated with transplant failure and clinical complications (Desai and Shea 2017). Retrievability is also an important issue for the regulatory approval processes (Matsumoto et al.

2016). Thereby, efforts are currently focused to generate alternative encapsulation systems readily scalable and conveniently retrievable, also allowing the delivery of a sufficient cell mass. Innovative approaches so far developed to generate easy retrievable devices consist in a cytocompatible enzymatic approach based on the addition of MMP (Amer et al. 2015). Alternatively, a highly wettable and Ca2 + −releasing nanoporous polymer, possessing particular mechanical properties which facilitates handling and retrieval of transplanted islets has been generated (An et al. 2018). Innovative retrievable encapsulation system has been also recently developed by An et al. (An et al. 2018). In particular, the thread-reinforced alginate fiber for islets encapsulation (TRAFFIC) consists in a one-step in situ cross-linking alginate hydrogel around nanoporous wettable, and Ca2 + −releasing polymer thread. TRAFFIC has the advantage to provide the essential fisical space and biocompatibility for islet transplantation, similar to conventional hydrogel capsules, but displaying mechanical strength enabling the easy handling, implantation, and retrieval (An et al. 2018). Furthermore, due to their particular design, this device may be extended to clinic (An et al. 2018). The therapeutic potential of TRAFFIC has been provided by the restoration of glucose control and inmuno protection of transplanted human or rat islets in diabetic mice, rats and dogs (An et al. 2018). Furthermore, rapid retrievability can be obtained through a simple laparoscopic procedure (An et al. 2018). The fast and minimally invasive retrievability makes TRAFFIC a promising encapsulation system with a great scale-up potential. Nevertheless, the feasible application in clinic, and particularly in T1DM, has to be evaluated.

4.6   ECM-Enriched Macroencapsulation

Islets in the pancreatic tissue are naturally surrounded by a capsule mainly composed by collagen type I, IV and laminins (Stendahl et al. 2009). Llacua et al. (2016) demonstrated the ability of specific ECM components (collagens or synthetic laminin peptides) to support survival and function of encapsulated human islets (Llacua et al. 2016). This technology was further applied to evaluate the immuno-regulatory effects of capsules covered with a mix of artificial ECM (A. ECM) and Layer-by-Layer (LbL) Pegylation (A. ECM + PEGylation) in non-human primate xenografts (Andrades et al. 2008; Haque et al. 2017; Llacua et al. 2018). The authors demonstrated that ECM-enriched encapsulation system acts as an effective immune barrier, translating in the increase of xenograft survival in the absence of immmunosuppresive therapy (Llacua et al. 2018). Although the ECM-enriched encapsulation system is also able to protect POs against the host immune response and to prolonge islet cell survival, its ability to induce islet re-vascularisation remains to be addressed. According with this concept, further studies have been performed to improve PO vascularization. In particular, vascularization improvement was obtained by incorporating heparin in nanofilms containing star-shaped polyethylene glycol (starPEG) (Lou et al. 2017) or silk fibroin macroporous (SF)-scaffolds (Mao et al. 2017). Additional anti-inflammatory and anti-coagulant properties along with the improvement of intra-islet vascularisation have been obtained by enriching capsules with glycosaminaglycans and heparin (Hep-PEG) (Lou et al. 2017). Heparin-mediated vascular endothelial growth factor (VEGF) binding, resulting in the activation of the endogenous VEGF/VEGFR2 pathway, has been reported to mediate endothelial proliferation and re-vascularisation (Lou et al. 2017). These mechanisms have been also previuosly described for both non-encapsulated (Cabric et al. 2007; Cabric et al. 2010) and encapsulated human islets (Marchioli et al. 2016). Given the pro-angiogenic, pro-survival and minimal post-transplantation inflammatory reaction, as well as the possibility to introduce different trophic factors, Hep-PEG-based encapsulation system is considered a very promising approach for clinical implementation.

4.7 Silk Fibroin-Based Macroencapsulation

The fibrin matrix, due to its good biocompatibility, rapid biodegradability, and easy production, is the most widely investigated natural biopolymeric material (Park and Woo 2018). In particular, the bioactive molecules conjugated with fibrinogen, can promote tissue morphogenesis, cell migration, proliferation, differentiation or maturation after cell adhesion on fibrin matrices (Park and Woo 2018). Thereby, fibrin matrices have been exploited as therapeutic strategies for tissue engineering applications (Park and Woo 2018). These concepts were used for the developmet of new materials such as autologous platelet-and plasma-derived fibrin scaffolds (Anitua et al. 2019). Synthetic scaffolds based on 3D silk matrices with ECM-derived motifs were also used to promote the aggregation of mouse and human primary cells into functional POs (Shalaly et al. 2016). Afterthought, POs generated by co-encapsulation of islets with MSCs in a silk hydrogel have been shown to increase insulin I, insulin II, glucagon and PDX-1 gene expression. This was associated with a better glucose-induced insulin response (Davis et al. 2012). Diabetic mice implanted with Silk fibroin hydrogel–islet–MSC system showed a prompt return to euglycaemia with a significative reduction of T-helper (Th)1-derived cytokines (Hamilton et al. 2017). Thereby, co-culture of islets and MSCs in a silk-hydrogen-islet system has been proposed to avoid the host immune-attack. However, additional studies are required to demonstrate the stability, the function and the biocompatibility before transfer of this approach to humans.

5 Oxygenated Devices

It has been postulated that chronic effects of non-immunologic factors such as hypoxia and the hyperglycemic milieu could damage the encapsulated islets, resulting in a gradual and short-term loss of efficacy (Safley et al. 2018b).

This prompted researchers to propose new preventing approaches. In particular, major efforts have been focused on generating new methodologies able to increase oxygen supply to POs. Oxygen supply is a crucial issue for cell survival in bioartificial pancreas (BAPs), since poor oxygen supply causes PO central necrosis. Implantation and removal of macrocapsules actually imply minimal risks, but the transport of oxygen and nutrients is still limited. Therefore, new strategies must be developed to improve oxygenation (Hwa and Weir 2018). At this regard, bioengineering research in diabetes mainly focused on the development of new 3D scaffolds able to supply sufficient oxygen and nutrients to POs (Iwata et al. 2018).

The generation of an oxygenated chamber system, based on the incorporation of a refillable oxygen tank (βAir device) into the immune-isolating alginate poly-membrane was the first technology developed to improve oxygenation of the encapsulated POs (Ludwig et al. 2012). Indeed, the feasibility to increase the oxygen supply in small and large animal models was demonstrated (Barkai et al. 2013; Neufeld et al. 2013). Based on these promising results, a phase 1 clinical trial was started to demonstrate the efficacy of the βAir (Ludwig et al. 2013). The authors demonstrated that βAir inhibits inflammation of human islets implanted in the absence of immunosuppression. Additionally, an increased vascularization and the enhancement of the oxygen supply were reported (Ludwig et al. 2013). A second phase 1 clinical trial demonstrated that βAir device was safe and able to enhance survival of allogeneic islets (Carlsson et al. 2018). Nevertheless, the real benefits obtained by the βAir device were limited, due to the low increase of circulating C-peptide and the lack of a real impact on the metabolic control (Carlsson et al. 2018). Additionally, the formation of fibrotic tissue, the presence of immune cells, and the deposition of amyloid were also detected in the endocrine tissue (Carlsson et al. 2018).

A new generation of 3D scaffolds containing oxygen generators have been proposed as alternative technology to improve PO oxygenation. Different approaches have been proposed. The new devices included, (1) polydimethylsiloxane encapsulated solid calcium peroxide (PDMS-CaO2) (Pedraza et al. 2012), (2) active microorganisms such as Synechococcus lividus, which photosynthetically generated oxygen (Evron et al. 2015), and (3) SPO and CPO (Lee et al. 2018). These studies demonstrated a successful enhancement of oxygenation that translated into a reduced hypoxia-induced cell loss during the precarious vascularization period (Pedraza et al. 2012). These beneficial effects resulted from the capability to supply oxygen at high cell loading densities (Pedraza et al. 2012), with an average of 0.026 mM/day oxygen, for more than 6 weeks (Evron et al. 2015). This prevented the development of detrimental oxygen gradients in the core of large implants. The beneficial effects of oxygen-generating scaffolds were further supported by the studies of Lee et al. (2018) who used neonatal pancreatic cell-clusters (NPCCs). The application of oxygen-generating microparticles (MP) included in a fibrin-conjugated heparin/VEGF and implanted in streptozotocin-induced diabetic NOD mice was also investigated (Montazeri et al. 2016). The MP improved the POs engraftment and function, the blood glucose control, body weight, glucose tolerance, serum C-peptide, and graft revascularization, by reducing the islet mass necessary to obtain this goal (Montazeri et al. 2016). It became evident that oxygen-generating materials such as sodium percarbonate (SPO) and calcium peroxide (CPO) could be efficient in supplementing oxygen to transplantable naked and encapsulated islets in diabetic patients (McQuilling et al. 2017). Indeed, it has been shown that SPO and CPO can be used to improve islet viability and function. Nevertheless, additional studies are required to control the oxygen generation rate able to provide the suitable oxygen supply over an extended period of time avoiding tissue damage resulting from oxidative stress (McQuilling et al. 2017). Of note, it has been suggested that this technology could be also applied, as a new animal-free experimental scaffold platform, to study the interactions between pig islets and human blood components *in vitro* (Lee et al. 2018).

5.1 Perfluorocarbons (PFCs) as Oxygen Carriers in POs

PFCs are obtained by fluorine replacement of hydrogen atoms in hydrocarbons (Hosgood and Nicholson 2010). The PFCs have a higher density than water and a high capacity to dissolve oxygen. Carbon-fluorine bonds are very long and render the molecule biologically and chemically inert (Hosgood and Nicholson 2010). Due to their lipophilic nature and their high oxygen solubility, PFCs have been largely explored to improve cell and tissue oxygenation. They can be also adapted and used as blood substitutes for the treatment of cardiac ischemia, anemia, and organ preservation (Hosgood and Nicholson 2010). However, controversal results have been reported. Bergert and collaborators (Hosgood and Nicholson 2010) have demonstrated that, compared to control rat islets, PFC-cultured islets showed a significant increase of DNA fragmentation and a reduced glucose sensitivity. On the contrary, other studies indicated that, although advantageous to transfer human harvested organs, the use of PFCs may be comparable to conventional islet culture protocols used for transplantation (Bergert et al. 2005). A different study demonstrated that perfluorotributylamine (PFTBA) emulsion co-encapsulated insulin secreting cells (βTC-tet insulinoma cells) did not have additional effects in terms of viability or function (Goh et al. 2010). Nevertheless, it was suggested that PFCs could be potentially applied as a culture method to improve islet preservation, in particular for pancreatic islets isolated from marginal pancreata (Ricordi et al. 2003). Additional studies are still required to demonstrate the biological benefits of PFCs on POs.

5.2 Generation of Hypoxia-Resistant Islets

It became evident that, while encapsulation clearly protects POs against the host immune attack, it impairs PO survival and long-term function. VEGF plays a crucial role in *β-cell* function and islet regeneration (Nikolova et al. 2006; Jabs et al. 2008). Moreover, it has been reported that a splice variant-1 of the GHRH receptor expressed in human pancreatic islets was able to generate antiapoptotic signals by modulating VEGF levels and by inducing angiogenesis (Ludwig et al. 2010). "Hypoxia-resistant islets" have been generated using implanted devices containing pancreatic islets and GHRH agonists (JI-36) (Ludwig et al. 2012). A significant hypervascularization and enhancement of the graft function, associated with a parallel improvement of glucose tolerance and increase of the *β-cell* insulin reserve were observed (Ludwig et al. 2012). This also allowed a reduction of the islet mass required for metabolic control (Ludwig et al. 2012). However, even if promising, the feasibility of this innovative technology as alternative xenotransplantation approach requires further validation.

5.3 Mathematical Models

As extensively documented, the main goal to improve biotechnological approaches is an efficient oxygenation, allowing a long-term survival and function of newly generated POs. One of the most relevant challenges to deal with relies on the the thickness of the cell layer (Wu et al. 1999). In encapsulated POs, the inner core mimics pancreatic islets while the peripheral shell acts as an inert defense. Oxygen consumption by the cells is a hurdle that follows two rules, (1) the limited diffusion, and (2) the limited consumption. Using mathematical models it has been suggested that hypoxia directly depends on the PO radius and would likely occur in high cell density conditions (King et al. 2019). According with this model, a better viability should be obtained with a PO radius around 142 μm or fewer, and with an encapsulating radius lower than 283 μm (King et al. 2019).

The effectiveness of supplementary oxygen suppliers in a *β-cell* 3D culture has been estimated by calculating the spatio-temporal distribution of oxygen concentrations inside the scaffolds (McReynolds et al. 2017). Two different types of simulations (constant cell density and cell growth simulation) used in this work, have lent a hand to determine the effectiveness of oxygen-releasing microbeads (McReynolds et al. 2017). Briefly, hydrogen peroxide was encapsulated into nontoxic PDMS disks which provided oxygenation to *β-cell* cultures (McReynolds et al. 2017). Detection of the spatial-temporal distribution of oxygen tension inside a scaffold would be a fascinating approach to be pursued in the future. MacReynolds and collaborators (McReynolds et al. 2017) have also suggested that optimization of the cell culture conditions, including cell seeding density and time of culture, are critical issues.

As widely described, the new 3D devices are mostly based on oxygen diffusion. Nonetheless, mathematical models indicate that convective transport should be a more efficient approach for the oxygen transfer (Iwata et al. 2018). Theoretical studies have described how oxygen supply in BAPs could predict the necrosis area in the islets (Iwata et al. 2018). Thereby, according with these models, to correct the difference in pressure between the device and the site of implantation, the BAPs should be directly connected to the blood vessels (Iwata et al. 2018). Platelet deposition usually takes place in implants subjected to blood flow, and eventually leads to thrombus formation (Affeld et al. n.d.). Thus, BAP coated with blood compatible materials could prevent thrombus formation in long-term BAPs.

In summary, despite enormous efforts devoted to the encapsulation technology, oxygenation remains the main challenge in diabetes bioengineering. Mathematical models, established that an oxygen-enabled 3D culture system should effectively provide an improved oxygen distribution within the scaffold, allowing an enhanced secretion of insulin from cells. Nevertheless, the application in humans is still missing.

6   Vascular Improvement in Transplants

Pancreatic islets are extremely vascularized structures by a capillary network wich is critical for glucose sensing and for providing nutrient supply and rapid secretion of hormones into the blood stream (Richards et al. 2010). Islets with an altered vascularization do not properly regulate glicemia (Lammert et al. 2003), indicating that engineered islets should permit an efficient mass transport and well-organized vascularization. This requires the presence of a porous implantable construct allowing vessel infiltration. Recent data indicate that the range of pore size that maximizes vascularization is narrow and corresponds to 30–40 μm, (Madden et al. 2010). This suggests that an accurate control of the pore size would significantly favor vascularization of implantable scaffolds (Buitinga et al. 2017). It has been clearly demonstrated that the lack of an adequate revascularization (Smink et al. 2017), the reccurrence of autoimmunity (Piemonti et al. 2013), the prompt blood-mediated inflammatory response (Kourtzelis et al. 2016; Samy et al. 2018), the ischemic injury (Faleo et al. 2017), and the activation of Natural Killer (NK)T cells (Saeki et al. 2017) are the main factors contributing to the engraftment failure.

Approaches exploiting angiogenic factors have been used in tissue engineering. Thus, encapsulation devices enriched in pro-angiogenic molecules such as VEGF, to improve PO vascularization is strongly encouraged (De Rosa et al. 2018; Schweicher et al. 2014). Trivedi et al. (2000) have proposed an alternative approach to solve the deficiency of vascularization in devices using a double layer capsule formed by three different layers combined with a transcutaneous infusion of VEGF. This approach leads to a reduced delay of the diffusion time due to the development of a new vascular network through large pores. The authors also showed that small pores could protect POs from the immune attack (Trivedi et al. 2000). However, even if a reduction in glycemia was detected, normoglycemia was never achieved in transplanted diabetic mice (Sweet et al. 2008). The authors have suggested that this unexpected result, rather than to the device itself, may be related to the number of implanted islets, corresponding to the half of islets usually transplanted into T1DM patients (Sweet et al. 2008).

Alternative studies applied a semi-interpenetrating polymer network (SIPN) as subcutaneous implantable carrier to deliver cells and to allow vascularization (Mahou et al. 2017). SIPN was generated by reacting a blend of vinyl sulfone-terminated polyethylene glycol (PEG-VS) and PMAA-Na with dithiothreitol (Mahou et al. 2017). This combination allowed the formation of vessels in the surrounding tissues, which is 2–3 times superior to that obtained with PEG alone (Mahou et al. 2017). The biological activity of the SIPN in glucose control was demonstrated in diabetic mice (Mahou et al. 2017).

Re-endothelization represents an innovative approach designed to solve the issue of blood supply. Endothelial progenitor cells (EPCs) have been used to generate pre-vascularized scaffolds. The main advantage identified was the direct connection between the pre-vascularized host blood circulation and the newly formed blood vessels (Guo et al. 2018). Very recently, Skrzypek et al. (2018a) developed a functional and not degradable flat polyethersulfone/polyvinylpyrrolidone (PES/PVP) porous membrane to create an early microvascular network without the addition of hydrogels (Skrzypek et al. 2018a). Applying this technology, pre-vascularized devices could be obtained by co-culturing different cells (Skrzypek et al. 2018a, b; Groot Nibbelink et al. 2018). It was demonstrated that the particular surface topography obtained with this method enhanced the formation of a stable vascular network in the membranes. Furthermore, the *in vitro* pre-vascularization allowed a faster connection of POs with the host vasculature, impacting on the engraftment, and on the long-term implant survival and function (Skrzypek et al. 2018b).

Oligomers have been used for tissue engineering approaches in T1DM pre-clinical models (Stephens et al. 2018). The feasibility of collagen I oligomeric-based macroencapsulation was already demonstrated (Stephens et al. 2018). Collagen-based macroencapsulation was able to improve mouse islet function and longevity (Stephens et al. 2018). Cytoarchitecture,

revascularization, and immunetolerance were also demonstrated (Stephens et al. 2018). Furthermore, STZ-induced diabetic mice showed a rapid glycemic control and a sustained normoglycemia until 40 days post-transplantation (Stephens et al. 2018).

7 Natural Extracellular Matrices as Biological Scaffolds

Undoubtedly, scaffolds dysplaing features similar to native islet extracellular matrix (ECM) would most likely succeed. The ECM is synthesized by cells of each tissue and provides a physical niche for cell attachment. ECM composition includes polysaccharides and proteins, such as collagen and elastin (Theocharis et al. 2016). One of the most recent technologies enabling the isolation of native ECM is the organ decellularization (Badylak et al. 2011). ECM derived from decellularized organs has been recently used for tissue engineering (Guruswamy Damodaran and Vermette 2018a). Decellularized scaffolds have been proposed as a source of native ECM exploitable in a wide-ranging regenerative medicine applications (Celikkin et al. 2017). ECM offers structural and function support and acts as a substrate for cell migration. The ECM mechanical behavior depends on its physical properties such as insolubility, porosity, rigidity, and its topography (Celikkin et al. 2017). The ECM components are very conserved through different species, and include fibers, collagens, proteoglycans, glycoproteins, and mucins (Brown and Badylak 2015). ECM-scaffolds respond and release growth factors, that modulate the immune response, and allow the recruitment of progenitor cells (Swinehart and Badylak 2016).

Islets embedded in the acinar tissue of the pancreas represent only the 1%–2% of pancreatic mass (Jansson et al. 2016; Aamodt and Powers 2017). They are spherical clusters of cells highly innervated and vascularized, containing ECM which coordinate and support cellular survival, proliferation and differentiation (Irving-Rodgers et al. 2014; Aamodt and Powers 2017; Cheng et al. 2011; Kuehn et al. 2014; Alismail and Jin 2014). In islets the ECM regulates multiple aspects of cell function, including insulin secretion, proliferation, survival, and participates to the preservation of spherical morphology (Stendahl et al. 2009). Pancreatic decellularized scaffolds play an essential role in regenerative medicine and represent a step toward the development of bioengineered pancreas. As other natural bioscaffolds, pancreatic decellularized scaffolds retain the native 3D architecture, the vasculature, the ductal channels, and the pancreatic ECM composition (Guruswamy Damodaran and Vermette 2018b). The interaction between cells and scaffolds improves pancreatic islet cell survival and insulin production potentially used for regenerative purposes (Goh et al. 2013).

Bioscaffold plasticity has been also demonstrated by using stem cells, which received and responded to the environmental signals. In this context mechanical forces play a central role in the regulation of multiple functions (Rana et al. 2017). Indeed, the adhesion of cells to ECM, the formation of cell-cell junctions, and of a mechanoresponsive cell cytoskeleton, are critical features in stem cell biology (Rana et al. 2017), particularly in cell fate determination (He et al. 2018). Plasticity and compatibility of ECM scaffolds derived from different species have been also demonstrated (Chaimov et al. 2017). Decellularized pancreatic scaffold biocompatibility in the absence of toxicity has been also confirmed (Chaimov et al. 2017). Particularly, Chaimov and collaborators (Chaimov et al. 2017) developed an innovative microencapsulation platform based on solubilized whole porcine pancreatic ECM mimicking the original decellularized ECM (Chaimov et al. 2017). This natural fibrous 3D niche was able to support the viability of pre-differentiated cells to promote differentiation and to improve insulin secretion (Chaimov et al. 2017). Moreover, these re-cellularized scaffolds were non-immunogenic and were able to improve the glycemic control in diabetic mice (Chaimov et al. 2017). It has been reported that growth factor-enriched decellularized pancreatic scaffolds induce differentiation of

mouse pancreatic stem cells (PSCs) into pancreatic β-like cells (Wan et al. 2017). Bioscaffold-induced differentiation, proliferation and insulin secretion, occur independently of scaffold-derived tissue (Zhou et al. 2016). This suggests the existence of a dual interplay between the ECM and the cell.

Decellularized pancreata show the characteristic homogeneous porous structure (Hashemi et al. 2018). Thus, to solve the problem of PO blood supply after *in vivo* transplantation, bioengineering has taken advantage of this valuable feature to enhance scaffold vascularization by co-seeding POs with EPCs (Guo et al. 2018). Newly formed blood vessels (Guo et al. 2018) and improvement in functional preservation characterized the re-endothelized scaffolds (Chaimov et al. 2017). Despite these promising results, vascularization still remains a challenge to generate a functional equivalent ECM-based pancreata for transplantation.

Generation of acellular pancreatic scaffolds from small animals represents a relevant issue for future pre-clinical studies. The next and most closely need is represented by the large-scale production. Some recent studies demonstrated the feasibility (Berman et al. 2016; Katsuki et al. 2016). Nonetheless, despite the improvement of islet metabolic function and the significant preservation of the islet architecture and the intrainsular vascularization (Berman et al. 2016; Katsuki et al. 2016), long-term studies and implantation *in vivo* are still needed. Moreover, it was recently described that signals promoting cell differentiation could be heterogeneous in nature, depending on the type of organ, and on the type of tissue from which they derive (Brown and Badylak 2015). For example, it has been shown that HLSC seeded in liver decellularized scaffolds can differentiate into three different cell types (hepatocyte-, endothelial- and epithelial-like cells) (Navarro-Tableros et al. 2015). This indicates the high ECM plasticity. Therefore, the use of ECM bioscaffold-based technology could be exploitable for hetero-organ approaches. This possibility has been recently validated by Vishwakarma and collaborators (Vishwakarma et al. 2018). Indeed, they were able to induce differentiation of human hepatic progenitor cells (hHPCs) into insulin-progenitor cells when hHCPs were seeded into xenogeneic rat acellular spleen in the presence of specific trophic factors and hyperglycemic environment (Vishwakarma et al. 2018). These data further support the hypothesis (Navarro-Tableros et al. 2015; Vishwakarma et al. 2018), that cell commitment is a complex process, modulated in part by the nature of the scaffold, but also by the type of hosted cells. In fact, the study of Vishwakarma et al. (2018) suggests that natural bioscaffold properties should be further investigated for regenerative approaches. Finally, pancreatic scaffolds, used in most if not all studies, have been generated from "healthy" tissues. Nevertheless, there are not definitive data on the feasibility to use "sick" organs for clinical applications. A very recent study, aimed to evaluate the feasibility at using diabetic pancreata to generate ECM for regenerative purposes (Huang et al. 2018a), demonstrated that similar to healthy scaffolds, those obtained from T1DM and T2DM pancreata preserved their ECM composition, 3D ultrastructure and released cytokines (Huang et al. 2018b). Nevertheless, the biological activity of this "diabetic scaffolds" has been not yet tested. These important results, even if preliminary, support the idea that marginal organs could serve as alternative sources of natural ECM scaffolds. Thereby, the application of marginal tissues could be implemented in future regenerative approaches in order to strongly support this hypothesis.

8   CRISPR/Cas9 Genome Editing Technology to Enhance Insulin-Producing Cell Function

The Clustered Regularly Interspaced Short Palindromic Repeats (CRISPR) combined with the DNA endonuclease Cas9 (CRISPR-Cas9) has been considered an innovative application which could completely change both medical and biotechnology approaches in our century. Using short guide target-specific sgRNAs, the Cas9 can be directed to any genomic location by inducing double strand breaks to allow non-homologous end joining or homologous recombination of the genome at specific sites (Kim et al. 2017). CRISPR works together with *Cas* gene to cleave genetic material, thus opening the possibility to use CRISPR-based approaches to induce β-cell differentiation (Gerace et al. 2017). Other approaches different from CRISPR, are mainly based on viral-mediated transfer of key pancreatic transcriptional factors such as *Pdx-1* (Karnieli et al. 2007), *NeuroD1* (Kojima et al. 2003) to either somatic or adult stem cells such as MSCs (Gerace et al. 2017). However, although CRISPR-Cas9 system is widely applied in a number of research fields (Kim et al. 2017), its application to generate or to enhance the function of insulin-producing cells is still limited. At this regard, Giménez et al. (2016) applied the CRISPR-ON system to activate the endogenous human insulin gene (INS), obtaining a significant upregulation of the insulin mRNA expression when the dCas9-VP160 construct and four sgRNAs (targeting the proximal INS promoter) were co-transfected in T1DM patients-derived skin fibroblasts (Giménez et al. 2016). More recently, one homozygous ATG > ATA mutation at codon 1 of the insulin gene was reverted to wild-type ATG in hiPSC by using CRISPR/Cas9 technology (Ma et al. 2018). The insulin mRNA expression and hormone secretion confirmed the functional correction (Ma et al. 2018). Moreover, the modified β-cells were able to reverse diabetes in STZ-diabetic mice (Ma et al. 2018). This work has opened the possibility to combine gene and cell therapy to restore glucose homeostasis in non-immune-mediated diabetic setting. As a future perspective, the combination of CRISPR-Cas9 and encapsulation technology may contribute to improve current methodologies to generate long-term functional POs to be applied in diabetes.

9   3D Bioreactor Systems to Produce and Maintain Insulin-Producing Cells

Although stem cells hold the promise to generate endless insulin-producing cells, most of the current preclinical protocols have been performed in lab-scale (Navarro-tableros et al. 2018; Kim et al. 2016a; Wu et al. 2007). Hence, the implementation of alternative culture systems, able to generate a large number of functional insulin-producing cells, is still needed. Cell culture in 3D bioreactors has been used for several decades by the biopharmaceutical industry (Abu-Absi et al. 2014), since they allow a tight control of relevant parameters such as pH, temperature and gas supply, and garantee a more efficient cell expansion and a higher cell viability and function (Petry et al. 2018). In this chapter, we will provide a general overview of different 3D bioreactors that have been employed to generate and/or maintain viable and functional POs obtained from different sources.

9.1   3D Bioreactors Used for Insulin-Producing Cell Cultures

Spinner Flasks (SF) SF, which are currently considered the simplest form of 3D bioreactors, consist in a magnetic stirrer-integrated flasks that provide a turbulent fluid flow environment to partly alleviate limitations of nutrient and oxygen diffusion (Ratcliffe and Niklason 2002) (Fig. 1a). However, SF suffer from similar drawbacks as classical 2D culture systems in terms of requirements of individual handling and scale-up (Ratcliffe and Niklason 2002), and to the best of our knowledge despite these limits, SF are the most commonly used bioreactors in the field of *β-cell* research.

In the earlies 2000s, Blyszczuk and collaborators (Blyszczuk et al. 2003) developed a 10 days culture protocol in SF (30 rpm agitation at 37°C and 5% of CO_2) to induce differentiation of mouse embryonic-derived nestin-positive progenitors into immature insulin-positive POs

(Blyszczuk et al. 2003). Trasplantation of these POs was able to ameliorate the glycemic control in immunocompromised STZ-diabetic mice (Blyszczuk et al. 2003). More recently, hESCs, initially maintained in 500 mL SF (70 rpm agitation at 37°C and 5% of CO_2) and subsequenly cultured in standard 2D plates in presence of specific growth factors, were found to successfully differentiate into POs able to restore normoglycemia in diabetic mice (Pagliuca et al. 2014). Long-term glycemic control was obtained in diabetic mice, when encapsulated hESCs-derived POs were pre-cultured for 24 h in SF, before their implantation (Vegas et al. 2016).

Chawla and collaborators (Chawla et al. 2006) developed a new SF system to generate POs in a large scale. To this end, porcine pancreatic neonatal endocrine tissues, at different densities, were seeded in SF (100 rpm agitation at 37°C and 5% of CO_2) in serum-free media with a specific differentiation-cocktail (Chawla et al. 2006). Using this approach, and after 9 days of culture, a large number of POs containing insulin-positive cells were produced. In addition, the POs responded to glucose and contained all islet cell types (Chawla et al. 2006). More recently, Lock et al. (2011) demonstrated the superiority of SF in the generation of mouse cell line (MIN6)-derived POs compared to the 2D standard culture. The POs generated in SF (60–100 rpm agitation at 37°C and 5% of CO_2) were associated with a 20% higher viability (64 vs 84%), a more regular shape morphology and the ability to grow for at least 2 weeks (Lock et al. 2011). In addition, the POs generated in SF, were more similar to native islets, in terms of ultrastructure, glucose-insulin secretion, incretin expression and were characterized by a lower necrosis grade (Lock et al. 2011).

Microgravity Bioreactor (MB) The original version of the MB was developed by NASA engineers and patented in 1990 under the form of rotating wall vessels (RWV), also known as high aspect ratio vessels (HARV) (Schwarz et al. 1992) (Fig. 1b). MB provided a continuous circular rotation of media and allowed cultured cells to

Fig. 1 Schematic representation of different bioreactor types used to generate and maintain insulin-producing cells. (**a**) Spinner Flasks (SF). (**b**) Rotating wall vessels (RWV). (**c**) Hollow fiber membrane bioreactor (HFB). (**d**) Fluidized-bed bioreactor (FBB)

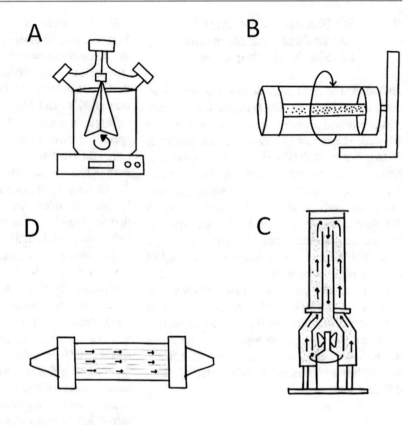

be suspended in a low gravitational field (free fall), characterized by a low turbulence and hydrodynamic shear stress coupled with a high mass of oxygen transfer (Daoud et al. 2010). The first study involving MB and islets was carried out by Cameron and collaborators (Cameron et al. 2001) in 2001. To facilitate islet engraftment after transplantation, they co-cultured porcine neonatal islets with Sertoli cells (isolated from rat testis) in MB for up to 14 days. After 4 days in MB culture, large Sertoli-islet POs were formed (0.5–3 mm diameter). However, although cellular viability (>90%) and glucose responsiveness was maintained, islets underwent disaggregation into single cells. Primary mouse islets cultured in MB were significantly less immunogenic compared to standard cultured islets (Rutzky et al. 2002). Moreover, to maintain euglycemia throughout 100 days in diabetic animals the number of MB-cultured POs required were 50% less than the non-MB counterpart (Rutzky et al. 2002). Furthermore, ultrastructure studies revealed that

islets appeared healthy and similar to fresh islets (Rutzky et al. 2002). Using MB (8 rpm at 37°C and 5% of CO_2) the improvement of human islet long-term function and viability was also demonstrated (Murray et al. 2005).

In an elegant work, Tanaka and collaborators (Tanaka et al. 2013) described the generation of mouse β-cell (MIN6 cell line)-derived POs. They loaded a two-compartment culture chamber onto the center of the 3D clinostat to generate microgravity-like conditions. Unlike the RWV, this new designed approach, allowed fresh media flow and gas exchange. In addition, POs had an average size of 250 μm and were glucose-responsive (Tanaka et al. 2013).

Hollow Fiber Membrane Bioreactor (HFB)

HFB utilizes specialized membranes to retain cells inside, and to allow an efficient exchange of nutrients and waste products by the flow of the medium through the lumen (Abu-Absi et al. 2014) (Fig. 1c). The main advantages of the membrane-based-bioreactors relied on the possibility to obtain an high cell capacity, high volumetric efficiency, and low shear stress (Abu-Absi et al. 2014). However, disadvantages such as a reduced cell viability, membrane fouling and clogging, the lack of product homogeneity have limited their large-scale application (Abu-Absi et al. 2014). Hoesli and collaborators (Hoesli et al. 2009) developed a procedure for immobilization of large-scale cell batches in alginate-filled hollow fiber bioreactors (AHFBRs). This process improved insulin secretion and cell viability of porcine neonatal pancreatic cells (Tanaka et al. 2013) and enhanced the expansion of the insulinoma cell line, INS-1 (Sharp and Vermette 2017; Gundersen et al. 2010).

Fluidized-Bed Bioreactor (FBB) In the FBB system, the cells are seeded on microcarriers (e.g Cytopore or Cytoline) or polymer-microencapsulated (e.g alginate, collagen) and loaded in a stirred tank which allows the flow of the oxygenated nutrient through the cells (Fig. 1d). This design avoids the formation of local high shear stress and provides an interface suitable for large-scale production (Abu-Absi et al. 2014). The feasibility of FBB for production of encapsulated-POs was previously assessed (Buchi-395 encapsulator). Indeed, it was demonstrated their ability to enhance function, viability and integrity as well as insulin responsivness of 1% alginate-encapsulated POs (range 180–220 μm) (Nikravesh et al. 2017).

10   *In-vitro* Expansion of β-Cells from Adult Human Pancreatic and Hepatic Tissues

hPSCs are a promising source of cells for tissue regeneration due to their unlimited proliferative potential and their capability to differentiate into three germ layers including: ectoderm, endoderm and mesoderm (Zuber and Grikscheit 2018). However, concerns still remain regarding cell differentiation commitment and scalability required for human transplantation approaches

(Jones and Zhang 2016). One of the major advantages of hPSCs compared to other stem cells, relies on their pancreatic origin. Indeed, they originated from the pancreatic islets, which could facilitate their commitment into mature POs (Nelson et al. 2009). Under physiological conditions, the endocrine pancreas (islets) has an extremely low turnover rate (Afelik and Rovira 2017), while under high metabolic demands (pregnancy or obesity), adaptation mechanisms such as cell hypertrophy, increased insulin synthesis and secretion, β-cell self-replication occur (Afelik and Rovira 2017). *In vivo*, the β-cell pool, is homogeneous (Brennand et al. 2007) and persist throughout the life as a "neogenic niche" within pancreatic islets (van der Meulen et al. 2017). Koop et al. (Kopp et al. 2011) demonstrated that after in vivo β-cell ablation, endocrine cells do not arise from the ducts, but from early pancreatic progenitors expressing Ptf1a, Nkx6.1, Pdx1 and Sox9 markers (Dor et al. 2004). Additionally, terminally differentiated adult β-cells retaining a significant proliferative capacity and accounting for pancreatic turnover and expansion throughout the life, have been identified (Wei et al. 2006). Pancreatic progenitors do not express hormones in basal conditions but can be induced to differentiate into hormone-expressing islet-like-cells (Dor et al. 2004). Although the hIPSC aggregation induces the expression of pancreatic endocrine markers such as Pdx1, MafA and hormones like insulin, glucagon or somatostatin, the mRNA and hormone levels are less expressed than in freshly isolated human islets. They fully differentiated only after implantation in mice (Kroon et al. 2008; Davani et al. 2007).

10.1 Pancreatic Islet-Derived Stem Cells

The presence of pancreatic precursor cells in adult human pancreas named pancreatic islet mesenchymal stromal cells (huPI-MSC) have been demonstrated and have been recently considered an alternative cell source to generate insulin-producing cells (Zhao et al. 2007; Joglekar and Hardikar 2012; Wang et al. 2013). These huPI-

MSCs exhibit all the characteristics of bone marrow MSC (BM-MSC), including the ability to suppress proliferation of lymphocyte stimulation. The number of harvested islet equivalents from a single donor varies from 300,000 to 600,000 (Kim et al. 2012). This implies that autologous huPI-MSC could be obtained from a small portion of adult pancreatic islets, including those which are discarded (Kim et al. 2012). These properties make huPI-MSC, an available and potentially exploitable source of adult human stem cells to generate *β-cell* substitutes for cell therapy in diabetes. From a therapeutic point of view, the use of adult stem cells from discarded human cadaveric islets represents an attractive perspective. It is well accepted that mature adult human *β-cells* could be significatively expanded *in vitro*, but complex de-differentiation and re-differentiation processes are required (Russ et al. 2008). Expanded human *β-cell*-derived (BCD) cells, constitute the ~40% of cells of the islet cell cultures, which can re-differentiated in response to a combination of soluble factors (Avnit-Sagi et al. 2009; Russ et al. 2011). However, only a part of these cells undergo re-differentiation (Russ et al. 2009). More recently, a different approach demonstrated that overexpression the miR-375, a miRNA highly expressed during human islet development, boosted adult pancreatic precursor cells to redifferentiate (Avnit-Sagi et al. 2009; Joglekar et al. 2009) into mature islets (Klein et al. 2013; Latreille et al. 2015).

Despite these promising results, further efforts are required for a successfully transability, and large scalability. In line with this concept, Carlotti and collaborators (Carlotti et al. 2010) have demonstrated the feasibility to use pancreatic-derived iPSCs in regenerative medicine. Recent studies, have demonstrated the presence of pancreatic stem/progenitor cell populations, named NIP/Nestin-positive Islet-derived Progenitors (Zulewski et al. 2001), PIDM/Pancreatic Islet-Derived Mesenchymal cells (Gallo et al. 2007) and PHID/Proliferating Human Islet-Derived cells (Ouziel-Yahalom et al. 2006) which are able to generate POs on tissue culture plates.

Very recently, Van der Meulen and collaborators (van der Meulen et al. 2017) identified a population of human immature *β-cells*, originating by trans-differentiation of *α-cells* on a specialized pancreatic "neogenic niche" located at the islet periphery (van der Meulen et al. 2017). These cells express insulin but not other *β-cell* markers. They are transcriptionally immature and do not sense glucose (van der Meulen et al. 2017). This cell population represents an intermediate stage of α-cell trans-differentiation into conventional *β-cell* (van der Meulen et al. 2017). These data suggest the possibility to use this cell population for generartion of functional POs. However, additional studies are crucial to demonstrate their potential application in clinic.

The presence of high aldehyde dehydrogenase activity (ALDHhi)-pancreatic stem cells, have been also reported by Loomans et al. (2018). The ALDHhi-derived POs contain insulin-producing cells that also express pancreatic progenitor markers such as PDX1, PTF1A, CPA1, and MYC. Nevertheless, these POs appeared immature and did not restore normoglycemia in diabetic mice (Loomans et al. 2018).

10.2 Human Pancreatic Ductal-Cells (PANC-1) and Non-endocrine Pancreatic Cells

It was recently demonstrated that de-differentiation of human pancreatic ductal-cells (PANC-1) promotes their differentiation into endocrine pancreatic cells (Donadel et al. 2017). Formation of POs expressing pancreatic markers such as C-peptide, insulin, pancreatic and duodenal homeobox 1 (PDX-1), Nkx2.2, Nkx6.1 was obtained from PANC-1 cells. The expression of glucagon, somatostatin, and Glut-2 was also observed. This was associated with a decrease in the expression of PANC-1 markers (Cytokeratin-19, MUC-1, CA19–9), supporting their pancreatic commitment (Donadel et al. 2017). However, additional studies are required to evaluate their potential use to generate functional and safe POs.

10.3 Biliary Tree-Derived Islet Progenitors

Recently, a new source of islet precursors has been identified in a niche located within the human biliary trees (Wang et al. 2013). This niche is characterized by a network of cells with overlapping commitment (Wang et al. 2013). The biliary tree-progenitor cells showed a high proliferation rate when cultured *in vitro*, under specific conditions (Wang et al. 2013). Moreover, these cells were able to aggregate in POs, resembling neo-islets generated from other cell sources (Wang et al. 2013). The POs generated from these progenitors showed ultrastructural, electrophysiological and functional characteristics, including the response to glucose, similar to other POs. Furthermore, the derived neo-islets were able to improve glucose control in immune-compromised diabetic mice (Wang et al. 2013).

10.4 Hepatic Stem Cell Derived Islet-Like Structures

We have recently found that adult human liver stem-like cells (HLSC, Fig. 2a), which are clonogenic and express several mesenchymal and embryonic markers, such as SSEA4, Oct4, SOX2 and nanog, were able to spontaneously generate insulin-producing organoids (Navarro-tableros et al. 2018). This is not surprising since liver and pancreas share common embryonic origins. Charge-induced 3D aggregation of HLSC (Fig. 2b) promoted differentiation into viable insulin expressing cells (Fig. 2c and d) (Navarro-tableros et al. 2018) which contained heterogeneous secretory granules (Fig. 2e). These HLSC-derived POs were able to secrete human C-peptide in response to high glucose concentrations both in static (Fig. 2f) and dynamic conditions (Navarro-tableros et al. 2018). *In vitro* differentiated islet-like structures showed an immature endocrine gene profile but were able to further differentiate after *in vivo* implantation (Navarro-tableros et al. 2018). In fact, when transplanted in non-immunocompetent diabetic mice hyperglycemia was reverted and the human C-petide was detectable in mice (Navarro-tableros et al. 2018). As a proof of concept, after removal of implanted islet-like structures diabetes was restablished, suggesting that diabetes reversal was dependent on the transplanted islet-like structures generated by HLSC differentiation (Navarro-tableros et al. 2018).

11 Clinical Trials

MSC derived from different sources such as bone marrow umbilical cord and adipose tissue have been used in clinical trials. Although MSC did not displayed curative properties it is quite astonishing that a partial improvement of glycemia has been reported (Päth et al. 2019). To date CinicalTrials.gov listed a number of therapeutic approaches using different cell types. Among them over 850 have used MSC to target a broad variety of diseases and 60 of them for T1DM and T2DM. A trial investigating the islet graft survival and function of MSC co-transplantation has been closed (Päth et al. 2019). It has been demonstrated that MSC mediate immune tolerance enabling a partial recovery of residual *β-cell* mass and/or delaying the *β-cell* damage during the onset of T1DM (Päth et al. 2019). Some of these clinical trials using co-transplanted MSC were found to reduce the loss of islet graft, the HbA1c percentage and to increase the release of C-peptide compared to control patients during 2 years follow-up (Päth et al. 2019). However, currently the small number of patients recruited for the studies dictate the requirement of a more prolonged observational study to elucidate whether MSC could really delay the development of T1DM (Päth et al. 2019). Furthermore different strategies using ESC iPSC and MSCs are still under investigation (Welsch et al. 2018). Moreover, since cellular debris and fragments of proteins may be sufficient to prime the immune system using the encapsulation devices an effective encapsulation device should be generated to prevent activation of the immune response leading to graft rejection. It has been shown that Encaptra devices are able to normalize blood glucose in spontaneous NOD

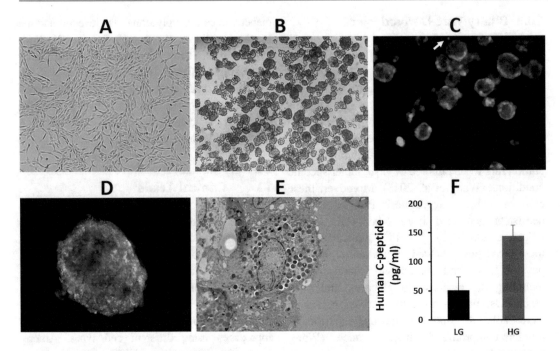

Fig. 2 Generation of POs from HLSC. (a) Non-differentiated HLSC showing their typical morphology. (b) Generation of HLSC-derived islet-like structures. (c) HLSC-ILS stained with fluorescein diacetate (FDA) and propidium iodide (PI) show viable (green) and dead cells (red, arrow). (d) Immnuofluorescence for human C-peptide (red) and Glucose transporter 2 (GLUT, green) in HLSC-ILS. (e) Transmission electron microscopy showing typical *β-cell* granules. (f) Glucose-stimulated C-peptide secretion in 2.8 mM glucose (LG) and 28 mM glucose (HG)

recipients (Faleo et al. 2016). In addition, in 2014 ViaCyte has started a phase I/II clinical trial (ClinicalTrials.gov identifier: Nbib2239354) currently ongoing. In 2017 a new Phase 1/2 clinical trial was started (STEP ONE) to test the PEC-Direct™ in patients with T1DM in San Diego (California USA) and Edmonton (Alberta Canada). More recently, the clinical trial was extended to Baltimore (Maryland USA) Minneapolis (Minnesota USA) Columbus (Ohio USA) and Vancouver (British Columbia Canada) to test the safety and efficacy of the ViaCyte's PEC-Encap (a.k.a. VC-01™) (ClinicalTrials.gov identifier: NCT03163511). The innovation consists in the use of a new open device that allows a direct vascularization of pancreatic progenitor cells (PEC-01) and avoids the use of immunosuppressive drugs. These studies are still ongoing but they could provide important information on safety and effectiveness for the use of stem cell-derived *β-cells* in humans. (Clinicaltrials.gov.)

The failure to generate feasible implantable devices to be applied in diabetic setting is still a challenge for many companies. This mainly depends on the fibrotic processes, the oxygenation and the diffusion of nutrients allowing a long-term PO viability.

12   Conclusions

In the last decades, the generation of insulin-secreting cells has gained particular attention. Different cell sources have been used and a number of studies have shown that functional insulin-secreting cells can be generated (Zhou and Melton 2018; Matsumoto et al. 2016; Ravassard et al. 2011; Boss et al. 2017; Clark et al. 1997; Scharfmann et al. 2014; Xie et al. 2016; Balboa et al. 2018; Wang et al. 2017; Russ et al. 2008; Avnit-Sagi et al. 2009; Russ et al. 2011; Russ et al. 2009; Joglekar et al. 2009; Klein et al. 2013;

Latreille et al. 2015; Carlotti et al. 2010; Zulewski et al. 2001; Gallo et al. 2007). Nevertheless, their clinical application is still missing.

Beeing organoids versatile for applications they have gained interest. In particular, the 3D conformation plays an important role in morphogenesis and tissue organization of pancreatic islets (Navarro-tableros et al. 2018; Kim et al. 2016a). Cell assembly in 3D structures facilitates cellular interactions, promotes morphogenesis and tissue organization, but also a fine modulation of hormone secretion (Orci et al. 1975; Kitasato et al. 1996; Kanno et al. 2002; Wojtusciszyn et al. 2008; Jo et al. 2009; Klee et al. 2011; Santos-Silva et al. 2012; Rorsman and Ashcroft 2018). Thereby, generation of POs, recapitulating the structure and function of native pancreatic islets, should be evaluated for long-term transplant viability. Moreover, organoid cultures are ideal to study stem cell-niche interactions in a 3D environment, and engineering POs could serve for drug studies (Scavuzzo et al. 2018).

Ideally, optimization would include feasible cell sources for a large production of functional POs. Thereby, generation of POs, mimicking the human native pancreatic islets in terms of morphology and function, is mandatory. Staminal cells isolated from a number of embryonic and adult tissues are currently used to generate functional POs (Zhou and Melton 2018; Balboa et al. 2018; Navarro-tableros et al. 2018; Kim et al. 2016a; Hinderer et al. 2016; Liu et al. 2018) (Fig. 3). Nevertheless, standardization of protocols for POs generation is still needed to achieve reproducibility and large scalability. Furthermore, efforts should be directed to generate mature POs, similar to native pancreatic islets, in self-regulation of glucose metabolism. The recent gene editing technology is a challenge to generate POs from different cell types and to educate cells obtained from T1DM patients in order to generate functional POs for autologous transplantation (Giménez et al. 2016; Ma et al. 2018).

The 3D bioreactor-based culture systems have demonstrated to improve the production of functional stem cell-derived β-like cells *in vitro* (Abu-Absi et al. 2014; Hoesli et al. 2009; Sharp and Vermette 2017; Gundersen et al. 2010)

(Fig. 3). Additionally, functional POs generated *in vitro*, were able to secrete insulin in response to glucose, and to restore the glycemic control when implanted in diabetic animals (Vaithilingam et al. 2011; Mason et al. 2009; Amer et al. 2015; Lee et al. 2006a; Jiang et al. 2017; An et al. 2018; Ludwig et al. 2012; Montazeri et al. 2016; Mahou et al. 2017; Faleo et al. 2016). Nevertheless, similarly to transplanted human pancreatic islets, oxygenation and activation of the host immune response are the main remaining hurdles.

To solve this problem, bioengineers and nanotechnologists have generated a number of new promising encapsulation approaches (Gálvez-Martín et al. 2017). Currently, nanotechnology research is focused on the generation of POs with a 3D morphology by using different polymers and natural ECM as encapsulating platforms (Gálvez-Martín et al. 2017) (Fig. 3). Pre-clinical studies have demonstrated their potential application to enhance PO vascularization (De Rosa et al. 2018; Schweicher et al. 2014; Trivedi et al. 2000; Sweet et al. 2008; Mahou et al. 2017; Guo et al. 2018; Skrzypek et al. 2018a, b; Groot Nibbelink et al. 2018; Stephens et al. 2018) and to improve the immune escape (Hinderer et al. 2016; Amer et al. 2015; Rios et al. 2016; Buitinga et al. 2013; Lee et al. 2004, 2006a, b, c, 2009; Jiang et al. 2017; Bowers et al. 2018; Liao et al. 2013; Andrades et al. 2008; Lou et al. 2017; Hamilton et al. 2017; Chaimov et al. 2017). Nevertheless, different clinical trials, many of which still ongoing, were not conclusive in confirming their safety for clinical application (CinicalTrials.gov).

A number of pre-clinical cell-based approaches have been also developed and each approach has contributed to solve the most relevant challenges. Indeed, the development of pre-vascularized or oxygenated POs represent the more promising and powerful tools for translational applications (Fig. 3). Finally, even though different clinical trials are currently ongoing to evaluate the potential application in diabetes, standardization of protocols for PO generation are still required to achieve reproducibility. Improvements should also include the feasibility and safety as well as the tuning of suitable protocols to obtain fully mature POs.

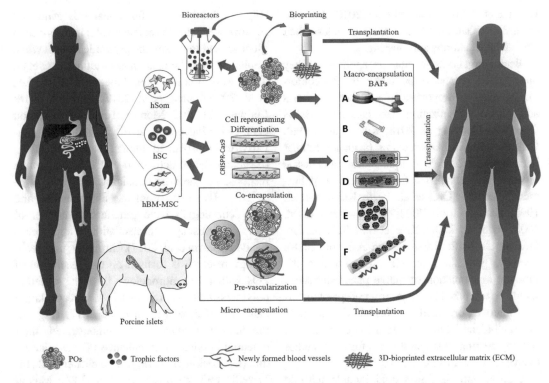

Bioreactors Bioprinting

Transplantation

hSom

hSC

hBM-MSC

CRISPR-Cas9

Cell reprograming Differentiation

Co-encapsulation

Pre-vascularization

Porcine islets

Micro-encapsulation

Macro-encapsulation BAPs

A

B

C

D

E

F

Transplantation

Transplantation

POs   Trophic factors   Newly formed blood vessels   3D-bioprinted extracellular matrix (ECM)

Fig. 3 Current strategies to generate 3D pancreatic organoids (POs)

A number of human cells including somatic (hSom), stem cells (hSC) and bone marrow-mesenchymal stem cells (hBM-MSC) as well as cells obtained from different embryonic or adult organs are currently used for their potential to generate *in vitro*, functional POs

In vitro cell commitment is promoted by the addition of specific trophic factors or by gene editing (CRISPR-Cas9)
Encapsulation allows oxygen, nutrients to pass through the porous membrane. Encapsulation also offers immuno-protection, promotes survival, differentiation, maturation and glucose-responsive insulin secretion. POs generated from human cells or from porcine islets (xenografts) can be macro-encapsulated and/or alginate-based micro-encapsulated to improve long-term survival.

Pre-vascularization or addition of oxygen generators are two potential approaches which could solve the reduced oxygen supply of transplanted POs

Bioartifical pancreas systems (BAPs) have been proposed to avoid the host immune attack against POs. The figure shows a schematic representation of some commercial BAP. βAir® Chamber System (**a**), Sernova's Cell Pouch System™ schematization (**b**), TheraCyte™ system schematization (**c**), VC-01TM (Pro-vascularized) System (**d**), Islets Sheet System (**e**) and Silk Fibroinc System (**f**)

3D bioprinting modalities are also used to generate ECM-containing POs. Natural or synthetic extracellular matrix (nECM, sECM) components are used for this purpose.

Bioreactors are proposed to large-scalability PO production.

Author Contributions V.N.T.: contributed to Manuscript writing and Figure preparation; Y.G.: contributed to manuscript writing and Figure preparation; M.F.B.: contributed to Manuscript editing; G.C.: contributed to Manuscript writing and editing.

Conflicts of Interest The Authors declare no conflict of interest. GC is a component of the Scientific Advisory Board of UNISYTE.

Funding This work has been supported by grants obtained by G.C. and M.F.B. from Unicyte AG.

References

Aamodt KI, Powers AC (2017) Signals in the pancreatic islet microenvironment influence β-cell proliferation. Diabetes Obes Metab. https://doi.org/10.1111/dom.13031

Abazari MF, Soleimanifar F, Nouri Aleagha M, Torabinejad S, Nasiri N, Khamisipour G, Amini Mahabadi J, Mahboudi H, Enderami SE, Saburi E, Hashemi J, Kehtari M (2018) PCL/PVA nanofibrous scaffold improve insulin-producing cells generation from human induced pluripotent stem cells. Gene. https://doi.org/10.1016/j.gene.2018.05.115

Abu-Absi S, Xu S, Graham H, Dalal N, Boyer M, Dave K (2014) Cell culture process operations for recombinant protein production. Adv Biochem Eng Biotechnol. https://doi.org/10.1007/10_2013_252

Afelik S, Rovira M (2017) Pancreatic β-cell regeneration: facultative or dedicated progenitors? Mol Cell Endocrinol. https://doi.org/10.1016/j.mce.2016.11.008

Affeld K, Schaller J, Wölken T, Krabatsch T, Kertzscher U (n.d.) Role of flow for the deposition of platelets. Biointerphases. https://doi.org/10.1116/1.4944383. Diventa 208 en la forma correctA

Agulnick AD, Ambruzs DM, Moorman MA, Bhoumik A, Cesario RM, Payne JK, Kelly JR, Haakmeester C, Srijemac R, Wilson AZ, Kerr J, Frazier MA, Kroon EJ, D'Amour KA (2015) Insulin-producing endocrine cells differentiated in vitro from human embryonic stem cells function in macroencapsulation devices in vivo. Stem Cells Transl Med. https://doi.org/10.5966/sctm.2015-0079

Alismail H, Jin S (2014) Microenvironmental stimuli for proliferation of functional islet β-cells. Cell Biosci. https://doi.org/10.1186/2045-3701-4-12

Amer LD, Holtzinger A, Keller G, Mahoney MJ, Bryant SJ (2015) Enzymatically degradable poly(ethylene glycol) hydrogels for the 3D culture and release of human embryonic stem cell derived pancreatic precursor cell aggregates. Acta Biomater. https://doi.org/10.1016/j.actbio.2015.04.013

An D, Chiu A, Flanders JA, Song W, Shou D, Lu Y-C, Grunnet LG, Winkel L, Ingvorsen C, Christophersen NS, Fels JJ, Sand FW, Ji Y, Qi L, Pardo Y, Luo D, Silberstein M, Fan J, Ma M (2018) Designing a retrievable and scalable cell encapsulation device for potential treatment of type 1 diabetes. Proc Natl Acad Sci. https://doi.org/10.1073/pnas.1708806115

Andrades P, Asiedu CK, Gansuvd B, Inusah S, Goodwin KJ, Deckard LA, Jargal U, Thomas JM (2008) Pancreatic islet isolation variables in non-human primates (rhesus macaques). Diabetologia. https://doi.org/10.1007/s00125-008-1030-z

Anitua E, Nurden P, Prado R, Nurden AT, Padilla S (2019) Autologous fibrin scaffolds: when platelet- and plasma-derived biomolecules meet fibrin. Biomaterials. https://doi.org/10.1016/j.biomaterials.2018.11.029

Avnit-Sagi T, Kantorovich L, Kredo-Russo S, Hornstein E, Walker MD (2009) The promoter of the pri-miR-375 gene directs expression selectively to the endocrine pancreas. PLoS One. https://doi.org/10.1371/journal.pone.0005033

Azadi SA, Vasheghani-Farahani E, Hashemi-Najafbabadi-S, Godini A (2016) Co-encapsulation of pancreatic islets and pentoxifylline in alginate-based microcapsules with enhanced immunosuppressive effects. Prog Biomater. https://doi.org/10.1007/s40204-016-0049-3

Badylak SF, Taylor D, Uygun K, Sabetkish S, Kajbafzadeh AM, Sabetkish N, Khorramirouz R, Akbarzadeh A, Seyedian SL, Pasalar P, Orangian S,

Beigi RSH, Aryan Z, Akbari H, Tavangar SM (2011) Whole-organ tissue engineering: decellularization and recellularization of three-dimensional matrix scaffolds. Annu Rev Biomed Eng. https://doi.org/10.1146/annurev-bioeng-071910-124743

Balboa D, Saarimäki-vire J, Otonkoski T (2018) Human pluripotent stem cells for the modelling of pancreatic beta-cell pathology. Stem Cells. https://doi.org/10.1002/stem.2913

Barati G, Nadri S, Hajian R, Rahmani A, Mostafavi H, Mortazavi Y, Taromchi AH (2018) Differentiation of microfluidic-encapsulated trabecular meshwork mesenchymal stem cells into insulin producing cells and their impact on diabetic rats. J Cell Physiol. https://doi.org/10.1002/jcp.27426

Barkai U, Weir GC, Colton CK, Ludwig B, Bornstein SR, Brendel MD, Neufeld T, Bremer C, Leon A, Evron Y, Yavriyants K, Azarov D, Zimermann B, Maimon S, Shabtay N, Balyura M, Rozenshtein T, Vardi P, Bloch K, De Vos P, Rotem A (2013) Enhanced oxygen supply improves islet viability in a new bioartificial pancreas. Cell Transplant. https://doi.org/10.3727/096368912X657341

Bellin MD, Barton FB, Heitman A, Harmon JV, Kandaswamy R, Balamurugan AN, Sutherland DER, Alejandro R, Hering BJ (2012) Potent induction immunotherapy promotes long-term insulin independence after islet transplantation in type 1 diabetes. Am J Transplant. https://doi.org/10.1111/j.1600-6143.2011.03977.x

Bergert H, Knoch KP, Meisterfeld R, Jäger M, Ouwendijk J, Kersting S, Saeger HD, Solimena M (2005) Effect of oxygenated perfluorocarbons on isolated rat pancreatic islets in culture. Cell Transplant. https://doi.org/10.3727/000000005783982873

Berman DM, Molano RD, Fotino C, Ulissi U, Gimeno J, Mendez AJ, Kenyon NM, Kenyon NS, Andrews DM, Ricordi C, Pileggi A (2016) Bioengineering the endocrine pancreas: intraomental islet transplantation within a biologic resorbable scaffold. Diabetes. https://doi.org/10.2337/db15-1525

Bettinger CJ, Langer R, Borenstein JT (2009) Engineering substrate topography at the micro- and nanoscale to control cell function. Angew Chem Int Ed Engl. https://doi.org/10.1002/anie.200805179

Blomeier H, Zhang X, Rives C, Brissova M, Hughes E, Baker M, Powers AC, Kaufman DB, Shea LD, Lowe WL (2006) Polymer scaffolds as synthetic microenvironments for extrahepatic islet transplantation. Transplantation. https://doi.org/10.1097/01.tp.0000231708.19937.21

Blyszczuk P, Czyz J, Kania G, Wagner M, Roll U, St-Onge L, Wobus AM (2003) Expression of Pax4 in embryonic stem cells promotes differentiation of nestin-positive progenitor and insulin-producing cells. Proc Natl Acad Sci U S A. https://doi.org/10.1073/pnas.0237371100

Boettler T, Schneider D, Cheng Y, Kadoya K, Brandon EP, Martinson L, Von Herrath M (2016) Pancreatic tissue transplanted in TheraCyte ™ encapsulation devices is protected and prevents hyperglycemia in a mouse model of immune-mediated diabetes. Cell Transplant. https://doi.org/10.3727/096368915X688939

Boss C, De Marchi U, Hermant A, Conrad M, Sizzano F, Palini A, Wiederkehr A, Bouche N (2017) Encapsulation of insulin-secreting cells expressing a genetically encoded fluorescent calcium Indicator for cell-based sensing in vivo. Adv Healthc Mater. https://doi.org/10.1002/adhm.201600869

Bowers DT, Olingy CE, Chhabra P, Langman L, Merrill PH, Linhart RS, Tanes ML, Lin D, Brayman KL, Botchwey EA (2018) An engineered macroencapsulation membrane releasing FTY720 to precondition pancreatic islet transplantation. J Biomed Mater Res B Appl Biomater. https://doi.org/10.1002/jbm.b.33862

Breger JC, Fisher B, Samy R, Pollack S, Wang NS, Isayeva I (2015) Synthesis of "click" alginate hydrogel capsules and comparison of their stability, water swelling, and diffusion properties with that of Ca^{+2} crosslinked alginate capsules. J Biomed Mater Res B Appl Biomater. https://doi.org/10.1002/jbm.b.33282

Brennan DC, Kopetskie HA, Sayre PH, Alejandro R, Cagliero E, Shapiro AMJ, Goldstein JS, DesMarais MR, Booher S, Bianchine PJ (2016) Long-term follow-up of the Edmonton protocol of islet transplantation in the United States. Am J Transplant. https://doi.org/10.1111/ajt.13458

Brennand K, Huangfu D, Melton D (2007) All β cells contribute equally to islet growth and maintenance. PLoS Biol. https://doi.org/10.1371/journal.pbio.0050163

Brown BN, Badylak SF (2015) Extracellular matrix as an inductive scaffold for functional tissue reconstruction. Transl Res. https://doi.org/10.1016/j.trsl.2013.11.003

Bruni A, Gala-Lopez B, Pepper AR, Abualhassan NS, James Shapiro AM (2014) Islet cell transplantation for the treatment of type 1 diabetes: recent advances and future challenges. Diabetes Metab Syndr Obes Targets Ther. https://doi.org/10.2147/DMSO.S50789

Buchwald P, Tamayo-Garcia A, Manzoli V, Tomei AA, Stabler CL (2018) Glucose-stimulated insulin release: parallel perifusion studies of free and hydrogel encapsulated human pancreatic islets. Biotechnol Bioeng. https://doi.org/10.1002/bit.26442

Buitinga M, Truckenmüller R, Engelse MA, Moroni L, Ten Hoopen HWM, van Blitterswijk CA, de Koning EJP, van Apeldoorn AA, Karperien M (2013) Microwell scaffolds for the Extrahepatic transplantation of islets of Langerhans. PLoS One. https://doi.org/10.1371/journal.pone.0064772

Buitinga M, Assen F, Hanegraaf M, Wieringa P, Hilderink J, Moroni L, Truckenmüller R, van Blitterswijk C, Römer GW, Carlotti F, de Koning E, Karperien M, van Apeldoorn A (2017) Microfabricated scaffolds lead to efficient remission of diabetes in mice. Biomaterials. https://doi.org/10.1016/j.biomaterials.2017.03.031

Cabric S, Sanchez J, Lundgren T, Foss A, Felldin M, Källen R, Salmela K, Tibell A, Tufveson G,

Larsson R, Korsgren O, Nilsson B (2007) Islet surface heparinization prevents the instant blood-mediated inflammatory reaction in islet transplantation. Diabetes. https://doi.org/10.2337/db07-0358

Cabric S, Sanchez J, Johansson U, Larsson R, Nilsson B, Korsgren O, Magnusson PU (2010) Anchoring of vascular endothelial growth factor to surface-immobilized heparin on pancreatic islets: implications for stimulating islet angiogenesis. Tissue Eng Part A. https://doi.org/10.1089/ten.tea.2009.0429

Calafiore R, Basta G, Luca G, Lemmi A, Montanucci MP, Calabrese G, Racanicchi L, Mancuso F, Brunetti P (2006) Microencapsulated pancreatic islet allografts into nonimmunosuppressed patients with type 1 diabetes. Diabetes Care. https://doi.org/10.2337/diacare.29.01.06.dc05-1270

Cameron D, Hushen J, Nazian S (2001) Formation of insulin-secreting, Sertoli-enriched tissue constructs by microgravity coculture of isolated pig islets and rat sertoli cells. In Vitr Cell Dev Biol Anim 37:490–498

Cardona K, Korbutt GS, Milas Z, Lyon J, Cano J, Jiang W, Bello-Laborn H, Hacquoil B, Strobert E, Gangappa S, Weber CJ, Pearson TC, Rajotte RV, Larsen CP (2006) Long-term survival of neonatal porcine islets in non-human primates by targeting costimulation pathways. Nat Med. https://doi.org/10.1038/nm1375

Carlotti F, Zaldumbide A, Loomans CJ, van Rossenberg E, Engelse M, de Koning EJ, Hoeben RC (2010) Isolated human islets contain a distinct population of mesenchymal stem cells. Islets. https://doi.org/10.4161/isl.2.3.11449

Carlsson PO, Palm F, Andersson A, Liss P (2001) Markedly decreased oxygen tension in transplanted rat pancreatic islets irrespective of the implantation site. Diabetes. https://doi.org/10.2337/diabetes.50.3.489

Carlsson PO, Espes D, Sedigh A, Rotem A, Zimerman B, Grinberg H, Goldman T, Barkai U, Avni Y, Westermark GT, Carlbom L, Ahlström H, Eriksson O, Olerud J, Korsgren O (2018) Transplantation of macroencapsulated human islets within the bioartificial pancreas βAir to patients with type 1 diabetes mellitus. Am J Transplant. https://doi.org/10.1111/ajt.14642

Celikkin N, Rinoldi C, Costantini M, Trombetta M, Rainer A, Święszkowski W (2017) Naturally derived proteins and glycosaminoglycan scaffolds for tissue engineering applications. Mater Sci Eng C Mater Biol Appl. https://doi.org/10.1016/j.msec.2017.04.016

Chaimov D, Baruch L, Krishtul S, Meivar-levy I, Ferber S, Machluf M (2017) Innovative encapsulation platform based on pancreatic extracellular matrix achieve substantial insulin delivery. J Control Release. https://doi.org/10.1016/j.jconrel.2016.07.045

Chawla M, Bodnar CA, Sen A, Kallos MS, Behie LA (2006) Production of islet-like structures from neonatal porcine pancreatic tissue in suspension bioreactors. Biotechnol Prog. https://doi.org/10.1021/bp050261i

Chawla K, Bin YT, Liao SW, Guan Z (2011) Biodegradable and biocompatible synthetic saccharide-peptide hydrogels for three-dimensional stem cell culture. Biomacromolecules. https://doi.org/10.1021/bm100980w

Chayosumrit M, Tuch B, Sidhu K (2010) Alginate microcapsule for propagation and directed differentiation of hESCs to definitive endoderm. Biomaterials. https://doi.org/10.1016/j.biomaterials.2009.09.071

Chen T, Yuan J, Duncanson S, Hibert ML, Kodish BC, Mylavaganam G, Maker M, Li H, Sremac M, Santosuosso M, Forbes B, Kashiwagi S, Cao J, Lei J, Thomas M, Hartono C, Sachs D, Markmann J, Sambanis A, Poznansky MC (2015) Alginate encapsulant incorporating CXCL12 supports long-term allo- and xenoislet transplantation without systemic immune suppression. Am J Transplant. https://doi.org/10.1111/ajt.13049

Cheng JYC, Raghunath M, Whitelock J, Poole-Warren L (2011) Matrix components and scaffolds for sustained islet function. Tissue Eng Part B Rev. https://doi.org/10.1089/ten.teb.2011.0004

Clark SA, Quaade C, Constandy H, Hansen P, Halban P, Ferber S, Newgard CB, Normington K (1997) Novel insulinoma cell lines produced by iterative engineering of GLUT2, glucokinase, and human insulin expression. Diabetes. https://doi.org/10.2337/diab.46.6.958

Croon AC, Karlsson R, Bergström C, Björklund E, Möller C, Tydén L, Tibell A (2003) Lack of donors limits the use of islet transplantation as treatment for diabetes. Transplant Proc 35(2):764

Crossan C, Mourad NI, Smith K, Gianello P, Scobie L (2018) Assessment of porcine endogenous retrovirus transmission across an alginate barrier used for the encapsulation of porcine islets. Xenotransplantation. https://doi.org/10.1111/xen.12409

Daoud J, Rosenberg L, Tabrizian M (2010) Pancreatic islet culture and preservation strategies: advances, challenges, and future outlook. Cell Transplant. https://doi.org/10.3727/096368910X515872

Datar A, Joshi P, Lee MY (2015) Biocompatible hydrogels for microarray cell printing and encapsulation. Biosensors (Basel). https://doi.org/10.3390/bios5040647

Davani B, Ikonomou L, Raaka BM, Geras-Raaka E, Morton RA, Marcus Samuels B, Gershengorn MC (2007) Human islet-derived precursor cells are mesenchymal stromal cells that differentiate and mature to hormone-expressing cells in vivo. Stem Cells. https://doi.org/10.1634/stemcells.2007-0323

Davis NE, Beenken-Rothkopf LN, Mirsoian A, Kojic N, Kaplan DL, Barron AE, Fontaine MJ (2012) Enhanced function of pancreatic islets co-encapsulated with ECM proteins and mesenchymal stromal cells in a silk hydrogel. Biomaterials. https://doi.org/10.1016/j.biomaterials.2012.06.015

De Mesmaeker I, Robert T, Suenens KG, Stangé GM, Van Hulle F, Ling Z, Tomme P, Jacobs-Tulleneers-Thevissen D, Keymeulen B, Pipeleers DG (2018) Increase functional β-cell mass in subcutaneous alginate capsules with porcine prenatal islet cells but loss with human adult islet cells. Diabetes. https://doi.org/10.2337/db18-0709

De Rosa L, Di Stasi R, D'Andrea LD (2018) Pro-angiogenic peptides in biomedicine. Arch Biochem Biophys. https://doi.org/10.1016/j.abb.2018.10.010

de Vos P, Faas MM, Strand B, Calafiore R (2006) Alginate-based microcapsules for immunoisolation of pancreatic islets. Biomaterials 27(32):5603–5617

De Vos P, Lazarjani HA, Poncelet D, Faas MM (2014) Polymers in cell encapsulation from an enveloped cell perspective. Adv Drug Deliv Rev. https://doi.org/10.1016/j.addr.2013.11.005

De Vos P, Smink AM, Paredes G, Lakey JRT, Kuipers J, Giepmans BNG, De Haan BJ, Faas MM (2016) Enzymes for pancreatic islet isolation impact chemokine-production and polarization of insulin-producing β-cells with reduced functional survival of immunoisolated rat islet-allografts as a consequence. PLoS One. https://doi.org/10.1371/journal.pone.0147992

Desai T, Shea LD (2017) Advances in islet encapsulation technologies. Nat Rev Drug Discov. https://doi.org/10.1038/nrd.2016.232

Desai NM, Goss JA, Deng S, Wolf BA, Markmann E, Palanjian M, Shock AP, Feliciano S, Brunicardi FC, Barker CF, Naji A, Markmann JF (2003) Elevated portal vein drug levels of sirolimus and tacrolimus in islet transplant recipients: local immunosuppression or islet toxicity. Transplantation 76:1623–1625. https://doi.org/10.1097/01.TP.0000081043.23751.81

Dimatteo R, Darling NJ, Segura T (2018) In situ forming injectable hydrogels for drug delivery and wound repair. Adv Drug Deliv Rev. https://doi.org/10.1016/j.addr.2018.03.007

Dolgin E (2016) Diabetes: encapsulating the problem. Nature. https://doi.org/10.1038/540S60a

Donadel G, Pastore D, Della-Morte D, Capuani B, Lombardo MF, Pacifici F, Bugliani M, Grieco FA, Marchetti P, Lauro D (2017) FGF-2b and h-PL transform duct and non-endocrine human pancreatic cells into endocrine insulin secreting cells by modulating differentiating genes. Int J Mol Sci. https://doi.org/10.3390/ijms18112234

Dor Y, Brown J, Martinez OI, Melton DA (2004) Adult pancreatic β-cells are formed by self-duplication rather than stem-cell differentiation. Nature. https://doi.org/10.1038/nature02520

Dufrane D, Goebbels RM, Saliez A, Guiot Y, Gianello P (2006) Six-month survival of microencapsulated pig islets and alginate biocompatibility in primates: proof of concept. Transplantation. https://doi.org/10.1097/01.tp.0000208610.75997.20

Dufrane D, Goebbels RM, Gianello P (2010) Alginate macroencapsulation of pig islets allows correction of streptozotocin-induced diabetes in primates up to 6 months without immunosuppression. Transplantation. https://doi.org/10.1097/TP.0b013e3181f6e267

Elliott RB, Escobar L, Tan PLJ, Garkavenko O, Calafiore R, Basta P, Vasconcellos AV, Emerich DF, Thanos C, Bambra C (2005) Intraperitoneal alginate-encapsulated neonatal porcine islets in a placebo-controlled study with 16 diabetic cynomolgus primates. Transplant Proc 37(8):3505–3508

Ellis C, Ramzy A, Kieffer TJ (2017) Regenerative medicine and cell-based approaches to restore pancreatic function. Nat Rev Gastroenterol Hepatol. https://doi.org/10.1038/nrgastro.2017.93

Enderami SE, Mortazavi Y, Soleimani M, Nadri S, Biglari A, Mansour RN (2017) Generation of insulin-producing cells from human-induced pluripotent stem cells using a stepwise differentiation protocol optimized with platelet-rich plasma. J Cell Physiol. https://doi.org/10.1002/jcp.25721

Evron Y, Zimermann B, Ludwig B, Barkai U, Colton CK, Weir GC, Arieli B, Maimon S, Shalev N, Yavriyants K, Goldman T, Gendler Z, Eizen L, Vardi P, Bloch K, Barthel A, Bornstein SR, Rotem A (2015) Oxygen supply by photosynthesis to an implantable islet cell device. Horm Metab Res. https://doi.org/10.1055/s-0034-1394375

Faleo G, Lee K, Nguyen V, Tang Q (2016) Assessment of immune isolation of allogeneic mouse pancreatic progenitor cells by a macroencapsulation device. Transplantation 100:1211–1218. https://doi.org/10.1097/TP.0000000000001146

Faleo G, Russ HA, Wisel S, Parent AV, Nguyen V, Nair GG, Freise JE, Villanueva KE, Szot GL, Hebrok M, Tang Q (2017) Mitigating ischemic injury of stem cell-derived insulin-producing cells after transplant. Stem Cell Rep. https://doi.org/10.1016/j.stemcr.2017.07.012

Farina M, Ballerini A, Fraga DW, Nicolov E, Hogan M, Demarchi D, Scaglione F, Sabek OM, Horner P, Thekkedath U, Gaber OA, Grattoni A (2017) 3D printed vascularized device for subcutaneous transplantation of human islets. Biotechnol J. https://doi.org/10.1002/biot.201700169

Farina M, Chua CYX, Ballerini A, Thekkedath U, Alexander JF, Rhudy JR, Torchio G, Fraga D, Pathak RR, Villanueva M, Shin CS, Niles JA, Sesana R, Demarchi D, Sikora AG, Acharya GS, Gaber AO, Nichols JE, Grattoni A (2018) Transcutaneously refillable, 3D-printed biopolymeric encapsulation system for the transplantation of endocrine cells. Biomaterials. https://doi.org/10.1016/j.biomaterials.2018.05.047

Formo K, Cho CHH, Vallier L, Strand BL (2015) Culture of hESC-derived pancreatic progenitors in alginate-based scaffolds. J Biomed Mater Res A. https://doi.org/10.1002/jbm.a.35507

Foster JL, Williams G, Williams LJ, Tuch BE (2007) Differentiation of transplanted microencapsulated fetal pancreatic cells. Transplantation. https://doi.org/10.1097/01.tp.0000264555.46417.7d

Gallo R, Gambelli F, Gava B, Sasdelli F, Tellone V, Masini M, Marchetti P, Dotta F, Sorrentino V (2007) Generation and expansion of multipotent mesenchymal progenitor cells from cultured human pancreatic islets. Cell Death Differ. https://doi.org/10.1038/sj.cdd.4402199

Gálvez-Martín P, Martin JM, Ruiz AM, Clares B (2017) Encapsulation in cell therapy: methodologies, materials, and clinical applications. Curr Pharm Biotechnol. https://doi.org/10.2174/1389201018666170502113252

Gerace D, Martiniello-Wilks R, Nassif NT, Lal S, Steptoe R, Simpson AM (2017) CRISPR-targeted genome editing of mesenchymal stem cell-derived therapies for type 1 diabetes: a path to clinical success? Stem Cell Res Ther. https://doi.org/10.1186/s13287-017-0511-8

Giménez CA, Ielpi M, Mutto A, Grosembacher L, Argibay P, Pereyra-Bonnet F (2016) CRISPR-on system for the activation of the endogenous human INS gene. Gene Ther. https://doi.org/10.1038/gt.2016.28

Goh F, Gross JD, Simpson NE, Sambanis A (2010) Limited beneficial effects of perfluorocarbon emulsions on encapsulated cells in culture: experimental and modeling studies. J Biotechnol. https://doi.org/10.1016/j.jbiotec.2010.08.013

Goh SK, Bertera S, Olsen P, Candiello JE, Halfter W, Uechi G, Balasubramani M, Johnson SA, Sicari BM, Kollar E, Badylak SF, Banerjee I (2013) Perfusion-decellularized pancreas as a natural 3D scaffold for pancreatic tissue and whole organ engineering. Biomaterials. https://doi.org/10.1016/j.biomaterials.2013.05.066

Groot Nibbelink M, Skrzypek K, Karbaat L, Both S, Plass J, Klomphaar B, van Lente J, Henke S, Karperien M, Stamatialis D, van Apeldoorn A (2018) An important step towards a prevascularized islet microencapsulation device: in vivo prevascularization by combination of mesenchymal stem cells on micropatterned membranes. J Mater Sci Mater Med. https://doi.org/10.1007/s10856-018-6178-6

Guilak F, Cohen DM, Estes BT, Gimble JM, Liedtke W, Chen CS (2009) Control of stem cell fate by physical interactions with the extracellular matrix. Cell Stem Cell. https://doi.org/10.1016/j.stem.2009.06.016

Gundersen SI, Chen G, Powell HM, Palmer AF (2010) Hemoglobin regulates the metabolic and synthetic function of rat insulinoma cells cultured in a hollow fiber bioreactor. Biotechnol Bioeng. https://doi.org/10.1002/bit.22830

Guo Y, Wu C, Xu L, Xu Y, Xiaohong L, Hui Z, Jingjing L, Lu Y, Wang Z (2018) Vascularization of pancreatic decellularized scaffold with endothelial progenitor cells. J Artif Organs. https://doi.org/10.1007/s10047-018-1017-6

Guruswamy Damodaran R, Vermette P (2018a) Tissue and organ decellularization in regenerative medicine. Biotechnol Prog. https://doi.org/10.1002/btpr.2699

Guruswamy Damodaran R, Vermette P (2018b) Decellularized pancreas as a native extracellular matrix scaffold for pancreatic islet seeding and culture. J Tissue Eng Regen Med. https://doi.org/10.1002/term.2655

Hamilton DC, Shih HH, Schubert RA, Michie SA, Staats PN, Kaplan DL, Fontaine MJ (2017) A silk-based encapsulation platform for pancreatic islet transplantation improves islet function in vivo. J Tissue Eng Regen Med. https://doi.org/10.1002/term.1990

Haque MR, Kim J, Park H, Lee HS, Lee KW, Al-Hilal TA, Jeong JH, Ahn CH, Lee DS, Kim SJ, Byun Y (2017) Xenotransplantation of layer-by-layer encapsulated non-human primate islets with a specified immunosuppressive drug protocol. J Control Release. https://doi.org/10.1016/j.jconrel.2017.04.021

Hashemi J, Pasalar P, Soleimani M, Arefian E, Khorramirouz R, Akbarzadeh A, Ghorbani F, Enderami SE, Kajbafzadeh AM (2018) Decellularized pancreas matrix scaffolds for tissue engineering using ductal or arterial catheterization. Cells Tissues Organs. https://doi.org/10.1159/000487230

He L, Ahmad M, Perrimon N (2018) Mechanosensitive channels and their functions in stem cell differentiation. Exp Cell Res. S0014482718311194

Henriksnäs J, Lau J, Zang G, Berggren PO, Köhler M, Carlsson PO (2012) Markedly decreased blood perfusion of pancreatic islets transplanted intraportally into the liver: disruption of islet integrity necessary for islet revascularization. Diabetes. https://doi.org/10.2337/db10-0895

Hering BJ, Wijkstrom M, Graham ML, Hårdstedt M, Aasheim TC, Jie T, Ansite JD, Nakano M, Cheng J, Li W, Moran K, Christians U, Finnegan C, Mills CD, Sutherland DE, Bansal-Pakala P, Murtaugh MP, Kirchhof N, Schuurman HJ (2006) Prolonged diabetes reversal after intraportal xenotransplantation of wild-type porcine islets in immunosuppressed nonhuman primates. Nat Med. https://doi.org/10.1038/nm1369

Hillberg AL, Oudshoorn M, Lam JBB, Kathirgamanathan K (2015) Encapsulation of porcine pancreatic islets within an immunoprotective capsule comprising methacrylated glycol chitosan and alginate. J Biomed Mater Res Appl Biomater. https://doi.org/10.1002/jbm.b.33185

Hinderer S, Layland SL, Schenke-Layland K (2016) ECM and ECM-like materials – biomaterials for applications in regenerative medicine and cancer therapy. Adv Drug Deliv Rev. https://doi.org/10.1016/j.addr.2015.11.019

Hoesli CA, Luu M, Piret JM (2009) A novel alginate hollow fiber bioreactor process for cellular therapy applications. Biotechnol Prog. https://doi.org/10.1002/btpr.260

Hosgood SA, Nicholson ML (2010) The role of perfluorocarbon in organ preservation. Transplantation. https://doi.org/10.1097/TP.0b013e3181da6064

Huang X, Zhang X, Wang X, Wang C, Tang B (2012) Microenvironment of alginate-based microcapsules for cell culture and tissue engineering. J Biosci Bioeng. https://doi.org/10.1016/j.jbiosc.2012.02.024

Huang YB, Mei J, Yu Y, Ding Y, Xia W, Yue T, Chen W, Zhou MT, Yang YJ (2018a) Comparative Decellularization and Recellularization of Normal versus Streptozotocin-induced diabetes mellitus rat pancreas. Artif Organs. https://doi.org/10.1111/aor.13353

Huang YB, Mei J, Yu Y, Ding Y, Xia W, Yue T, Chen W, Zhou MT, Yang YJ (2018b) Comparative decellularization and recellularization of normal versus streptozotocin-induced diabetes mellitus rat pancreas. Artif Organs. https://doi.org/10.1111/aor.13353

Hwa AJ, Weir GC (2018) Transplantation of macroencapsulated insulin-producing cells. Curr Diab Rep. https://doi.org/10.1007/s11892-018-1028-y

Im B, Jeong J, Haque MR, Yun D, Ahn C, Eun J, Byun Y (2013) Biomaterials the effects of 8-arm-PEG-catechol/heparin shielding system and immunosuppressive drug, FK506 on the survival of intraportally allotransplanted islets. Biomaterials. https://doi.org/10.1016/j.biomaterials.2012.11.028

Irving-Rodgers HF, Choong FJ, Hummitzsch K, Parish CR, Rodgers RJ, Simeonovic CJ (2014) Pancreatic Islet basement membrane loss and remodeling after mouse islet isolation and transplantation: impact for allograft rejection. Cell Transplant. https://doi.org/10.3727/096368912X659880

Itkin-Ansari P, Geron I, Hao E, Demeterco C, Tyrberg B, Levine F (2003) Cell-based therapies for diabetes: progress towards a transplantable human β cell line. In: Ann N Y Acad Sci 1005(1): 138–147

Itoh T, Iwahashi S, Shimoda M, Chujo D, Takita M, Sorelle JA, Naziruddin B, Levy MF, Matsumoto S (2011) High-mobility group box 1 expressions in hypoxia-induced damaged mouse islets. Transplant Proc. https://doi.org/10.1016/j.transproceed.2011.09.100

Iwata H, Arima Y, Tsutsui Y (2018) Design of bioartificial pancreases from the standpoint of oxygen supply. Artif Organs. https://doi.org/10.1111/aor.13106

Jabs N, Franklin L, Brenner MB, Gromada J, Ferrara N, Wollheim CB, Lammert E (2008) Reduced insulin secretion and content in VEGF-A deficient mouse pancreatic islets. Exp Clin Endocrinol Diabetes. https://doi.org/10.1055/s-2008-1081486

Jacobs-Tulleneers-Thevissen D, Chintinne M, Ling Z, Gillard P, Schoonjans L, Delvaux G, Strand BL, Gorus F, Keymeulen B, Pipeleers D (2013) Sustained function of alginate-encapsulated human islet cell implants in the peritoneal cavity of mice leading to a pilot study in a type 1 diabetic patient. Diabetologia. https://doi.org/10.1007/s00125-013-2906-0

Jansson L, Barbu A, Bodin B, Drott CJ, Espes D, Gao X, Grapensparr L, Källskog Ö, Lau J, Liljebäck H, Palm F, Quach M, Sandberg M, Strömberg V, Ullsten S, Carlsson PO (2016) Pancreatic islet blood flow and its measurement. Ups J Med Sci. https://doi.org/10.3109/03009734.2016.1164769

Jeong JH, Yook S, Hwang JW, Jung MJ, Moon HT, Lee DY, Byun Y (2013) Synergistic effect of surface modification with poly(ethylene glycol) and immunosuppressants on repetitive pancreatic islet transplantation into antecedently sensitized rat. Transplant Proc. https://doi.org/10.1016/j.transproceed.2012.02.028

Jiang K, Weaver JD, Li Y, Chen X, Liang J, Stabler CL (2017) Local release of dexamethasone from macroporous scaffolds accelerates islet transplant engraftment by promotion of anti-inflammatory M2 macrophages. Biomaterials. https://doi.org/10.1016/j.biomaterials.2016.11.004

Jo J, Choi MY, Koh DS (2009) Beneficial effects of intercellular interactions between pancreatic islet cells in blood glucose regulation. J Theor Biol. https://doi.org/10.1016/j.jtbi.2008.12.005

Jo EH, Hwang YH, Lee DY (2015) Encapsulation of pancreatic islet with HMGB1 fragment for attenuating inflammation. Biomater Res. https://doi.org/10.1186/s40824-015-0042-2

Joglekar MV, Hardikar AA (2012) Isolation, expansion, and characterization of human islet-derived progenitor cells. Methods Mol Biol. https://doi.org/10.1007/978-1-61779-815-3_21

Joglekar MV, Joglekar VM, Hardikar AA (2009) Expression of islet-specific microRNAs during human pancreatic development. Gene Expr Patterns. https://doi.org/10.1016/j.gep.2008.10.001

Jones JR, Zhang SC (2016) Engineering human cells and tissues through pluripotent stem cells. Curr Opin Biotechnol. https://doi.org/10.1016/j.copbio.2016.03.010

Kang AR, Park JS, Ju J, Jeong GS, Lee SH (2014) Cell encapsulation via microtechnologies. Biomaterials. https://doi.org/10.1016/j.biomaterials.2013.12.073

Kanno T, Göpel SO, Rorsman P, Wakui M (2002) Cellular function in multicellular system for hormone-secretion: electrophysiological aspect of studies on α-, β- and δ-cells of the pancreatic islet. Neurosci Res 42(2):79–90

Karnieli O, Izhar-Prato Y, Bulvik S, Efrat S (2007) Generation of insulin-producing cells from human bone marrow mesenchymal stem cells by genetic manipulation. Stem Cells. https://doi.org/10.1634/stemcells.2007-0164

Katsuki Y, Yagi H, Okitsu T, Kitago M, Tajima K, Kadota Y, Hibi T, Abe Y, Shinoda M, Itano O, Takeuchi S, Kitagawa Y (2016) Endocrine pancreas engineered using porcine islets and partial pancreatic scaffolds. Pancreatology. https://doi.org/10.1016/j.pan.2016.06.007

Kilimnik G, Zhao B, Jo J, Periwal V, Witkowski P, Misawa R, Hara M (2011) Altered islet composition and disproportionate loss of large islets in patients with type 2 diabetes. PLoS One. https://doi.org/10.1371/journal.pone.0027445

Kim J, Breunig MJ, Escalante LE, Bhatia N, Denu RA, Dollar BA, Stein AP, Hanson SE, Naderi N, Radek J, Haughy D, Bloom DD, Assadi-Porter FM, Hematti P (2012) Biologic and immunomodulatory properties of mesenchymal stromal cells derived from human pancreatic islets. Cytotherapy. https://doi.org/10.3109/14653249.2012.684376

Kim Y, Kim H, Ko UH, Oh Y, Lim A, Sohn JW, Shin JH, Kim H, Han YM (2016a) Islet-like organoids derived from human pluripotent stem cells efficiently function in the glucose responsiveness in vitro and in vivo. Sci Rep. https://doi.org/10.1038/srep35145

Kim JH, Kim HW, Cha KJ, Han J, Jang YJ, Kim DS, Kim JH (2016b) Nanotopography promotes pancreatic differentiation of human embryonic stem cells and induced pluripotent stem cells. ACS Nano. https://doi.org/10.1021/acsnano.5b06985

Kim EJ, Kang KH, Ju JH (2017) Crispr-cas9: a promising tool for gene editing on induced pluripotent stem cells. Korean J Intern Med. https://doi.org/10.3904/kjim.2016.198

King CC, Brown AA, Sargin I, Bratlie KM, Beckman SP (2019) Modeling of reaction-diffusion transport into a core-shell geometry. J Theor Biol. https://doi.org/10.1016/j.jtbi.2018.09.026

Kirk K, Hao E, Lahmy R, Itkin-Ansari P (2014) Human embryonic stem cell derived islet progenitors mature inside an encapsulation device without evidence of increased biomass or cell escape. Stem Cell Res. https://doi.org/10.1016/j.scr.2014.03.003

Kitasato H, Kai R, Ding WG, Omatsu-Kanbe M (1996) The intrinsic rhythmicity of spike-burst generation in pancreatic beta- cells and intercellular interaction within an islet. Jpn J Physiol. https://doi.org/10.2170/jjphysiol.46.363

Klee P, Allagnat F, Pontes H, Cederroth M, Charollais A, Caille D, Britan A, Haefliger JA, Meda P (2011) Connexins protect mouse pancreatic β cells against apoptosis. J Clin Invest. https://doi.org/10.1172/JCI40509

Klein D, Misawa R, Bravo-Egana V, Vargas N, Rosero S, Piroso J, Ichii H, Umland O, Zhijie J, Tsinoremas N, Ricordi C, Inverardi L, Domínguez-Bendala J, Pastori RL (2013) MicroRNA expression in alpha and Beta cells of human pancreatic islets. PLoS One. https://doi.org/10.1371/journal.pone.0055064

Kojima H, Fujimiya M, Matsumura K, Younan P, Imaeda H, Maeda M, Chan L (2003) NeuroD-betacellulin gene therapy induces islet neogenesis in the liver and reverses diabetes in mice. Nat Med. https://doi.org/10.1038/nm867

Kopp JL, Dubois CL, Hao E, Thorel F, Herrera PL, Sander M (2011) Progenitor cell domains in the developing and adult pancreas. Cell Cycle 15;10(12):1921–1927

Korbutt GS, Elliott JF, Ao Z, Smith DK, Warnock GL, Rajotte R (1996) Large scale isolation, growth, and function of porcine neonatal islet cells. J Clin Invest. https://doi.org/10.1172/JCI118649

Kourtzelis I, Kotlabova K, Lim JH, Mitroulis I, Ferreira A, Chen LS, Gercken B, Steffen A, Kemter E, von Ameln AK, Waskow C, Hosur K, Chatzigeorgiou A, Ludwig B, Wolf E, Hajishengallis G, Chavakis T (2016) Developmental endothelial locus-1 modulates platelet-monocyte interactions and instant blood-mediated inflammatory reaction in islet transplantation. Thromb Haemost. https://doi.org/10.1160/TH15-05-0429

Kroon E, Martinson LA, Kadoya K, Bang AG, Kelly OG, Eliazer S, Young H, Richardson M, Smart NG, Cunningham J, Agulnick AD, D'Amour KA, Carpenter MK, Baetge EE (2008) Pancreatic endoderm derived from human embryonic stem cells generates glucose-responsive insulin-secreting cells in vivo. Nat Biotechnol. https://doi.org/10.1038/nbt1393

Kuehn C, Vermette P, Fülöp T (2014) Cross talk between the extracellular matrix and the immune system in the context of endocrine pancreatic islet transplantation. A review article. Pathol Biol. https://doi.org/10.1016/j.patbio.2014.01.001

Lammert E, Gu G, McLaughlin M, Brown D, Brekken R, Murtaugh LC, Gerber HP, Ferrara N, Melton DA (2003) Role of VEGF-A in vascularization of pancreatic islets. Curr Biol. https://doi.org/10.1016/S0960-9822(03)00378-6

Latreille M, Herrmanns K, Renwick N, Tuschl T, Malecki MT, McCarthy MI, Owen KR, Rülicke T, Stoffel M (2015) miR-375 gene dosage in pancreatic β-cells: implications for regulation of β-cell mass and biomarker development. J Mol Med. https://doi.org/10.1007/s00109-015-1296-9

Lavallard V, Armanet M, Parnaud G, Meyer J, Barbieux C, Montanari E, Meier R, Morel P, Berney T, Bosco D (2016) Cell rearrangement in transplanted human islets. FASEB J. https://doi.org/10.1096/fj.15-273805

Lee DY, Yang K, Lee S, Chae SY, Kim KW, Lee MK, Han DJ, Byun Y (2002) Optimization of monomethoxy-polyethylene glycol grafting on the pancreatic islet capsules. J Biomed Mater Res. https://doi.org/10.1002/jbm.10246

Lee DY, Nam JH, Byun Y (2004) Effect of polyethylene glycol grafted onto islet capsules on prevention of splenocyte and cytokine attacks. J Biomater Sci Polym Ed. https://doi.org/10.1163/156856204774196144

Lee DY, Park SJ, Nam JH, Byun Y (2006a) A combination therapy of PEGylation and immunosuppressive agent for successful islet transplantation. J Control Release. https://doi.org/10.1016/j.jconrel.2005.10.023

Lee DY, Lee S, Nam JH, Byun Y (2006b) Minimization of immunosuppressive therapy after islet transplantation: combined action of heme oxygenase-1 and PEGylation to islet. Am J Transplant. https://doi.org/10.1111/j.1600-6143.2006.01414.x

Lee DY, Park SJ, Nam JH, Byun Y (2006c) A new strategy toward improving immunoprotection in cell therapy for diabetes mellitus: long-functioning PEGylated islets in vivo. Tissue Eng. https://doi.org/10.1089/ten.2006.12.ft-55

Lee SH, Hao E, Savinov AY, Geron I, Strongin AY, Itkin-Ansari P (2009) Human β-cell precursors mature into functional insulin-producing cells in an immunoisolation device: implications for diabetes cell therapies. Transplantation. https://doi.org/10.1097/TP.0b013e31819c86ea

Lee KH, No DY, Kim SH, Ryoo JH, Wong SF, Lee SH (2011) Diffusion-mediated in situ alginate encapsulation of cell spheroids using microscale concave well and nanoporous membrane. Lab Chip. https://doi.org/10.1039/c0lc00540a

Lee EM, Jung JI, Alam Z, Yi HG, Kim H, Choi JW, Hurh S, Kim YJ, Jeong JC, Yang J, Oh KH, Kim HC, Lee BC, Choi I, Cho DW, Ahn C (2018) Effect of an oxygen-generating scaffold on the viability and insulin secretion function of porcine neonatal pancreatic cell clusters. Xenotransplantation. https://doi.org/10.1111/xen.12378

Li N, Sun G, Wang S, Wang Y, Xiu Z, Sun D, Guo X, Zhang Y, Ma X (2017) Engineering islet for improved performance by optimized reaggregation in alginate gel beads. Biotechnol Appl Biochem. https://doi.org/10.1002/bab.1489

Liao SW, Rawson J, Omori K, Ishiyama K, Mozhdehi D, Oancea AR, Ito T, Guan Z, Mullen Y (2013) Maintaining functional islets through encapsulation in an injectable saccharide-peptide hydrogel. Biomaterials. https://doi.org/10.1016/j.biomaterials.2013.02.007

Liu X, Li X, Zhang N, Wen X (2015) Engineering β-cell islets or islet-like structures for type 1 diabetes treatment. Med Hypotheses. https://doi.org/10.1016/j.mehy.2015.04.005

Liu Z, Hu W, He T, Dai Y, Hara H, Bottino R, Cooper DKC, Cai Z, Mou L (2017) Pig-to-primate islet xenotransplantation: past, present, and future. Cell Transplant. https://doi.org/10.3727/096368917X694859

Liu Z, Tang M, Zhao J, Chai R, Kang J (2018) Looking into the future: toward advanced 3D biomaterials for stem-cell-based regenerative medicine. Adv Mater. https://doi.org/10.1002/adma.201705388

Llacua A, De Haan BJ, Smink SA, De Vos P (2016) Extracellular matrix components supporting human islet function in alginate-based immunoprotective microcapsules for treatment of diabetes. J Biomed Mater Res A. https://doi.org/10.1002/jbm.a.35706

Llacua LA, de Haan BJ, de Vos P (2018) Laminin and collagen IV inclusion in immunoisolating microcapsules reduces cytokine-mediated cell death in human pancreatic islets. J Tissue Eng Regen Med. https://doi.org/10.1002/term.2472

Lock LT, Laychock SG, Tzanakakis ES (2011) Pseudoislets in stirred-suspension culture exhibit enhanced cell survival, propagation and insulin secretion. J Biotechnol. https://doi.org/10.1016/j.jbiotec.2010.12.015

Loomans CJM, Williams Giuliani N, Balak J, Ringnalda F, van Gurp L, Huch M, Boj SF, Sato T, Kester L, de Sousa Lopes SMC, Roost MS, Bonner-Weir S, Engelse MA, Rabelink TJ, Heimberg H, Vries RGJ, van Oudenaarden A, Carlotti F, Clevers H, de Koning EJP (2018) Expansion of adult human pancreatic tissue yields Organoids harboring progenitor cells with endocrine differentiation potential. Stem Cell Rep. https://doi.org/10.1016/j.stemcr.2018.02.005

Lou S, Zhang X, Zhang J, Deng J, Kong D, Li C (2017) Pancreatic islet surface bioengineering with a heparin-incorporated starPEG nanofilm. Mater Sci Eng C. https://doi.org/10.1016/j.msec.2017.03.295

Ludwig B, Ziegler CG, Schally AV, Richter C, Steffen A, Jabs N, Funk RH, Brendel MD, Block NL, Ehrhart-Bornstein M, Bornstein SR (2010) Agonist of growth hormone-releasing hormone as a potential effector for survival and proliferation of pancreatic islets. Proc Natl Acad Sci. https://doi.org/10.1073/pnas.1005098107

Ludwig B, Rotem A, Schmid J, Weir GC, Colton CK, Brendel MD, Neufeld T, Block NL, Yavriyants K, Steffen A, Ludwig S, Chavakis T, Reichel A, Azarov D, Zimermann B, Maimon S, Balyura M, Rozenshtein T, Shabtay N, Vardi P, Bloch K, de Vos P, Schally AV, Bornstein SR, Barkai U (2012) Improvement of islet function in a bioartificial pancreas by enhanced oxygen supply and growth hormone releasing hormone agonist. Proc Natl Acad Sci. https://doi.org/10.1073/pnas.1201868109

Ludwig B, Reichel A, Steffen A, Zimerman B, Schally AV, Block NL, Colton CK, Ludwig S, Kersting S, Bonifacio E, Solimena M, Gendler Z, Rotem A, Barkai U, Bornstein SR (2013) Transplantation of human islets without immunosuppression. Proc Natl Acad Sci. https://doi.org/10.1073/pnas.1317561110

Lutolf MP, Hubbell JA (2005) Synthetic biomaterials as instructive extracellular microenvironments for morphogenesis in tissue engineering. Nat Biotechnol 23 (1):47–55

Ma S, Viola R, Sui L, Cherubini V, Barbetti F, Egli D (2018) β cell replacement after gene editing of a neonatal diabetes-causing mutation at the insulin locus. Stem Cell Rep. https://doi.org/10.1016/j.stemcr.2018.11.006

Madden LR, Mortisen DJ, Sussman EM, Dupras SK, Fugate JA, Cuy JL, Hauch KD, Laflamme MA, Murry CE, Ratner BD (2010) Proangiogenic scaffolds as functional templates for cardiac tissue engineering. Proc Natl Acad Sci. https://doi.org/10.1073/pnas.1006442107

Mahboudi H, Kazemi B, Soleimani M, Hanaee-Ahvaz H, Ghanbarian H, Bandehpour M, Enderami SE, Kehtari M, Barati G (2018) Enhanced chondrogenesis of human bone marrow mesenchymal stem cell (BMSC) on nanofiber-based polyethersulfone (PES) scaffold. Gene. https://doi.org/10.1016/j.gene.2017.11.073

Mahou R, Zhang DKY, Vlahos AE, Sefton MV (2017) Injectable and inherently vascularizing semi-interpenetrating polymer network for delivering cells to the subcutaneous space. Biomaterials. https://doi.org/10.1016/j.biomaterials.2017.03.032

Mao GH, Chen GA, Bai HY, Song TR, Wang YX (2009) The reversal of hyperglycaemia in diabetic mice using PLGA scaffolds seeded with islet-like cells derived from human embryonic stem cells. Biomaterials. https://doi.org/10.1016/j.biomaterials.2008.12.030

Mao D, Zhu M, Zhang X, Ma R, Yang X, Ke T, Wang L, Li Z, Kong D, Li C (2017) A macroporous heparin-releasing silk fibroin scaffold improves islet transplantation outcome by promoting islet revascularisation and survival. Acta Biomater. https://doi.org/10.1016/j.actbio.2017.06.039

Marchioli G, Di LA, de Koning E, Engelse M, Van Blitterswijk CA, Karperien M, Van Apeldoorn AA, Moroni L (2016) Hybrid Polycaprolactone/alginate scaffolds functionalized with VEGF to promote de novo vessel formation for the transplantation of islets of Langerhans. Adv Healthc Mater. https://doi.org/10.1002/adhm.201600058

Mason MN, Arnold CA, Mahoney MJ (2009) Entrapped collagen type 1 promotes differentiation of embryonic pancreatic precursor cells into glucose-responsive b. Cells Tissue Eng Part A. https://doi.org/10.1089/ten.TEA.2009.0148

Matsumoto S (2010) Islet cell transplantation for type 1 diabetes. J Diabetes. https://doi.org/10.1111/j.1753-0407.2009.00048.x

Matsumoto S, Tan P, Baker J, Durbin K, Tomiya M, Azuma K, Doi M, Elliott RB (2014) Clinical porcine islet xenotransplantation under comprehensive regulation. Transplant Proc. https://doi.org/10.1016/j.transproceed.2014.06.008

Matsumoto S, Tomiya M, Sawamoto O (2016) Current status and future of clinical islet xenotransplantation. 8:483–493. doi: https://doi.org/10.1111/1753-0407.12395

McQuilling JP, Sittadjody S, Pendergraft S, Farney AC, Opara EC (2017) Applications of particulate oxygen-generating substances (POGS) in the bioartificial pancreas. Biomater Sci. https://doi.org/10.1039/c7bm00790f

McReynolds J, Wen Y, Li X, Guan J, Jin S (2017) Modeling spatial distribution of oxygen in 3d culture of islet beta-cells. Biotechnol Prog. https://doi.org/10.1002/btpr.2395

Mironov AV, Grigoryev AM, Krotova LI, Skaletsky NN, Popov VK, Sevastianov VI (2017) 3D printing of PLGA scaffolds for tissue engineering. J Biomed Mater Res A. https://doi.org/10.1002/jbm.a.35871

Mobarra N, Soleimani M, Pakzad R, Enderami SE, Pasalar P (2018) Three-dimensional nanofiberous PLLA/PCL scaffold improved biochemical and molecular markers hiPS cell-derived insulin-producing islet-like cells. Artif Cells Nanomed Biotechnol. https://doi.org/10.1080/21691401.2018.1505747

Moberg L, Johansson H, Lukinius A, Berne C, Foss A, Källen R, Østraat SK, Tibell A, Tufveson G, Elgue G, Nilsson Ekdahl K, Korsgren O, Nilsson B (2002) Production of tissue factor by pancreatic islet cells as a trigger of detrimental thrombotic reactions in clinical islet transplantation. Lancet. https://doi.org/10.1016/S0140-6736(02)12020-4

Montanucci P, Pennoni I, Pescara T, Basta G, Calafiore R (2013) Treatment of diabetes mellitus with microencapsulated fetal human liver (FH-B-TPN) engineered cells. Biomaterials. https://doi.org/10.1016/j.biomaterials.2013.02.026

Montazeri L, Hojjati-Emami S, Bonakdar S, Tahamtani Y, Hajizadeh-Saffar E, Noori-Keshtkar M, Najar-Asl M, Ashtiani MK, Baharvand H (2016) Improvement of islet engrafts by enhanced angiogenesis and microparticle-mediated oxygenation. Biomaterials. https://doi.org/10.1016/j.biomaterials.2016.02.043

Mooranian A, Negrulj R, Arfuso F, Al-Salami H (2016) Multicompartmental, multilayered probucol microcapsules for diabetes mellitus: formulation characterization and effects on production of insulin and inflammation in a pancreatic β-cell line. Artif Cells Nanomed Biotechnol. https://doi.org/10.3109/21691401.2015.1069299

Mooranian A, Zamani N, Mikov M, Goločorbin-Kon S, Stojanovic G, Arfuso F, Al-Salami H (2018) Novel nano-encapsulation of probucol in microgels: scanning electron micrograph characterizations, buoyancy profiling, and antioxidant assay analyses. Artif Cells Nanomed Biotechnol:1–7. https://doi.org/10.1080/21691401.2018.1511571

Morini S, Braun M, Onori P, Cicalese L, Elias G, Gaudio E, Rastellini C (2006) Morphological changes of isolated rat pancreatic islets: a structural, ultrastructural and morphometric study. J Anat. https://doi.org/10.1111/j.1469-7580.2006.00620.x

Morozov VA, Wynyard S, Matsumoto S, Abalovich A, Denner J, Elliott R (2017) No PERV transmission during a clinical trial of pig islet cell transplantation. Virus Res. https://doi.org/10.1016/j.virusres.2016.08.012

Motte E, Szepessy E, Suenens K, Stange G, Bomans M, Jacobs-Tulleneers-Thevissen D, Ling Z, Kroon E, Pipeleers D (2014) Composition and function of macroencapsulated human embryonic stem cell-derived implants: comparison with clinical human islet cell grafts. Am J Physiol Endocrinol Metab. https://doi.org/10.1152/ajpendo.00219.2014

Murray HE, Paget MB, Downing R (2005) Preservation of glucose responsiveness in human islets maintained in a rotational cell culture system. Mol Cell Endocrinol. https://doi.org/10.1016/j.mce.2005.03.014

Nabavimanesh MM, Hashemi-Najafabadi S, Vasheghani-Farahani E (2015) Islets immunoisolation using encapsulation and PEGylation, simultaneously, as a novel design. J Biosci Bioeng. https://doi.org/10.1016/j.jbiosc.2014.09.023

Nadri S, Barati G, Mostafavi H, Esmaeilzadeh A, Enderami SE (2017) Differentiation of conjunctiva mesenchymal stem cells into secreting islet beta cells on plasma treated electrospun nanofibrous scaffold. Artif Cells Nanomed Biotechnol. https://doi.org/10.1080/21691401.2017.1416391

Nassiri Mansour R, Barati G, Soleimani M, Ghoraeian P, Nouri Aleagha N, Kehtari M, Mahboudi H, Hosseini F, Hassannia H, Abazari MF, Enderami SE (2018) Generation of high-yield insulin producing cells from human-induced pluripotent stem cells on polyethersulfone nanofibrous scaffold. Artif Cells Nanomed Biotechnol. https://doi.org/10.1080/21691401.2018.1434663

Navarro-Tableros V, Herrera Sanchez MB, Figliolini F, Romagnoli R, Tetta C, Camussi G (2015) Recellularization of rat liver scaffolds by human liver stem cells. Tissue Eng Part A. https://doi.org/10.1089/ten.tea.2014.0573

Navarro-tableros V, Gai C, Gomez Y, Giunti S, Pasquino C, Deregibus MC, Tapparo M, Pitino A, Tetta C, Brizzi MF, Ricordi C, Camussi G (2018) Islet-like structures generated in vitro from adult human liver stem cells revert hyperglycemia in diabetic SCID mice. Stem Cell Rev Rep. https://doi.org/10.1007/s12015-018-9845-6

Nelson TJ, Behfar A, Yamada S, Martinez-Fernandez A, Terzic A (2009) Stem cell platforms for regenerative medicine. Clin Transl Sci. https://doi.org/10.1111/j.1752-8062.2009.00096.x

Neufeld T, Ludwig B, Barkai U, Weir GC, Colton CK, Evron Y, Balyura M, Yavriyants K, Zimermann B, Azarov D, Maimon S, Shabtay N, Rozenshtein T, Lorber D, Steffen A, Willenz U, Bloch K, Vardi P, Taube R, de Vos P, Lewis EC, Bornstein SR, Rotem A (2013) The efficacy of an Immunoisolating membrane system for islet xenotransplantation in minipigs. PLoS One. https://doi.org/10.1371/journal.pone.0070150

Nikolova G, Jabs N, Konstantinova I, Domogatskaya A, Tryggvason K, Sorokin L, Fässler R, Gu G, Gerber HP, Ferrara N, Melton DA, Lammert E (2006) The vascular basement membrane: a niche for insulin gene expression and β cell proliferation. Dev Cell. https://doi.org/10.1016/j.devcel.2006.01.015

Nikravesh N, Cox SC, Ellis MJ, Grover LM (2017) Encapsulation and fluidization maintains the viability and glucose sensitivity of Beta-cells. ACS Biomater Sci Eng. https://doi.org/10.1021/acsbiomaterials.7b00191

Nilsson B, Ekdahl KN, Korsgren O (2011) Control of instant blood-mediated inflammatory reaction to improve islets of Langerhans engraftment. Curr Opin Organ Transplant. https://doi.org/10.1097/MOT.0b013e32834c2393

Nir T, Melton DA, Dor Y (2007) Recovery from diabetes in mice by β cell regeneration. J Clin Invest. https://doi.org/10.1172/JCI32959

Olsson R, Olerud J, Pettersson U, Carlsson PO (2011) Increased numbers of low-oxygenated pancreatic islets after intraportal islet transplantation. Diabetes 60:2350–2353. https://doi.org/10.2337/db09-0490

Onoe H, Okitsu T, Itou A, Kato-Negishi M, Gojo R, Kiriya D, Sato K, Miura S, Iwanaga S, Kuribayashi-Shigetomi K, Matsunaga YT, Shimoyama Y, Takeuchi S (2013) Metre-long cell-laden microfibres exhibit tissue morphologies and functions. Nat Mater. https://doi.org/10.1038/nmat3606

Orci L, Malaisse-Lagae F, Ravazzola M, Rouiller D, Renold AE, Perrelet A, Unger R (1975) A morphological basis for intercellular communication between α and β cells in the endocrine pancreas. J Clin Invest. https://doi.org/10.1172/JCI108154

Orive G, Santos E, Poncelet D, Hernández RM, Pedraz JL, Wahlberg LU, De Vos P, Emerich D (2015) Cell encapsulation: technical and clinical advances. Trends Pharmacol Sci 36:537–546. https://doi.org/10.1016/j.tips.2015.05.003

Ouziel-Yahalom L, Zalzman M, Anker-Kitai L, Knoller S, Bar Y, Glandt M, Herold K, Efrat S (2006) Expansion and redifferentiation of adult human pancreatic islet cells. Biochem Biophys Res Commun. https://doi.org/10.1016/j.bbrc.2005.12.187

Pagliuca FW, Melton DA (2013) How to make a functional β-cell. Development 140:2472–2483. https://doi.org/10.1242/dev.093187

Pagliuca FW, Millman JR, Gürtler M, Segel M, Van Dervort A, Ryu JH, Peterson QP, Greiner D, Melton DA (2014) Generation of functional human pancreatic β cells in vitro. Cell. https://doi.org/10.1016/j.cell.2014.09.040

Paraskevas S, Maysinger D, Wang R, Duguid WP, Rosenberg L (2000) Cell loss in isolated human islets occurs by apoptosis. Pancreas. https://doi.org/10.1097/00006676-200004000-00008

Park CH, Woo KM (2018) Fibrin-based biomaterial applications in tissue engineering and regenerative medicine. Adv Exp Med Biol. https://doi.org/10.1007/978-981-13-0445-3_16.doi:10.1007/978-981-13-0445-3_16

Park HS, Kim JW, Lee SH, Yang HK, Ham DS, Sun CL, Hong TH, Khang G, Park CG, Yoon KH (2017) Antifibrotic effect of rapamycin containing polyethylene glycol-coated alginate microcapsule in islet xenotransplantation. J Tissue Eng Regen Med. https://doi.org/10.1002/term.2029

Päth G, Perakakis N, Mantzoros CS, Seufert J (2019) Stem cells in the treatment of diabetes mellitus-focus on mesenchymal stem cells. Metabolism. https://doi.org/10.1016/j.metabol.2018.10.005

Pedraza E, Coronel MM, Fraker CA, Ricordi C, Stabler CL (2012) Preventing hypoxia-induced cell death in beta cells and islets via hydrolytically activated, oxygen-generating biomaterials. Proc Natl Acad Sci. https://doi.org/10.1073/pnas.1113560109

Pepper AR, Gala-Lopez B, Pawlick R, Merani S, Kin T, Shapiro AMJ (2015) A prevascularized subcutaneous device-less site for islet and cellular transplantation. Nat Biotechnol. https://doi.org/10.1038/nbt.3211

Pepper AR, Bruni A, Shapiro AMJ (2018) Clinical islet transplantation: is the future finally now? Curr Opin Organ Transplant. https://doi.org/10.1097/MOT.0000000000000546

Perignon C, Ongmayeb G, Neufeld R, Frere Y, Poncelet D (2015) Microencapsulation by interfacial polymerisation: membrane formation and structure. J Microencapsul. https://doi.org/10.3109/02652048.2014.950711

Petry F, Weidner T, Czermak P, Salzig D (2018) Three-dimensional bioreactor technologies for the cocultivation of human mesenchymal stem/stromal cells and beta cells. Stem Cells Int. https://doi.org/10.1155/2018/2547098

Piemonti L, Everly MJ, Maffi P, Scavini M, Poli F, Nano R, Cardillo M, Melzi R, Mercalli A, Sordi V, Lampasona V, De Arias AE, Scalamogna M, Bosi E, Bonifacio E, Secchi A, Terasaki PI (2013) Alloantibody and autoantibody monitoring predicts islet transplantation outcome in human type 1 diabetes. Diabetes. https://doi.org/10.2337/db12-1258

Pipeleers D, Chintinne M, Denys B, Martens G, Keymeulen B, Gorus F (2008) Restoring a functional β-cell mass in diabetes. Diabetes Obes Metab. https://doi.org/10.1111/j.1463-1326.2008.00941.x

Rana D, Zreiqat H, Benkirane-Jessel N, Ramakrishna S, Ramalingam M (2017) Development of decellularized scaffolds for stem cell-driven tissue engineering. J Tissue Eng Regen Med. https://doi.org/10.1002/term.2061

Ratcliffe A, Niklason LE (2002) Bioreactors and bioprocessing for tissue engineering. Ann N Y Acad Sci 961(1):210–215

Ravassard P, Hazhouz Y, Pechberty S, Bricout-Neveu E, Armanet M, Czernichow P, Scharfmann R (2011) A genetically engineered human pancreatic β cell line exhibiting glucose-inducible insulin secretion. J Clin Invest. https://doi.org/10.1172/JCI58447

Rengifo HR, Giraldo JA, Labrada I, Stabler CL (2014) Long-term survival of allograft murine islets coated via covalently stabilized polymers. Adv Healthc Mater. https://doi.org/10.1002/adhm.201300573

Richards OC, Raines SM, Attie AD (2010) The role of blood vessels, endothelial cells, and vascular pericytes in insulin secretion and peripheral insulin action. Endocr Rev. https://doi.org/10.1210/er.2009-0035

Ricordi C, Fraker C, Szust J, Al-Abdullah I, Poggioli R, Kirlew T, Khan A, Alejandro R (2003) Improved human islet isolation outcome from marginal donors following addition of oxygenated perfluorocarbon to the cold-storage solution. Transplantation. https://doi.org/10.1097/01.TP.0000058813.95063.7A

Ricordi C, Goldstein JS, Balamurugan AN, Szot GL, Kin T, Liu C, Czarniecki CW, Barbaro B, Bridges ND, Cano J, Clarke WR, Eggerman TL, Hunsicker LG, Kaufman DB, Khan A, Lafontant DE, Linetsky E, Luo X, Markmann JF, Naji A, Korsgren O, Oberholzer J, Turgeon NA, Brandhorst D, Friberg AS, Lei J, Wang LJ, Wilhelm JJ, Willits J, Zhang X, Hering BJ, Posselt AM, Stock PG, Shapiro AMJ (2016) National institutes of health-sponsored clinical islet transplantation consortium phase 3 trial: manufacture of a complex cellular product at eight processing facilities. Diabetes. https://doi.org/10.2337/db16-0234

Rios PD, Zhang X, Luo X, Shea LD (2016) Mold-casted non-degradable, islet macro-encapsulating hydrogel devices for restoration of normoglycemia in diabetic mice. Biotechnol Bioeng. https://doi.org/10.1002/bit.26005

Rios PD, Skoumal M, Liu J, Youngblood R, Kniazeva E, Garcia AJ, Shea LD (2018) Evaluation of encapsulating and microporous nondegradable hydrogel scaffold designs on islet engraftment in rodent models of diabetes. Biotechnol Bioeng. https://doi.org/10.1002/bit.26741

Rogers SA, Tripathi P, Mohanakumar T, Liapis H, Chen F, Talcott MR, Faulkner C, Hammerman MR (2011) Engraftment of cells from porcine islets of langerhans following transplantation of pig pancreatic primordia in non-immunosuppressed diabetic rhesus macaques. Organogenesis. https://doi.org/10.4161/org.7.3.16522

Rorsman P, Ashcroft FM (2018) Pancreatic β-cell electrical activity and insulin secretion: of mice and men. Physiol Rev. https://doi.org/10.1152/physrev.00008.2017

Russ HA, Bar Y, Ravassard P, Efrat S (2008) In vitro proliferation of cells derived from adult human β-cells revealed by cell-lineage tracing. Diabetes. https://doi.org/10.2337/db07-1283

Russ HA, Ravassard P, Kerr-Conte J, Pattou F, Efrat S (2009) Epithelial-mesenchymal transition in cells expanded in vitro from lineage-traced adult human pancreatic beta cells. PLoS One. https://doi.org/10.1371/journal.pone.0006417

Russ HA, Sintov E, Anker-Kitai L, Friedman O, Lenz A, Toren G, Farhy C, Pasmanik-Chor M, Oron-Karni V, Ravassard P, Efrat S (2011) Insulin-producing cells generated from dedifferentiated human pancreatic beta cells expanded in vitro. PLoS One. https://doi.org/10.1371/journal.pone.0025566

Rutter GA, Hodson DJ (2015) Beta cell connectivity in pancreatic islets: a type 2 diabetes target? Cell Mol Life Sci. https://doi.org/10.1007/s00018-014-1755-4

Rutzky LP, Bilinski S, Kloc M, Phan T, Zhang H, Katz SM, Stepkowski SM (2002) Microgravity culture condition reduces immunogenicity and improves function of pancreatic islets. Transplantation. https://doi.org/10.1097/00007890-200207150-00004

Ryan AJ, O'Neill HS, Duffy GP, O'Brien FJ (2017) Advances in polymeric islet cell encapsulation technologies to limit the foreign body response and provide immunoisolation. Curr Opin Pharmacol. https://doi.org/10.1016/j.coph.2017.07.013

Saeki Y, Ishiyama K, Ishida N, Tanaka Y, Ohdan H (2017) Role of natural killer cells in the innate immune system after intraportal islet transplantation in mice. Transplant Proc. https://doi.org/10.1016/j.transproceed.2016.10.010

Safley SA, Kenyon NS, Berman DM, Barber GF, Willman M, Duncanson S, Iwakoshi N, Holdcraft R, Gazda L, Thompson P, Badell IR, Sambanis A, Ricordi C, Weber CJ (2018a) Microencapsulated adult porcine islets transplanted intraperitoneally in streptozotocin-diabetic non-human primates. Xenotransplantation. https://doi.org/10.1111/xen.12095

Safley SA, Kenyon NS, Berman DM, Barber GF, Willman M, Duncanson S, Iwakoshi N, Holdcraft R, Gazda L, Thompson P, Badell IR, Sambanis A, Ricordi C, Weber CJ (2018b) Microencapsulated adult porcine islets transplanted intraperitoneally in streptozotocin-diabetic non-human primates. Xenotransplantation. https://doi.org/10.1111/xen.12450

Salvay DM, Rives CB, Zhang X, Chen F, Kaufman DB, Lowe WL, Shea LD (2008) Extracellular matrix protein-coated scaffolds promote the reversal of diabetes after extrahepatic islet transplantation. Transplantation. https://doi.org/10.1097/TP.0b013e31816fc0ea

Sama AE, D'Amore J, Ward MF, Chen G, Wang H (2004) Bench to bedside: HMGB1 – a novel proinflammatory cytokine and potential therapeutic target for septic patients in the emergency department. Acad Emerg Med

Samy KP, Martin BM, Turgeon NA, Kirk AD (2014) Islet cell xenotransplantation: a serious look toward the clinic. Xenotransplantation. https://doi.org/10.1111/xen.12095

Samy KP, Davis RP, Gao Q, Martin BM, Song M, Cano J, Farris AB, McDonald A, Gall EK, Dove CR, Leopardi FV, How T, Williams KD, Devi GR, Collins BH, Kirk AD (2018) Early barriers to neonatal porcine islet engraftment in a dual transplant model. Am J Transplant. https://doi.org/10.1111/ajt.14601

Santos-Silva JC, Carvalho CP de F, de Oliveira RB, Boschero AC, Collares-Buzato CB (2012) Cell-to-cell contact dependence and junctional protein content are correlated with in vivo maturation of pancreatic beta cells. Can J Physiol Pharmacol. https://doi.org/10.1139/y2012-064

Scavuzzo MA, Teaw J, Yang DBM (2018) Generation of scaffold-free, three-dimensional insulin expressing pancreatoids from mouse pancreatic progenitors in vitro. J Vis Exp. https://doi.org/10.3791/57599

Scharfmann R, Pechberty S, Hazhouz Y, Von Bülow M, Bricout-Neveu E, Grenier-Godard M, Guez F, Rachdi L, Lohmann M, Czernichow P, Ravassard P (2014) Development of a conditionally immortalized human pancreatic β cell line. J Clin Invest. https://doi.org/10.1172/JCI72674

Scharp DW, Marchetti P (2014) Encapsulated islets for diabetes therapy: history, current progress, and critical issues requiring solution. Adv Drug Deliv Rev. https://doi.org/10.1016/j.addr.2013.07.018

Schuetz C, Anazawa T, Cross SE, Labriola L, Meier RPH, Redfield RR, Scholz H, Stock PG, Zammit NW (2018) β cell replacement therapy: the next 10 years. Transplantation. https://doi.org/10.1097/TP.0000000000001937

Schwarz RP, Goodwin TJ, Wolf DA (1992) Cell culture for three-dimensional modeling in rotating-wall vessels: an application of simulated microgravity. J Tissue Cult Methods. https://doi.org/10.1007/BF01404744

Schweicher J, Nyitray C, Desai TA (2014) Membranes to achieve immunoprotection of transplanted islets. Front Biosci 19:49–76

Shalaly ND, Ria M, Johansson U, Åvall K, Berggren PO, Hedhammar M (2016) Silk matrices promote formation of insulin-secreting islet-like clusters. Biomaterials. https://doi.org/10.1016/j.biomaterials.2016.03.006

Shapiro AMJ, Ricordi C, Hering BJ, Auchincloss H, Lindblad R, Robertson RP, Secchi A, Brendel MD, Berney T, Brennan DC, Cagliero E, Alejandro R, Ryan EA, DiMercurio B, Morel P, Polonsky KS, Reems J-A, Bretzel RG, Bertuzzi F, Froud T, Kandaswamy R, Sutherland DER, Eisenbarth G, Segal M, Preiksaitis J, Korbutt GS, Barton FB, Viviano L, Seyfert-Margolis V, Bluestone J, Lakey JRT (2006) International trial of the Edmonton protocol for islet transplantation. N Engl J Med. https://doi.org/10.1056/NEJMoa061267

Shapiro AMJ, Pokrywczynska M, Ricordi C (2017) Clinical pancreatic islet transplantation. Nat Rev Endocrinol. https://doi.org/10.1038/nrendo.2016.178

Sharp J, Vermette P (2017) An in-situ glucose-stimulated insulin secretion assay under perfusion bioreactor conditions. Biotechnol Prog. https://doi.org/10.1002/btpr.2407

Shen Y, Hou Y, Yao S, Huang P, Yobas L (2015) In vitro epithelial Organoid generation induced by substrate nanotopography. Sci Rep. https://doi.org/10.1038/srep09293

Skrzypek K, Nibbelink MG, Karbaat LP, Karperien M, van Apeldoorn A, Stamatialis D (2018a) An important step towards a prevascularized islet macroencapsulation device—effect of micropatterned membranes on development of endothelial cell network. J Mater Sci Mater Med. https://doi.org/10.1007/s10856-018-6102-0

Skrzypek K, Barrera YB, Groth T, Stamatialis D (2018b) Endothelial and beta cell composite aggregates for improved function of a bioartificial pancreas encapsulation device. Int J Artif Organs. https://doi.org/10.1177/0391398817752295

Smink AM, Li S, Hertsig DT, De Haan BJ, Schwab L, Van Apeldoorn AA, De Koning E, Faas MM, Lakey JRT, De Vos P (2017) The efficacy of a Prevascularized, retrievable poly(D,L,-lactide-co-ε-caprolactone) subcutaneous scaffold as transplantation site for pancreatic islets. Transplantation. https://doi.org/10.1097/TP.0000000000001663

Song S, Roy S (2016) Progress and challenges in macroencapsulation approaches for type 1 diabetes (T1D) treatment: cells, biomaterials, and devices. Biotechnol Bioeng. https://doi.org/10.1002/bit.25895

Soon-Shiong P, Heintz RE, Merideth N, Yao QX, Yao Z, Zheng T, Murphy M, Moloney MK, Schmehl M, Harris M, Mendez R, Mendez R, Sandford PA (1994) Insulin independence in a type 1 diabetic patient after encapsulated islet transplantation. Lancet. https://doi.org/10.1016/S0140-6736(94)90067-1

Sörenby AK, Kumagai-Braesch M, Sharma A, Hultenby KR, Wernerson AM, Tibell AB (2008) Preimplantation of an immunoprotective device can lower the curative dose of islets to that of free islet transplantation-studies in a rodent model. Transplantation. https://doi.org/10.1097/TP.0b013e31817efc78

Steele JAM, Hallé JP, Poncelet D, Neufeld RJ (2014) Therapeutic cell encapsulation techniques and applications in diabetes. Adv Drug Deliv Rev. https://doi.org/10.1016/j.addr.2013.09.015

Steffens D, Braghirolli DI, Maurmann N, Pranke P (2018) Update on the main use of biomaterials and techniques associated with tissue engineering. Drug Discov Today. https://doi.org/10.1016/j.drudis.2018.03.013

Stendahl JC, Kaufman DB, Stupp SI (2009) Extracellular matrix in pancreatic islets: relevance to scaffold design and transplantation. Cell Transplant. https://doi.org/10.3727/096368909788237195

Stephens CH, Orr KS, Acton AJ, Tersey SA, Mirmira RG, Considine RV, Voytik-Harbin SL (2018) In situ type I oligomeric collagen macroencapsulation promotes islet longevity and function in vitro and in vivo. Am J Physiol Endocrinol Metab. https://doi.org/10.1152/ajpendo.00073.2018

Sun Y, Ma X, Zhou D, Vacek I, Sun AM (1996) Normalization of diabetes in spontaneously diabetic cynomolgus monkeys by xenografts of microencapsulated porcine islets without immunosuppression. J Clin Invest. https://doi.org/10.1172/JCI118929

Sweet IR, Yanay O, Waldron L, Gilbert M, Fuller JM, Tupling T, Lernmark A, Osborne WRA (2008) Treatment of diabetic rats with encapsulated islets. J Cell Mol Med 1. https://doi.org/10.1111/j.1582-4934.2008.00322.x

Swinehart IT, Badylak SF (2016) Extracellular matrix bioscaffolds in tissue remodeling and morphogenesis. Dev Dyn. https://doi.org/10.1002/dvdy.24379

Szot GL, Yadav M, Lang J, Kroon E, Kerr J, Kadoya K, Brandon EP, Baetge EE, Bour-Jordan H, Bluestone JA (2015) Tolerance induction and reversal of diabetes in mice transplanted with human embryonic stem cell-derived pancreatic endoderm. Cell Stem Cell. https://doi.org/10.1016/j.stem.2014.12.001

Tanaka H, Tanaka S, Sekine K, Kita S, Okamura A, Takebe T, Zheng YW, Ueno Y, Tanaka J, Taniguchi H (2013) The generation of pancreatic β-cell spheroids in a simulated microgravity culture system. Biomaterials. https://doi.org/10.1016/j.biomaterials.2013.04.003

Theocharis AD, Skandalis SS, Gialeli C, Karamanos NK (2016) Extracellular matrix structure. Adv Drug Deliv Rev. https://doi.org/10.1016/j.addr.2015.11.001

Thompson P, Badell IR, Lowe M, Cano J, Song M, Leopardi F, Avila J, Ruhil R, Strobert E, Korbutt G, Rayat G, Rajotte R, Iwakoshi N, Larsen CP, Kirk AD (2011) Islet xenotransplantation using gal-deficient neonatal donors improves engraftment and function. Am J Transplant. https://doi.org/10.1111/j.1600-6143.2011.03720.x

Tibell A, Rafael E, Wennberg L, Nordenström J, Bergström M, Geller RL, Loudovaris T, Johnson RC, Brauker JH, Neuenfeldt S, Wernerson A (2001)

Survival of macroencapsulated allogeneic parathyroid tissue one year after transplantation in nonimmunosuppressed humans. Cell Transplant

Tomei AA, Villa C, Ricordi C (2015) Development of an encapsulated stem cell-based therapy for diabetes. Expert Opin Biol Ther. https://doi.org/10.1517/14712598.2015.1055242

Trivedi N, Steil GM, Colton CK, Bonner-Weir S, Weir GC (2000) Improved vascularization of planar membrane diffusion devices following continuous infusion of vascular endothelial growth factor. Cell Transplant. https://doi.org/10.1177/096368970000900114

Tuch BE, Keogh GW, Williams LJ, Wu W, Foster JL, Vaithilingam V, Philips R (2009) Safety and viability of microencapsulated human islets transplanted into diabetic humans. Diabetes Care. https://doi.org/10.2337/dc09-0744

Vaithilingam V, Kollarikova G, Qi M, Lacik I, Oberholzer J, Guillemin GJ, Tuch BE (2011) Effect of prolonged gelling time on the intrinsic properties of barium alginate microcapsules and its biocompatibility. J Microencapsul. https://doi.org/10.3109/02652048.2011.586067

Van Der Laan LJW, Lockey C, Griffeth BC, Frasler FS, Wilson CA, Onlons DE, Hering BJ, Long Z, Otto E, Torbett BE, Salomon DR (2000) Infection by porcine endogenous retrovirus after islet xenotransplantation in SCID mice. Nature. https://doi.org/10.1038/35024089

van der Meulen T, Mawla AM, DiGruccio MR, Adams MW, Nies V, Dólleman S, Liu S, Ackermann AM, Cáceres E, Hunter AE, Kaestner KH, Donaldson CJ, Huising MO (2017) Virgin beta cells persist throughout life at a Neogenic niche within pancreatic islets. Cell Metab. https://doi.org/10.1016/j.cmet.2017.03.017

Vegas AJ, Veiseh O, Gürtler M, Millman JR, Pagliuca FW, Bader AR, Doloff JC, Li J, Chen M, Olejnik K, Tam HH, Jhunjhunwala S, Langan E, Aresta-Dasilva S, Gandham S, McGarrigle JJ, Bochenek MA, Hollister-Lock J, Oberholzer J, Greiner DL, Weir GC, Melton DA, Langer R, Anderson DG (2016) Long-term glycemic control using polymer-encapsulated human stem cell-derived beta cells in immune-competent mice. Nat Med. https://doi.org/10.1038/nm.4030

Veiseh O, Doloff JC, Ma M, Vegas AJ, Tam HH, Bader AR, Li J, Langan E, Wyckoff J, Loo WS, Jhunjhunwala S, Chiu A, Siebert S, Tang K, Hollister-Lock J, Aresta-Dasilva S, Bochenek M, Mendoza-Elias J, Wang Y, Qi M, Lavin DM, Chen M, Dholakia N, Thakrar R, Lacík I, Weir GC, Oberholzer J, Greiner DL, Langer R, Anderson DG (2015) Size- and shape-dependent foreign body immune response to materials implanted in rodents and non-human primates. Nat Mater. https://doi.org/10.1038/nmat4290

Vériter S, Gianello P, Igarashi Y, Beaurin G, Ghyselinck A, Aouassar N, Jordan B, Gallez B, Dufrane D (2014) Improvement of subcutaneous bioartificial pancreas vascularization and function by coencapsulation of pig islets and mesenchymal stem cells in primates. Cell Transplant. https://doi.org/10.3727/096368913X663550

Vishwakarma SK, Lakkireddy C, Bardia A, Raju N, Paspala SAB, Habeeb MA, Khan AA (2018) Molecular dynamics of pancreatic transcription factors in bioengineered humanized insulin producing neoorgan. Gene. https://doi.org/10.1016/j.gene.2018.07.006

Wan J, Huang Y, Zhou P, Guo Y, Wu C, Zhu S, Wang Y, Wang L, Lu Y, Wang Z (2017) Culture of iPSCs derived pancreatic β -like cells in vitro using decellularized pancreatic scaffolds: a preliminary trial. Biomed Res Int. https://doi.org/10.1155/2017/4276928

Wang Q, Li S, Xie Y, Yu W, Xiong Y, Ma X, Yuan Q (2006) Cytoskeletal reorganization and repolarization of hepatocarcinoma cells in APA microcapsule to mimic native tumor characteristics. Hepatol Res. https://doi.org/10.1016/j.hepres.2006.03.003

Wang Y, Lanzoni G, Carpino G, Bin CC, Dominguez-Bendala J, Wauthier E, Cardinale V, Oikawa T, Pileggi A, Gerber D, Furth ME, Alvaro D, Gaudio E, Inverardi L, Reid LM (2013) Biliary tree stem cells, precursors to pancreatic committed progenitors: evidence for possible life-long pancreatic organogenesis. Stem Cells. https://doi.org/10.1002/stem.1460

Wang Y, Zhong J, Zhang X, Liu Z, Yang Y, Gong Q, Ren B (2016) The role of HMGB1 in the pathogenesis of type 2 diabetes. J Diabetes Res 2016:2543268. https://doi.org/10.1155/2016/2543268

Wang W, Jin S, Ye K (2017) Development of islet organoids from H9 human embryonic stem cells in biomimetic 3D scaffolds. Stem Cells Dev. https://doi.org/10.1089/scd.2016.0115

Weaver JD, Headen DM, Hunckler MD, Coronel MM, Stabler CL, García AJ (2018) Design of a vascularized synthetic poly(ethylene glycol) macroencapsulation device for islet transplantation. Biomaterials. https://doi.org/10.1016/j.biomaterials.2018.04.047

Wei C, Geras-Raaka E, Marcus-Samuels B, Oron Y, Gershengorn MC (2006) Trypsin and thrombin accelerate aggregation of human endocrine pancreas precursor cells. J Cell Physiol. https://doi.org/10.1002/jcp.20459

Welsch CA, Rust WL, Csete M1 (2018) Concise review : lessons learned from islet transplant clinical trials in developing stem cell therapies for type 1 diabetes. Stem Cells Transl Med. https://doi.org/10.1002/sctm.18-0156

Wilson JT, Cui W, Chaikof EL (2008) Layer-by-layer assembly of a conformal nanothin PEG coating for intraportal islet transplantation. Nano Lett. https://doi.org/10.1021/nl080694q

Wojtusciszyn A, Armanet M, Morel P, Berney T, Bosco D (2008) Insulin secretion from human beta cells is heterogeneous and dependent on cell-to-cell contacts. Diabetologia. https://doi.org/10.1007/s00125-008-1103-z

Wu H, Avgoustiniatos ES, Swette L, Bonner-Weir S, Weir GC, Colton CK (1999) In situ electrochemical oxygen generation with an immunoisolation device. Ann N Y Acad Sci 875(1):105–125

Wu XH, Liu CP, Xu KF, Mao XD, Zhu J, Jiang JJ, Cui D, Zhang M, Xu Y, Liu C (2007) Reversal of hyperglycemia in diabetic rats by portal vein transplantation of islet-like cells generated from bone marrow mesenchymal stem cells. World J Gastroenterol. https://doi.org/10.3748/wjg.v13.i24.3342

Wynyard S, Nathu D, Garkavenko O, Denner J, Elliott R (2014) Microbiological safety of the first clinical pig islet xenotransplantation trial in New Zealand. Xenotransplantation. https://doi.org/10.1111/xen.12102

Xie M, Ye H, Wang H, Charpin-El Hamri G, Lormeau C, Saxena P, Stelling J, Fussenegger M (2016) β-cell-mimetic designer cells provide closed-loop glycemic control. Science. https://doi.org/10.1126/science.aaf4006

Yim EKF, Darling EM, Kulangara K, Guilak F, Leong KW (2010) Nanotopography-induced changes in focal adhesions, cytoskeletal organization, and mechanical properties of human mesenchymal stem cells. Biomaterials. https://doi.org/10.1016/j.biomaterials.2009.10.037

Zarekhalili Z, Bahrami SH, Ranjbar-Mohammadi M, Milan PB (2017) Fabrication and characterization of PVA/Gum tragacanth/PCL hybrid nanofibrous scaffolds for skin substitutes. Int J Biol Macromol. https://doi.org/10.1016/j.ijbiomac.2016.10.042

Zhang X, Wang W, Xie Y, Zhang Y, Wang X, Guo X, Ma X (2006) Proliferation, viability, and metabolism of human tumor and normal cells cultured in microcapsule. Appl Biochem Biotechnol. https://doi.org/10.1385/ABAB:134:1:61

Zhao M, Amiel SA, Christie MR, Muiesan P, Srinivasan P, Littlejohn W, Rela M, Arno M, Heaton N, Huang GC (2007) Evidence for the presence of stem cell-like progenitor cells in human adult pancreas. J Endocrinol. https://doi.org/10.1677/JOE-07-0436

Zhou Q, Melton DA (2018) Pancreas regeneration. Nature. https://doi.org/10.1038/s41586-018-0088-0

Zhou P, Guo Y, Huang Y, Zhu M, Fan X, Wang L, Wang Y, Zhu S, Xu T, Wu D, Lu Y, Wang Z (2016) The dynamic three-dimensional culture of islet-like clusters in decellularized liver scaffolds. Cell Tissue Res. https://doi.org/10.1007/s00441-015-2356-8

Zimmermann H, Shirley SG, Zimmermann U (2007) Alginate-based encapsulation of cells: past, present, and future. Curr Diab Rep

Zuber SM, Grikscheit TC (2018) Stem cells for babies and their surgeons: the future is now. J Pediatr Surg. https://doi.org/10.1016/j.jpedsurg.2018.10.027

Zulewski H, Abraham EJ, Gerlach MJ, Daniel PB, Moritz W, Müller B, Vallejo M, Thomas MK, Habener JF (2001) Multipotential nestin-positive stem cells isolated from adult pancreatic islets differentiate ex vivo into pancreatic endocrine, exocrine, and hepatic phenotypes. Diabetes. https://doi.org/10.2337/diabetes.50.3.521

Adv Exp Med Biol – Cell Biology and Translational Medicine (2019) 6: 221–223
https://doi.org/10.1007/978-3-030-32823-8
© Springer Nature Switzerland AG 2020

Index

Printed in the United States
By Bookmasters